ENZYME ASSAYS FOR FOOD SCIENTISTS

ENZYME ASSAYS FOR FOOD SCIENTISTS

Clyde E. Stauffer, Ph.D.
Technical Food Consultants
Cincinnati, Ohio

AN AVI BOOK
Published by Van Nostrand Reinhold
New York

AN AVI BOOK
(AVI is an imprint of Van Nostrand Reinhold)

Copyright © 1989 by Van Nostrand Reinhold
Library of Congress Catalog Card Number 88–26189
ISBN 0-442-20765-4

Printed in the United States of America

Van Nostrand Reinhold
115 Fifth Avenue
New York, New York 10003

Van Nostrand Reinhold International Company Limited
11 New Fetter Lane
London EC4P 4EE, England

Van Nostrand Reinhold
480 La Trobe Street
Melbourne, Victoria 3000, Australia

Macmillan of Canada
Division of Canada Publishing Corporation
164 Commander Boulevard
Agincourt, Ontario M1S 3C7, Canada

16 15 14 13 12 11 10 9 8 7 6 5 4 3 2 1

Library of Congress Cataloging-in-Publication Data

Stauffer, Clyde E.
    Enzyme assays for food scientists / Clyde E. Stauffer.
        p.   cm.
    Bibliography: p.
    Includes index.
    ISBN 0-442-20765-4
    1. Enzymes—Analysis.   2. Enzyme kinetics.   3. Food—Analysis.
I. Title
    [DNLM: 1. Enzymes—analysis.   2. Food Technology.   QU 135 S798e]
QP601.S5664 1989
574.19'25'028—dc19
DNLM/DLC
for Library of Congress   88-26189
                                                                      CIP

Contents

PART II. PRACTICAL

APPENDIXES

ENZYME ASSAYS FOR FOOD SCIENTISTS

Part I   THEORETICAL

1   Introduction

No apologia is needed for considering the food industry as a major influence in the use of enzymes, or for recognizing the importance of enzymes to the processing and preserving of foods. The importance of enzymes to food scientists is well attested by the frequent appearance of enzymological papers in food-oriented research journals. To help deal with this flood of information, several authors have written or edited books on the topic of food enzymology. Libraries in food science departments have many such books on their shelves. Some of the more familiar ones are the texts by Reed (1975), Whitaker (1972) and Schwimmer (1981) as well as the symposium proceedings edited by Whitaker (1974). In addition, every year or two sees the publication of many books addressing some aspect of enzymes and the food industry, for example those edited by Linko and Larinkari (1980), Birch, Blakebrough and Parker (1981), Godfrey and Reichelt (1983) and Kruger, Lineback and Stauffer (1987).

The basic discipline of enzymology is well covered in numerous textbooks (Dixon and Webb 1979), multi-volume sets (Barman 1969; Boyer 1970), and frequent updates by series such as *Methods in Enzymology* or *Advances in Enzymology*. Some of the articles in those volumes are applicable to food enzymes, although many of them tend to be more oriented towards work in medical areas. And of course, there are several research journals which frequently publish articles on the isolation and study of enzymes in foods.

Given this plethora of information, the obvious question is, What need is there for another book on food enzymes? The present volume is intended to lie somewhere between the broad coverage of a text such as Schwimmer's (1981) and the detail of an article in *Methods in Enzymology*. The central focus of this book is on accurate, reliable methods for measuring the amount of enzymes in foods and food-related systems. It is not meant to be a cookbook, although the later chapters present experimental details which will be helpful in designing and testing an assay for an unfamiliar enzyme. It is not an in-depth theoretical treatise; that has already been done, and is readily available. However, the discussions in the earlier chapters should make clear the kind of requirements applicable to a good enzyme assay, and demonstrate the limitations of many assays commonly used in the food industry. Finally, it is not a handbook, although a number of tables of useful data are found in various chapters.

This book is written as a tool for the working scientist who has some background in enzymes but is not an in-depth expert enzymologist. The job demands that he/she does a competent, professional study on the properties and uses of an enzyme, but does not allow the luxury of several months browsing in the library and performing trial-and-error experiments. This book will answer most of the questions regarding experimental design and data analysis and interpretation which arise during the course of the study of a food enzyme. Hopefully it will be regarded as a working manual, and not just another reference book on the shelf.

STRATEGY OF AN ENZYME STUDY

A comprehensive study of a particular enzyme includes the following steps.

1. Identify the catalytic activity of interest. Develop an assay which will allow the quantification of that activity. At this stage, simplicity and speed are more important attributes of an assay than are sophistication and accuracy. For instance, the hydrolysis of p-nitrophenyl-acetate in a rack full of test tubes may be very useful for testing fractions from a chromatographic separation of esterase, although the accuracy may be no better than $\pm$ 20%.
2. Purify the enzyme. This may be only partial isolation, not the achievement of complete purity. However, it should be carried to the point at which there is only one enzyme present (not two or more with similar activities) and endogenous inhibitors or activators are removed or identified.
3. Characterize the enzymatic activity. This step deals with the kinetics of the enzyme. At this stage more care must be taken in choosing an assay which will give appropriate experimental data. In investigating different properties (e.g., the pH-dependence of activity and the rate of heat-denaturation under various conditions), different substrates and assay conditions may be needed. Other characterizing parameters might be specificity (rate of reaction with different substrates) and the effect of inhibitors and activators.
4. Determine the enzyme's chemical and physical properties. Once a pure (single molecular species) material is obtained, there are a number of factors to be found: composition (amino acids, sugars, metal ions, and any other prosthetic groups), molecular weight (quaternary structure, if any), amino acid sequence, secondary and tertiary structure. Of practical importance is the stability of the enzyme against denaturation under conditions which might be found in use.

5. Integrate the protein characteristics and enzymatic nature of the molecule. The amino acid residues which make up the active site are identified by chemical and kinetic studies. By crystallographic studies the orientation of the substrate at the active site is determined. If the source of the enzyme is amenable to gene engineering, it may be possible to tailor the amino acids around the active site to alter activity or specificity in ways which are technologically desirable.

Different readers are undoubtedly involved with enzymes at different points in this broad, overall scheme. A researcher for a company which produces enzymes will spend much time in the early stages of purification, and a fair amount of effort in understanding the factors which lead to stability (or instability) of the enzyme during use. An applications scientist using the enzyme in food processing is perhaps more interested in the ways that pH, inhibitors and activators affect the enzyme. A university graduate student might be concerned with the first step of developing a workable assay for an enzyme in a foodstuff, or with the last step, depending upon the interests and funding available in the department.

In choosing an enzyme assay, the first point to establish is the level of information which is needed. An assay for quickly screening hundreds of fractions from a chromatographic column needs speed and convenience, not accuracy. For following the rate of heat denaturation of an enzyme less speed but a higher degree of precision is required. In standardizing enzyme preparations for food processing applications moderately high accuracy is needed, and the question of substrate specificity comes into play. For investigating the effect of pH on enzyme action a sensitive, accurate assay is required which will allow the determination of V_{max} and K_M. A similar sort of sophisticated assay is needed to properly study inhibitors and activators of the enzyme.

The first question the researcher must ask is, What do I really need to know from the assay? and secondly, What kind of assay will give me that level of information? An assay will give one of four kinds of information: 1. the number of moles of catalytic center present; 2. the rate of product formation under given reaction conditions; 3. the extent of product formation under conditions of use; or 4. the amount of substrate present, inferred from the rate or amount of product formation under defined assay conditions.

The first kind of information is of most importance to researchers engaged in enzyme purification or to bioengineers who are trying to find optimum conditions for producing an enzyme. In fact, most assays as normally conducted yield this information only by inference. The assay result (v) is related to the molar amount of enzyme present (E_t) via the appropriate rate equation. If no other factors (inhibitors, activators, multiple enzymes) vary, then the proportionality factor is in fact relatively constant, and the inference is correct. Balanced against this uncertainty is the fact that this assay is probably much faster and easier to perform than an assay which would give an unequivocal measure of enzyme molarity. The researcher is usually willing to trade off reliability for speed; the challenge is to always be aware of the compromise and its possible consequences.

The second kind of information represents the vast majority of assay results as performed by food enzymologists and product developers today. Such assays are often used for standardizing enzyme preparations. The difficulty is that they give a false sense of comparability between different enzymes. The problem is usually one of specificity; the assay substrate is different from the actual substrate to which the enzymes are applied. Thus "100,000 H.U. (hemoglobin units)" of each of three proteases may, upon application to bread doughs, have quite different effects because they are now hydrolyzing a different protein substrate (Petit 1974).

The third type of information, performance under practical applications condition, is of extreme importance and also difficult to obtain. For example, an assay to measure the effectiveness of a protease in reducing chill haze formation in beer is highly desirable. The substrate is ill defined (some mixture of malt proteins and polyphenols), the conditions for activity (refrigerator temperatures) are less than optimum for the enzyme, and the time to reach the endpoint (days to weeks) is greater than the patience of the enzymologist. Hence the tendency is to use some quick, familiar measure of protease activity, and correlate it with the ability to prevent chill haze. Again, the benefits of the quick assay are considered to outweigh the risks inherent in depending on a correlation. If the enzyme is being used as an immobilized enzyme, an accurate assay will require that the configuration of the applications equipment be reproduced, because such factors as flow rate, turbulence, and viscosity can have a major impact on the amount of product formed per unit time per unit of enzyme protein present.

The fourth kind of information requires the use of enzymes as analytical tools. They may be utilized to measure residual glucose in egg whites, lactic acid in a sourdough culture, ethanol in wine or beer, pectin in a fruit extract; the applications are legion. Often in developing the analytical method full use is not made of the insights available from enzyme kinetics. Rather, an existing assay method for the enzyme is turned on its head, and amounts of enzyme, cofactors (if any), temperature, time and pH are adjusted semi-empirically until answers obtained by the enzymatic method agree with answers obtained from another, more traditional analytical method. The resulting method is usually wasteful of biochemical reagents, and the inherent experimental error is determined from making many replicate determinations rather than from considerations of the theoretical basis for the reaction.

At this point a word about the units in which we express the results of enzyme assays is appropriate. For certain kinds of assays, e.g., the titration assay for serine proteases, the results are expressed in fundamental chemical units, moles per gram of protein. This corresponds to the first kind of information. Far more common are units correlating with the second kind of information, namely the rate, in moles per time unit, of appearance of product or disappearance of substrate. During the 1970s the Enzyme Commission established an International Unit (IU) as that amount of enzyme which causes the formation of 1 μmole of product per minute under defined assay conditions. Thus one may express enzyme activity as IU per some amount of enzyme preparation (mg protein, ml solution, etc.). More recently, in order to bring enzymology into line with the CGS system, enzyme catalytic activity is expressed as the katal, which is that amount of enzyme which leads to the formation of 1 mole of product per second. One katal is equivalent to 60×10^6 IU, or 1 IU equals 16.67 nanokatals (a more convenient magnitude for ordinary work). At this time, most activities such as specifications in biochemical supply house catalogs are still expressed in International Units. Finally, note should be made of the practice of defining enzyme units in terms of rate of change of the instrumental signal (e.g., 1 unit = a change of 0.001 Absorbance unit per second). This is often found where the instrument signal cannot be readily connected to the molecular events which comprise the enzyme action (turbidimetric measurement of lysozyme activity), in which case we just have to live with it. But too often the publication of work based on this sort of "enzyme unit" merely indicates laziness on the part of the researcher who has not gone the extra step of translating instrument signal into product molarity. This greatly decreases the value of the work to other scientists.

REFLECTIONS ON ENZYME ASSAY DESIGN

"Why do we never have time to do it right, but we always have time to do it over?" This wry comment on the pressures which we all feel has been copied and posted over the desks of many product development personnel, graduate students, and analytical service laboratory technicians. Unfortunately it is all too often apropos. In the beginning of an enzyme project an assay is put together which detects the activity and measures it during the steps of purification. The plan is that once the enzyme is purified the question of an accurate, powerful assay will be addressed and the shortcomings of the current one will be corrected. Other more pressing questions seem to intervene, and suddenly it is realized that several notebooks have been filled with data on studies of the properties of the enzyme, all based on this admittedly inadequate assay. If the assay system is changed at this point, much of that data will be irrelevant and the experiments will have to be repeated, a task for which the researcher doesn't seem to have the time. So we do the best we can with the existing data, and write up a publication (a paper, a technical data sheet, a method for the analytical control laboratory) which is based on a weak assay, and hope the weaknesses don't compromise subsequent research based upon our work.

It is inefficient to try to develop a powerful, accurate assay for an enzyme which is available only as a minor component of a crude homogenate or extract. The presence of many other complicating factors (inhibitors, activators, stabilizing and destabilizing species) make the interpretation of results from a sophisticated kinetic experiment problematic at best. Nevertheless, even at the early stages, as purification is proceeding, monitored by the inelegant but rapid assay, thought should be given to the design of an ideal assay. As some small amounts of relatively pure enzyme become available this design may be tested and refined. The ultimate goal is that at the point where purification of the enzyme is routine and adequate amounts are available for studies of its catalytic characteristics, the assay will also be available which will enable those studies to be done in a scientifically elegant manner. Then notebooks #2, 3 . . . will contain data which will stand up under the scrutiny of your scientific peers, and will form a firm foundation for further work by other researchers.

As an analogy, consider the measurement of pH. No one would make a precise study of the effect of pH upon an enzyme reaction based on measurement of pH using Universal Indicator Paper. At the least a pH-meter which can accurately measure to the nearest 0.05 pH unit would be used, and perhaps an expanded-scale meter if the rest of the experiment warranted the expense. Nonetheless Universal Indicator Paper has a definite place in the biochemical laboratory; it is excellent for adjusting a homogenate to neutral pH, say 6.5 to 7.5, where the expanded-scale meter would definitely qualify as overkill. In the same way, the rough-and-ready assay is fine for choosing the time at which to terminate a biofermentation for production of an enzyme. But when a detailed study of the effect of certain inhibitors upon that enzyme is initiated, the accuracy of the assay needs to be an order of magnitude better.

This analogy points up the real nature of an enzyme assay: it is a tool. For the most part biochemists do not study the theory of pH meter operation or spectrophotometry or chromatography for their own sake. The purpose of many laboratory courses is to make sure that students learn enough theory to operate the tools properly, recognizing both their power and their limitations. The object of most enzymatic research is to learn more about the enzyme, not about the tools used in the research. Nevertheless, having more powerful tools (assays) available, and knowing how to use them to the utmost, enhances the possibilities in exploring enzyme properties. It is a truism that for the most part advances in science during the last 50 years have been initiated by advances in the tools available for doing that science. On a personal note, when I began my graduate research 30 years ago I felt lucky to have a Coleman Jr. colorimeter at my disposal. Before I finished, the availability of a Beckman DU UV-VIS spectrophotometer broadened the scope of the project, and a continuous-recording UV-VIS at my first research laboratory made even more sophisticated research possible.

Modern instrumentation does not decrease the work load; it enables us to obtain more penetrating insights with the same amount of work. The scientist who worked 60 hours per week before getting a personal computer still works 60 hours per week, but the nature of the output has changed due to the added analytical power of the computer. But if the PC is only used to add, subtract, multiply and divide, it is not much better than a simple pocket calculator. For many fixed-time assays the Coleman Jr. is equivalent to the most modern spectrophotometer. Taking full advantage of the capabilities inherent in the modern instrument requires an understanding of the kinetics underlying the experiments we would like to do, just as making good use of a PC requires some understanding and use of programming language. This book might be called a BASIC primer for enzyme assays; mastering and applying the contents will enable you to achieve improved insights into the nature of the enzyme you are working with, through better application of the fundamental tool of the enzymologist, the assay.

While a cursory glance at the literature concerned with food enzymes gives the impression that most of the studies would have benefitted from a better application of kinetic theory, a caveat must still be made. The justification for a new, more penetrating study of an enzyme should not simply be "Now we can do it." The aimless application of any tool without some goal in mind is a waste of resources. As stated in the next chapter, first decide what level of sophistication is required to meet the goal, then work at that level with neither compromise downwards nor overkill. Don't try to shave with an axe, but don't split firewood with a straight razor. Fit the tool to the task; choose an assay which gives the information you need.

2 Kinetics

The description of enzymatic characteristics involves kinetic analysis. In some cases this might be rather superficial, while in other instances many details of the rate of reaction will be cataloged. The researcher should decide what level of information is required before embarking on large-scale experimentation. The enzyme of interest might be an α-amylase for desizing fabric; the desired information might be no more than a surface response map of pH, temperature and enzyme quantity, finding the time to produce a desired level of modification. In this case a detailed analysis of kinetic rate parameters would not be justified. On the other hand, a study might be ordered to measure all the parameters which would allow the prediction of the optimum temperature and pH for operation of a glucose isomerase column. Now a rather more detailed experimental design is needed, to obtain rate constants for both the forward and reverse reactions, the influence of pH and temperature on these rate constants, and also the rate constant for heat-inactivation of enzyme in the presence of substrate and product. Such a project would obviously entail much more in the way of data-gathering and analysis than the former example; nevertheless, both studies are kinetic in nature, and the appropriate level of planning must be applied in order to obtain good results with reasonable expenditure of resources.

Determining rate constants in the first example would probably be over-kill, using more time and money than is warranted for the application's needs. In the second example, using a response surface methodology approach might solve an immediate problem, but would not provide the insights and detailed information needed to quickly solve future problems, where some variation in the process is desired. As a general observation, the laboratory researcher and the applications engineer often find themselves arguing about how much effort to spend on characterizing the enzyme. If instead they work together to define how much information is needed and how it will be used, they will both be doing a better, more professional job.

Once the level of sophistication of information is established, the design of experiments to obtain that data requires knowledge and some insights into the kinetics of enzyme action. These chapters on kinetics, inhibitors, pH and temperature will present the requisite knowledge. The insight comes only from experience in working with enzymes, both in the laboratory and in the plant. If more detailed discussion is wanted, there are many excellent texts which deal with enzyme kinetics (Cornish-Bowden 1976; Fromm 1975; Reiner 1969; Segel 1975).

THE MICHAELIS-MENTEN RATE EQUATION

One Enzyme, One Substrate Systems

Rapid Equilibrium Model. In early investigations of enzyme reactions it was found that the rate of product formation was a hyperbolic function of the concentration of substrate (Figure 2-1). At low substrate levels doubling the substrate concentration led to doubling the rate of the reaction, while at high substrate concentrations the rate became independent of substrate concentrations. The key to this behavior (transition from a rate which was first order in substrate to one which was zero order in substrate) was first proposed by V. Henri in 1902 and elaborated by L. Michaelis and M. L. Menten in 1913.

The basic concept is that an enzyme-catalyzed reaction involves two steps: 1. the reversible, rapid combination of enzyme (E) and substrate (S) to form a complex (ES); and 2. a somewhat slower breakdown of ES to give product (P) and regenerate free enzyme. The key idea is that E, S, and ES are in rapid equilibrium characterized by an equilibrium *dissociation* constant K_m. The rate of formation of P (i.e., v, the rate of the reaction) is directly proportional to the concentration of ES at any given time. The kinetic model is given as:

$$E + S \underset{k_{-1}}{\overset{k_1}{\rightleftharpoons}} ES \qquad K_m = \frac{[E][S]}{[ES]} = \frac{k_{-1}}{k_1} \qquad (2\text{-}1a)$$

$$ES \overset{k_2}{\longrightarrow} E + P \qquad v = \frac{dP}{dt} = k_2[ES] \qquad (2\text{-}1b)$$

The basic Michaelis-Menten (M-M) equation is readily derived from this model by remembering that the total enzyme added, E_t is equal to $E + ES$, and defining a parameter $V_{\max}$, the velocity at high $[S]$, as equal to $k_2[E_t]$. The derivation is shown in Appendix A, and the M-M equation is:

$$v = \frac{V_{\max}[S]}{K_m + [S]} = \frac{dP}{dt} = \frac{k_2[E_t]}{k_{-1}/k_1 + [S]} \qquad (2\text{-}2)$$

Steady-State Model. In about 1925 G. E. Briggs and J. B. S. Haldane pointed out that when enzyme and substrate are initially combined the only reaction occurring is the combination of $E + S$ to give ES. After some short period of time ES begins to break down to form P, as well as dissociating back to $E + S$. In less than a second the three rates attain a balance in which $k_1[E][S] = (k_{-1} + k_2)[ES]$ and the concentration of ES reaches a steady state, $d[ES]/dt = 0$. The existence of the pre-steady state stage has been demonstrated for a large number of enzymes using stopped-flow experimental techniques. Its existence is usually assumed in developing any kinetic schemes for an enzyme under investigation. It has implications for interpreting certain kinds of data, particularly those pertaining to the effect of pH or temperature on enzyme activity. It has also been used to study the association of high-affinity inhibitors with enzymes.

The derivation of the steady state M-M equation is also presented in Appendix A. The form of the rate equation is identical; all that changes is the significance of the equilibrium constant. For the steady state the equilibrium dissociation constant $K_s = (k_{-1} + k_2)/k_1$. If the rate constant for the breakdown to product k_2 is much less than the rate constant for dissociation k_{-1} then $K_s \approx K_m$. If k_2 is much greater than k_{-1} then the reaction looks (kinetically) like two sequential irreversible steps. This difference may become significant when one is interpreting the effects of pH and temperature upon the fundamental kinetic parameters. These nuances will be mentioned at the appropriate points later in this book.

For most practical purposes the equilibrium and steady state models give identical results. Operationally, either K_m or K_s represent that value of $[S]$ at which $v = V_{max}/2$ (Figure 2-1), and will be designated K_M. In those few instances where it makes a difference, the equilibrium or steady state parameter will be given as K_m or K_s respectively.

Intermediate Reaction Species. The kinetic model presented is the simplest possible one which represents the rate equation for most one enzyme-one substrate systems. The first part of equation 2-2, $v = V_{max}[S]/(K_M + [S])$, is useful for most of our discussions. This model implies the presence of only four molecular species, E, S, ES, and P. Most one enzyme-one substrate systems entail many intermediate species; in some cases the recognition of these species is important for gathering certain kinds of information. A prime example of this is the group of serine proteases (chymotrypsin, trypsin, subtilisin, etc.), hydrolyzing esters of amino acids, and certain organic acids. The ES complex first liberates P_1, the alcohol moiety of the substrate, and forms an acylated enzyme EA. This then reacts with a water molecule to give the acid product P_2 and E.

$$E+S \underset{k_{-1}}{\overset{k_1}{\rightleftharpoons}} ES \overset{k_2}{\longrightarrow} EA+P_1 \qquad (2\text{-}3a)$$

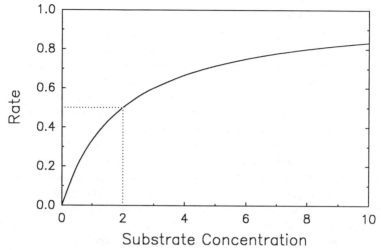

FIGURE 2-1. Hyperbolic dependence of enzyme rate on substrate concentration according to the Michaelis-Menten equation.

$$EA + H_2O \xrightarrow{k_3} E + P_2 \qquad (2\text{-}3b)$$

The overall kinetics of this system is still described by the basic M-M equation, but V_{max} and K_M have a somewhat different significance.

$$V_{max} = [E_t]k_{cat} = [E_t]k_2\left(\frac{k_3}{k_2 + k_3}\right) \qquad (2\text{-}4a)$$

$$K_M = K_s = \frac{(k_{-1} + k_2)/k_1}{k_3/(k_2 + k_3)} \qquad (2\text{-}4b)$$

These equations are important in the operation of many of the "burst" or active site titration methods for assaying absolute moles of those enzymes which react in this way. Briefly, enzyme is mixed with a substrate such that either EA or P_1 can be measured, under conditions where $k_2 \gg k_3$. Then the molar concentration of EA or P_1 is essentially equal to E_t. The first demonstration of this was by Hartley and Kilby (1954) who reacted p-nitrophenyl acetate with milligram quantities of chymotrypsin. They observed a rapid release of p-nitrophenol (P_1) which they were able to relate to the number of moles of chymotrypsin present. The kinetics of this situation are discussed more fully in Appendix G, the application to enzyme assays is given in Chapter 6, and some of the details of these "titration" assays are given in Chapter 8.

One Enzyme, Two Substrate Systems

Many enzymatic reactions involve more than one substrate. Common examples are the oxidases and oxygenases where one substrate is O_2, various kinases and phosphorylases where a high-energy phosphate compound such as ATP contributes a phosphate group, and carboxylases which combine CO_2 with a receptor such as ribulose diphosphate, phospho-enol pyruvate, or acetyl CoA to generate a molecule needed in the metabolism of the organism. The kinetic analysis of two substrate systems can become exceedingly complex. Do the substrates combine with the enzyme in a specific order, or is the addition random? Are product molecules released at intermediate stages of the reaction? Are any or all of the steps reversible? How is the reaction affected by inhibitors and activators? The inclusion of these possibilities in setting up the initial model can lead to some frightful-looking rate equations. Fortunately a number of workers have reduced these factors to manageable proportions and systematized the approach to analyzing them experimentally (Cleland 1970; Segel 1975).

In this chapter the discussion will be limited to the cases most often seen in practical food enzymes application work, namely the two-substrate, random- or ordered-addition reaction. For simplicity the rapid equilibrium equations will be used. The steady state equations are more exact but also more cumbersome than is needed for present purposes. The kinetic model for the random-addition case with two substrates A and B is:

$$E + A \overset{K_A}{\rightleftharpoons} EA + B \overset{\alpha K_B}{\rightleftharpoons} EAB \qquad (2\text{-}5a)$$

$$E + B \overset{K_B}{\rightleftharpoons} EB + A \overset{\alpha K_A}{\rightleftharpoons} EAB \qquad (2\text{-}5b)$$

$$EAB \overset{k_p}{\longrightarrow} E + P \qquad (2\text{-}5c)$$

Random Order of Substrate Binding.   The order of addition of two substrates may be random, that is either 2-5a or 2-5b may occur first on the path to EAB and thence to P. If the factor α (see Appendix A for an explanation of this) is less than 1, the binding of the first substrate *increases* the binding of the second substrate (the dissociation constant is decreased). If α is greater than 1, the binding of the second substrate is decreased; while if α equals 1, there is no effect. The rate equation obtained from this model is:

$$v = \frac{V_{\max}[A][B]}{\alpha K_A K_B + \alpha K_B[A] + \alpha K_A[B] + [A][B]} \qquad (2\text{-}6)$$

If [B] is held constant and [A] is varied, for example in a run made with lipoxygenase where the buffer is saturated with O_2 and linoleate is varied, then equation 2-6 is manipulated by dividing the denominator and numerator by [B] and rearranging, to get:

$$v = \frac{V_{max}[A]}{\alpha K_A(1 + K_B/[B]) + [A](1 + \alpha K_B/[B])} \tag{2-7}$$

Equation 2-7 is symmetrical; if [A] is held constant and [B] is varied the concentration terms and equilibrium constants are reversed.

Both the expressions in parentheses in 2-7 are constant. Dividing the numerator and the denominator by the second of these expressions gives the familiar hyperbolic rate expression $v = X[A]/(Y + [A])$, where X is the rate at high [A] and Y is the concentration [A] giving half that rate. X and Y are complex terms involving all four reaction parameters V_{max}, K_A, K_B, and α. The method for finding these individual parameters from experimental data will be discussed later in this chapter.

Specific Order of Substrate Binding. If the addition of the two substrates to the enzyme active site is ordered, that is substrate A must bind before substrate B, then the rate equation is simplified. Equation 2-5b does not exist, and the factor α is dropped in 2-5a. The resulting rate is given by:

$$v = \frac{V_{max}[A][B]}{K_A K_B + K_B[A] + [A][B]} \tag{2-8}$$

This can be rearranged to show the dependence of v on [A] or [B] as before, but now the two rate expressions are not simple reversals.

$$v = \frac{V_{max}[A]}{K_A K_B/[B] + [A](1 + K_B/[B])} \tag{2-9a}$$

$$v = \frac{V_{max}[B]}{K_B(1 + K_A/[A]) + [B]} \tag{2-9b}$$

For the random-addition two substrate system both substrates must be present at saturating concentrations in order to measure a true V_{max}. In the ordered-addition case the true V_{max} is found with saturating concentrations of B even if [A] is lower than K_A. The purpose of the proper analysis of the data is to determine whether the system under investigation is ordered or random in nature, and if it is ordered, which of the two substrates is B. This is a topic in the data analysis section below. Beyond that, if a reversible reaction system is suspected, the more detailed work by Cleland (1970) or Segel (1975) should be consulted.

Sequential Enzyme Reactions

In applying enzymes to a food process usually only one or a few related enzymes are investigated. In assaying for an enzyme activity of interest sometimes the primary product(s) of the reaction are not readily measurable. The experimenter may then have recourse to auxiliary enzymes, and set up a so-called coupled enzyme assay. As an example, the enzyme under investigation might be amyloglucosidase, and the product is glucose. The usual chemical methods for measuring glucose are not particularly convenient, so some glucose oxidase is added to the assay mixture and the disappearance of O_2 is followed using a gas electrode. This gives a quick, continuous measure of amyloglucosidase action. The question of the accuracy of the assay, however, is still to be determined. Understanding the kinetics of such a coupled enzyme assay can lend confidence in the accuracy of the results. McClure (1969) analyzed the kinetics of such a system, and also the system in which two auxiliary enzymes are used. His recommendations provide a basis for the rational design of a coupled enzyme assay system (Rudolph et al. 1979).

One Auxiliary Enzyme.   The simple coupled enzyme assay is as described above, where the first enzyme catalyzes the reaction of A to B, and the auxiliary enzyme catalyzes a reaction B to C.

$$A \xrightarrow{k_1} B \xrightarrow{k_2} C$$

We assume that the first reaction is irreversible (since the product B is continuously removed) and the rate k_1 does not change. This means that either $[S]$ is large enough that the first reaction rate is V_{max}, or else the fraction of A which is converted to B is very small. The second reaction is also considered to be irreversible, and first order with respect to B. If $[B] << K_B$ then the M-M equation $v = V_b[B]/(K_B + [B])$ reduces to an equation which is first order in $[B]$, $v = -dB/dt = (V_b/K_B)[B]$, and the rate constant k_2 is equal to V_b/K_B.

The system may be analyzed mathematically, to arrive at an expression which specifies the amount of auxiliary enzyme required to give a reliable assay (see Appendix B). After a lag period of t^* seconds, $[B]$ equals some fraction F of the steady-state concentration, at which the rate of formation of C equals the rate of reaction of A to form B. Then

$$\frac{V_b}{K_B} = k_2 = \frac{-\ln(1 - F)}{t^*} \tag{2-10}$$

The first order rate constant V_b/K_B for the auxiliary enzyme is usually easy to determine under the conditions of the assay. If F is taken as 0.99 (i.e. the rate of formation of C is only 1% less than the rate of reaction of A) and a lag period of 5 seconds is chosen (the amount of time to mix the assay ingredients and put them in the detection device) then an amount of auxiliary enzyme to give a first order rate constant of 0.92 sec^{-1} is used per assay. Some numerical examples are given in Chapter 6 to demonstrate this calculation.

Two Auxiliary Enzymes.   In those cases where the product of the reaction of the auxiliary enzyme is not easily measurable, a second reaction may be necessary. To extend the earlier example, glucose formed by the action of amyloglucosidase is oxidized by glucose oxidase to give H_2O_2 (product C) which then reacts with pyrogallol in a reaction catalyzed by peroxidase to give a colored product D which is monitored spectrophotometrically.

$$A \overset{k_1}{\longrightarrow} B \overset{k_2}{\longrightarrow} C \overset{k_3}{\longrightarrow} D$$

Unlike the case with one auxiliary enzyme, a unique analytical expression such as equation 2-10 cannot be developed for this case. McClure (1969) worked out the relationship between k_2 and k_3 for any specified lag time in the attainment of 99% of the steady state concentration of C, and showed that with $F = 0.99$, $1/k_2 + 1/k_3 = 0.21715t^*$. If the two first-order constants are equal, then to obtain an overall lag time of five seconds in this system, $k_2 = k_3 = 1.83$ sec^{-1}. If the two enzymes for the sequence are quite different in price and/or availability, the respective amounts may be adjusted. A numerical example is given in Chapter 6.

THE HENRI INTEGRATED RATE EQUATION

One Enzyme, One Substrate Systems

Non-denaturing, Non-inhibiting Conditions.   The M-M equation is a differential equation, that is, it gives the rate of change of concentration of product. There are many enzyme applications where it is useful to know how much product has accumulated, given a certain initial substrate concentration $[S_0]$. Moreover there are certain situations (e.g. inhibition by the product of the reaction or progressive denaturation of enzyme during the course of the reaction) in which experimentation using the integrated rate equation is the easiest or only way to get the desired inhibition or denaturation rate constants. Rearranging equation 2-2, noting that at any time, t, $[S] = [S_0] - [P]$, we have:

$$dP\left(\frac{K_M + [S_0] - [P]}{[S_0] - [P]}\right) = V_{max}dt$$

Gathering terms, the equation becomes:

$$\left(\frac{K_M}{[S_0] - [P]}\right)dP + dP = V_{max}dt$$

Integrating between limits $t = (0, t)$ and $P = (0, P)$ yields the Henri equation:

$$[P] = V_{max}t - K_M \ln\left(\frac{[S_0]}{[S_0] - [P]}\right) \qquad (2\text{-}11)$$

Enzyme Denatures during the Reaction.   The integrated rate equation for product inhibition will be presented in Chapter 3 in connection with a general discussion on inhibitors. However, the progressive denaturation of enzyme during the course of the reaction with substrate will be treated here. In addition to combining with substrate (equation 2-1a) the enzyme also undergoes irreversible inactivation with a first order rate constant k:

$$E \xrightarrow{k} E_{inact}$$

At any time t the amount of total active enzyme present $[E_t]_t$ is related to the original amount of enzyme $[E_t]_0$ by $[E_t]_t = [E_t]_0(e^{-kt})$, and the maximum rate at time t is given by $V_{max}(e^{-kt})$. Substituting this for V_{max} in the rearranged equation above and gathering terms, we have:

$$\left(\frac{K_M}{[S_0] - [P]}\right)dP + dP = V_{max}(e^{-kt})dt$$

Integration between limits $t = (0, t)$ and $P = (0, P)$ gives the integrated "denaturing" equation:

$$[P] = \left(\frac{V_{max}}{k}\right)(1 - e^{-kt}) - K_M \ln\left(\frac{[S_0]}{[S_0] - [P]}\right) \qquad (2\text{-}12a)$$

The term e^{-kt} is expanded to $(1 - kt + k^2t^2/2 - k^3t^3/6 \ldots)$. Substituting this into equation 2-12a gives:

$$[P] = V_{\text{max}}t \left(1 - \frac{kt}{2} + \frac{k^2t^2}{6} \cdots \right) - K_M \ln \left(\frac{[S_0]}{[S_0] - [P]}\right) \qquad (2\text{-}12b)$$

The difference between this and the "undenatured" equation 2-11 is the parenthetical factor in the V_{max} term. The way in which this is used to measure the denaturation rate constant k will be discussed later in this chapter.

DATA ANALYSIS

The raw data from an enzyme kinetic experiment consist of some numbers related to product or substrate concentration determined at different times after the initiation of the reaction. How these numbers are obtained is covered in Chapters 6 and 7. The present discussion is concerned with the ways in which the data are manipulated mathematically to arrive at the kinetic parameters of interest (i.e., V_{max} and K_M, inhibitor binding constants, denaturation rate constants, etc.). The two aims of data manipulatory methods are to minimize the effort required to perform them, and to maximize the reliability of the derived parameters. While the ease of performance is related to the method itself, I must emphasize that the main determinant of parameter reliability is the quality of the experimental data. Good mathematical analytical methods cannot transform poor data into reliable derived constants. GIGO(Garbage In, Garbage Out) is still valid.

First-Order Kinetics

At low $[S]$, when $[S] << K_M$, the M-M equation describes a reaction which is kinetically first order in $[S]$:

$$v = \frac{dP}{dt} = -\frac{dS}{dt} = \left(\frac{V_{\text{max}}}{K_M}\right)[S] \qquad (2\text{-}13)$$

At any time t the concentration of substrate $[S]$ is related to the initial concentration $[S_0]$ by $[S] = [S_0](e^{-k't})$ where the first order rate constant $k' = V_{\text{max}}/K_M$. Taking logarithms, $\ln[S] = \ln[S_0] - k't$, so a plot of $\ln[S]$ versus t is a straight line with a slope $-k'$ (Figure 2-2).

If the experimental procedure is detecting $[S]$ this works out conveniently. Most enzyme reaction procedures tend to follow the appearance of P which makes data analysis more complicated. When the reaction goes to completion, the product concentration is $[P_\infty]$ which equals $[S_0]$. At any time t,

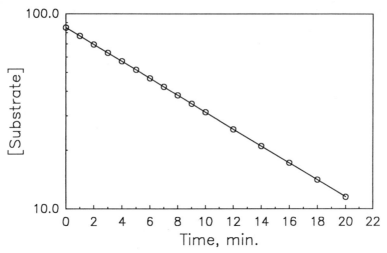

FIGURE 2-2. Semi-logarithmic plot of disappearance of substrate with time.

$[S]$ equals $[P_\infty] - [P]$, so the plot to find $-k'$ is $\ln([P_\infty] - [P])$ versus t. The difficulty arises in obtaining an accurate value for $[P_\infty]$. There are three ways to do this.

Complete Reaction Method.   The first method is to let the reaction run long enough to approach 100% completion. To get 99% completion requires 7 half-lives, where the half-life $t_{1/2} = 0.697/k'$ is the interval required to reduce $[S]$ by 50%. In a run with $t_{1/2}$ of 10 minutes, most of the usable data will be obtained during the first 30 minutes of the reaction, but then it is necessary to wait an additional 40 minutes to get an estimate of $[P_\infty]$. If $[P]$ is being detected directly, say in a spectrophotometer, this ties up the equipment for an inordinate amount of time, as well as running the risk of baseline drift. In some indirect reactions (e.g. amide hydrolysis by aminopeptidase followed by ninhydrin analysis of an aliquot, see Chapter 7) this procedure works quite well.

Swinbourne Method.   A number of methods for graphical evaluation of first-order data without having to know $[P_\infty]$ have been suggested, but probably the easiest to use is that suggested by Swinbourne (1960). Appendix C shows the mathematics of this method. In practice a listing is made of $[P]$ versus t at regular intervals. A value T is chosen which is between one-half and one times the roughly estimated $t_{1/2}$ for the reaction. If p is the value of $[P]$ at time t, then p' is $[P]$ at time $t + T$; a plot is made of p on the y-axis versus p' on the x-axis for each data pair of $t, t + T$ (note that a given

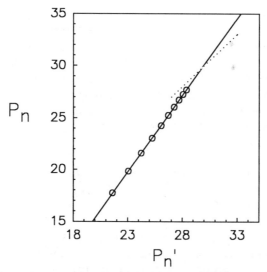

FIGURE 2-3. Swinbourne plot for finding P_x for a first-order reaction.

[P] may be p' for one data pair and p for a subsequent data pair). A plot of the first-order data given by Swinbourne is shown in Figure 2-3. While the straight-line fit may be made by eye, it is simpler to calculate the Linear Least Squares (LLS) fit (Appendix D) to the line. Then the first-order rate constant $k = \ln(\text{slope})/T$, and the value of $[P_\infty]$ is equal to intercept/(1 − slope). The Swinbourne plot is rather insensitive to slight deviations from strict first-order kinetics, so conformity of the reaction to the first-order rate law must be checked by other means if there is any question.

Computer-Assisted Curve Fitting.   A third method of obtaining $[P_\infty]$ and k' is to use a personal computer and a program to arrive at a value of $[P_\infty]$ which gives the best straight line on the semi-logarithmic plot. Given a set of $[P], t$ data, an estimate for the rate constant is made, then by Gauss-Newton iteration, corrected values for k', P_∞ and P_0 are found. Standard error estimates for each parameter are also obtained. The program EST1ST included in Appendix E is written in QuickBasic to do this. When applied to Swinbourne's data, the program gave an estimate of 29.77 ± 0.06 for P_∞, 15.10 ± 0.04 for P_0, and $1.629 \pm 0.081 \times 10^{-3}$ per second for k'. These compare well with the values of 30.0, 15.1, and 1.63×10^{-3} found by Swinbourne.

Two Substrate Systems.   In the one enzyme-one substrate system the reaction will inevitably become first-order if allowed to proceed to completion, even if $[S_0]$ is large enough so that initial conditions are in the intermediate kinetic part of the M-M curve. While this is also true for two substrate systems, somewhat more care must be taken in selecting the initial concentrations for A and for B so that the reaction is first-order with respect to the substrate of choice but the change in the second substrate does not significantly perturb the calculations for finding k'.

To analyze the random-addition situation note that equation 2-7, giving the dependence of v on $[A]$ with constant $[B]$, may be rearranged to isolate $[A]$ in the denominator:

$$v = \frac{[A]V_{\max}[B]/(1 + \alpha K_B)}{\alpha k_A[B] + K_B/([B] + \alpha K_B) + [A]} \tag{2-7'}$$

The key point in establishing first-order reaction conditions is that the substrate concentration $[A]$ is much less than the K_M term, i.e. $[A] <<$ $\alpha K_A\{([B] + K_B)/([B] + \alpha K_B)\}$. The question is how changes in $[B]$ affect this inequality. A second condition in the two-substrate system is that $[B] >> [A]$, so that upon complete reaction of $[A]$ only a negligible fraction of $[B]$ is changed.

To assess the $[A]$, K_M inequality, let us evaluate the K_M term for two values of α and two values of $[B]$. When $\alpha = 0.1$ and $[B] = K_B$, the K_M term equals $0.18K_A$. With $\alpha = 10$ and $[B] = K_B$, the term equals $1.8K_A$. With $\alpha = 0.1$ and $[B] = 10K_B$, $K_M = K_A$, and at $\alpha = 10$, $[B] = 10K_B$, then $K_M = 10K_A$. Within this region of reasonable values for α and $[B]$, if $[A] << 0.2K_A$ there should be no difficulty in obtaining first-order kinetic conditions. Since the random-addition case is symmetrical, $[A]$ and $[B]$ may be interchanged in the above discussion.

The ordered-addition situation is somewhat different. Rearranging equation 2-9a to get $[A]$ alone in the denominator gives:

$$v = \frac{[A]V_{\max}[B]/([B] + K_B)}{K_A K_B/([B] + K_B) + [A]} \tag{2-9a'}$$

The denominator inequality for first-order kinetics is: $[A] << K_A K_B/([B] + K_B)$. If $[B] = K_B$ then $[A] << K_A/2$, and if $[B] = 10K_B$ then $[A] << K_A/11$. In other words, it is not desirable to increase $[B]$ greatly, since this decreases the range of $[A]$ which can be used for determining k'. A value of $[B]$ between K_B and $3K_B$ seems most suitable.

If $[A]$ is the constant substrate, then equation 2-9b gives:

$$v = \frac{[B]V_{\max}}{K_B(1 + K_A/[A]) + [B]} \tag{2-9b'}$$

The required denominator inequality is: $[B] << K_B(1 + K_A/[A])$. If $[A] = K_A$ then $[B] << 2K_B$ is needed, and if $[A] = 10K_A$ the inequality is $[B] << 1.1K_B$. This situation is much less sensitive to the concentration of the constant substrate than is the reverse; almost any reasonable concentration of $[A]$ should yield a good value of k'.

The first-order rate constants for random addition (k'_r), ordered addition in $[A]$ with $[B]$ constant (k'_a) and ordered addition in $[B]$ with $[A]$ constant (k'_b) are:

$$k'_r = \frac{V_{max}}{\alpha K_A([B] + K_B)/([B] + \alpha K_B)} \qquad (2\text{-}14a)$$

$$k'_a = \frac{V_{max}}{K_A K_B/[B]} \qquad (2\text{-}14b)$$

$$k'_b = \frac{V_{max}}{K_B(1 + K_A/[A])} \qquad (2\text{-}14c)$$

Initial Velocity Data

Most enzyme rate experiments are performed on the assumption that the observed rate $v = (dP/dt)$ is correlated with $[S_0]$, the concentration of substrate present in the assay mixture at zero time. The correlation factor is the appropriate rate equation, and while some period of time may be required to get a value which is interpreted as the rate, that value is still usually considered to be the rate at zero time, the initial velocity v_0. It is recognized that this assumption is not always correct, and much effort is expended to overcome these facts of life (see the initial rate measurements section in Chapter 6). In this discussion it is assumed that the researcher has made the appropriate corrections, and that the $v,[S]$ data pairs in hand do indeed represent the instantaneous values observed at time zero. The methods for manipulating the data pairs to arrive at V_{max} and the equilibrium constants in the rate equations are the topic of this section.

Reciprocal Plots. The hyperbolic M-M equation does not allow accurate estimates of V_{max} and K_M from a direct plot of v versus $[S]$. However there are three transformations of the M-M equation which give linear forms. These forms are:

$$\frac{1}{v} = \frac{1}{V_{max}} + \left(\frac{K_M}{V_{max}}\right)\frac{1}{[S]} \qquad \text{(Lineweaver-Burk)} \qquad (2\text{-}15a)$$

$$\frac{[S]}{v} = \frac{K_M}{V_{max}} + \left(\frac{1}{V_{max}}\right)[S] \qquad \text{(Hanes)} \qquad (2\text{-}15b)$$

$$\frac{v}{[S]} = \frac{V_{max}}{K_M} - \left(\frac{1}{K_M}\right)v \qquad \text{(Eadie)} \qquad (2\text{-}15c)$$

The algebraic manipulations to obtain these forms are straightforward. Simple inversion of the M-M equation gives 2-15a. Multiplying both sides of 2-15a by [S] gives 2-15b. For 2-15c both sides of the M-M equation are multiplied by $(K_M + [S])/K_M[S]$, then the terms are separated and reordered. While quite a few different names have been associated with these transformations over the years, the ones given above seem to be most common in recent literature and will be used in this book.

Which of the above transformations is the best one to use? The Lineweaver-Burk plot (often referred to as the double reciprocal plot) is the most commonly applied, yet it is by far the worst choice. Experimental measurements of v are usually made at approximately equal increments in [S]. When the results are plotted with $1/[S]$ on the x-axis the data points tend to crowd together near the y-axis with a few points widely spaced towards the right side. This can be overcome by choosing [S] so that the reciprocals are more evenly spaced, i.e., 1, 1.25, 1.67, 2.5, 5, etc. This approach magnifies the second difficulty which is that rate measurements are less accurate at the lower values of [S]. There is usually some inherent methodological error in enzyme assays which might amount to as much as 25% of the rate at the lowest [S], but is a more acceptable 2% to 5% at higher [S]. The points towards the right side of the Lineweaver-Burk plot have a large vertical error component. In making a LLS (see Appendix D) fit to find the best straight line for the plot, these less accurate points have a much larger influence than do those points nearer the origin which are more accurate.

The Hanes and the Eadie transformations are less prone to these difficulties. The x-axis of the Hanes plot spaces the points according to the experimental spacing of [S]. If the error in determining v is inversely related to [S] then this plot is most reliable; the points containing higher experimental error are closer to the origin and have a reduced influence on the LLS best-fit line. The Eadie plot is more reliable when inherent experimental difficulties result in rather large relative errors in v (i.e. ±25% at any v) regardless of [S]. Since v is used directly on both axes the magnification of relative error introduced by taking the reciprocal does not appear.

Figure 2-4 shows the M-M plot and the three transformation plots of a set of theoretical data. The theoretical rate v at each [S] has been altered by an amount equal to a randomly generated error which is proportional

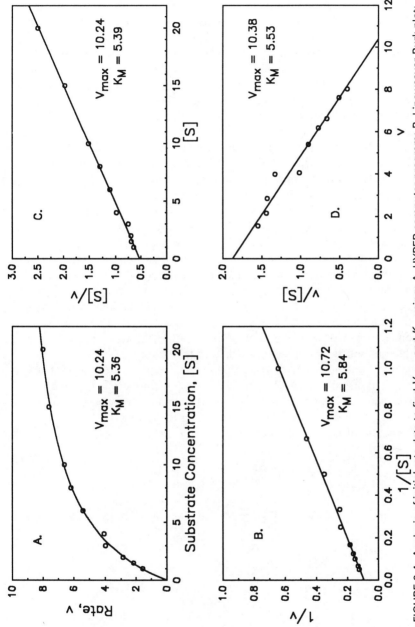

FIGURE 2-4. Analysis of initial rate data to find V_{max} and K_M using: A. HYPER computer program; B. Lineweaver-Burk plot; C. Hanes plot; D. Eadie plot.

to $1/[S]$ and represents a relative standard deviation of 20% in the actual rate. The spread of points around the theoretical line in Figure 2-4A is representative of what would be considered "good" experimental data from a laboratory run. Values of V_{max} and K_M from the Hanes plot are closest to the theoretical numbers from which the original rate data were derived; similar results are found when actual experimental data are used in such a comparative analysis. Atkins and Nimmo (1980), in a theoretical analysis of error propagation in the three methods, concluded that the Hanes transformation was preferable to the Eadie plot in most situations, and was superior to the Lineweaver-Burk plot in all situations. All the linear transformation plots in this book are based upon the Hanes plot; it is hoped that the use of the Lineweaver-Burk plot by enzyme researchers will soon cease.

The analysis of an actual set of $v, [S]$ data is rather straightforward. A table of $[S]/v$ versus $[S]$ is made and the best straight line fit is found using a LLS fit program (Appendix D). Then V_{max} equals 1/slope and K_M is intercept/slope. While all this can be done strictly mathematically, it is advisable to also make the actual plot; if one data point is found to be quite far off the line, an undetected experimental error should be suspected and the data point discarded (or at least v should be remeasured at that $[S]$). More importantly, the plot may curve, indicating the presence of complicating factors: substrate inhibition, two enzymes catalyzing the same reaction, inhibitor or activator contaminants in the substrate, etc. Some of these factors will be discussed elsewhere in this book.

The above methods can also be applied to the two-substrate systems, although it is usually necessary to do a replot. Considering first the random addition situation (equation 2-6), experimentally v is measured as a function of $[A]$ at one value for $[B]$, according to equation 2-7. The Hanes transformation of this equation gives:

$$\frac{[A]}{v} = \left(\frac{1}{V_{max}}\right)\left\{\alpha K_A\left(1 + \frac{K_B}{[B]}\right)\right\} + \left(\frac{1}{V_{max}}\right)\left(1 + \frac{\alpha K_B}{[B]}\right)[A] \qquad (2\text{-}16a)$$

On the Hanes plot shown in Figure 2-5 the slope and intercept of the line is given by:

$$\text{Slope} = \frac{1 + \alpha K_B/[B]}{V_{max}} = \frac{1}{V_{max}} + \left(\frac{\alpha K_B}{V_{max}}\right)\left(\frac{1}{[B]}\right) \qquad (2\text{-}16b)$$

$$\text{Intercept} = \alpha K_A\left(\frac{1 + K_B/[B]}{V_{max}}\right) = \alpha K_A\left(\frac{1}{V_{max}}\right) + \left(\frac{\alpha K_A K_B}{V_{max}}\right)\left(\frac{1}{[B]}\right) \qquad (2\text{-}16c)$$

This process is then repeated at several other concentrations $[B]$, resulting in a set of slopes and intercepts, S and I which are functions of $[B]$. These are then replotted as shown in Figures 2-5B and 2-5C. The slope and intercept of the replotted slopes (Figure 2-5B) are designated S_s and I_s respectively. The slope and intercept of the replotted intercepts (Figure 2-5C) are designated S_i and I_i. The significance of these four values are:

$$S_s = \frac{\alpha K_B}{V_{max}} \qquad I_s = \frac{1}{V_{max}}$$

$$S_i = \frac{\alpha K_A K_B}{V_{max}} \qquad I_i = \frac{\alpha K_A}{V_{max}}$$

The four parameters are then easily calculated: $V_{max} = 1/I_s$, $K_A = S_i/S_s$, $K_B + S_i/I_i$, and $\alpha = S_s V_{max}/K_B$.

It is immaterial which of the two substrates used is A and which is B; the plots will have the same form in either sequence of experimental runs. The quality of the numbers obtained from the replots will depend upon the quality of the original data as well as the reliability of the method used for determining the original slopes and intercepts. The reliability of these values is enhanced by using the LLS method for both the original set of plots and for the replots. A test of the internal consistency of the calculations is indicated in Figure 2-5A. The lines of the $[A]/v$ versus $[A]$ plots will intersect in the third quadrant at a point equal to $-K_A$ (on the x-

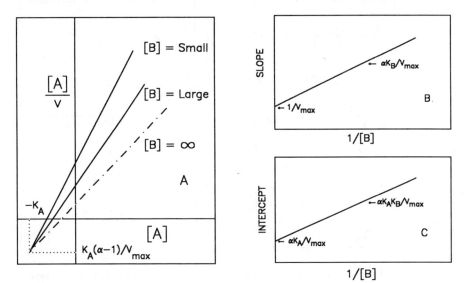

FIGURE 2-5. Random-addition two-substrate system. A. Hanes plot of experimental rate data. B. Replot of slopes from A. C. Replot of intercepts from A.

axis) and $K_A(\alpha - 1)/V_{max}$ (on the y-axis). A similar plot of $[B]/v$ versus $[B]$ at constant $[A]$ will intersect at $-K_B$, $K_B(\alpha - 1)/V_{max}$. The intersections are not as accurate as calculations but do provide confirmatory evidence for the calculated values of the parameters.

The rate equation for the ordered addition of two substrates is given by equation 2-8. At constant $[B]$, equation 2-9a is obtained. The Hanes transformation results in:

$$\frac{[A]}{v} = \left(\frac{K_A K_B}{V_{max}[B]} \right) + \left(\frac{[B] + K_B}{V_{max}[B]} \right)[A] \qquad (2\text{-}17a)$$

As before, the slope S and intercept I are obtained at several values of $[B]$. $S = 1/V_{max} + (K_B/V_{max})(1/[B])$ and $I = (K_A K_B/V_{max})(1/[B])$. The replot of S versus $1/[B]$ gives both a slope S_s and an intercept I_s, but the replot of I versus $1/[B]$ passes through the origin. The three replot values are:

$$S_s = \frac{K_B}{V_{max}} \qquad I_s = \frac{1}{V_{max}} \qquad S_i = \frac{K_A K_B}{V_{max}}$$

It is readily apparent how the three parameters are calculated from these three values. The Hanes plot and the replots are depicted in Figure 2-6.

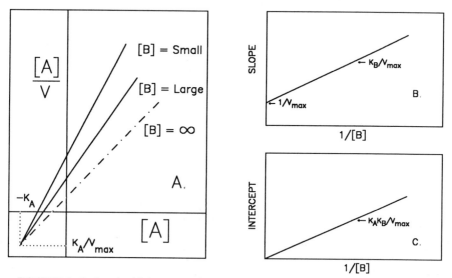

FIGURE 2-6. Ordered-addition two-substrate system, B is the second substrate to bind to the enzyme. A. Hanes plot of experimental rate data. B. Replot of slopes from A. C. Replot of intercepts from A.

If by chance the experimenter should choose to hold A constant because of not knowing that this is an ordered-addition system, nor which is the first substrate to add, the appropriate rate equation is 2-9b. The Hanes transformation results in:

$$\frac{[B]}{v} = K_B\left(1 + \frac{K_A}{[A]}\right)\left(\frac{1}{V_{max}}\right) + \left(\frac{1}{V_{max}}\right)[B]$$

The Hanes plots at a series of $[A]$ will give parallel lines all having the slope $1/V_{max}$. This is the clue that the constant substrate was the first one in an ordered addition situation. The intercept I will vary as $[A]$ varies:

$$I = \frac{K_B}{V_{max}} + \left(\frac{K_B K_A}{V_{max}}\right)\left(\frac{1}{[A]}\right)$$

The replot of I versus $1/[A]$ has the intercept K_B/V_{max} and slope $K_B K_A/V_{max}$ which, along with V_{max} from the Hanes plot allows the calculation of all three parameters. The Hanes plot and intercept replot are shown in Figure 2-7.

These two models will cover the two-substrate systems most likely to occur in food enzyme systems. The more complex multi-substrate systems are best treated using the steady-state approaches of Cleland (1970) or Fromm (1975). A full discussion of the graphical methods for analyzing

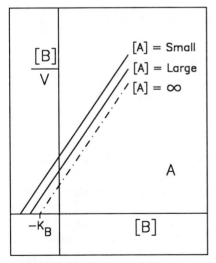
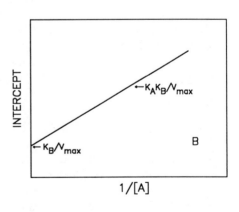

FIGURE 2-7. Ordered-addition two-substrate system, A is the first substrate to bind to the enzyme. A. Hanes plot of experimental rate data. B. Replot of intercepts from A.

these systems is also given by Segel (1975). Most practical systems can be adequately characterized using the approaches explained here.

Direct Linear Plots.   The Hanes plot transforms the M-M equation into a linear equation $y = a + bx$. The statistical method of LLS is a way of finding the values of a and b which define the straight line through a set of data points such that the deviation of the actual y values from the line is minimized. While LLS deals with the effect of normally distributed random errors in rate data, it is less successful in handling the occasional non-random error which gives rise to an "outlier," the point unexpectedly far-off the line of the linear plot. The direct linear plot method described by Cornish-Bowden and Eisenthal (1974; 1978) circumvents this problem by using a different approach to minimizing deviations.

Another way of writing an equation describing a straight line in Cartesian coordinates is:

$$\frac{y}{b} + \frac{x}{a} = 1 \qquad (2\text{-}18\text{a})$$

Point a (on the x-axis) and point b (on the y-axis) will define a straight line. A second pair of points, a' and b', will define a second straight line. The point of intersection of the two lines defines unique values of X and Y such that the (rearranged) equation 2-18a is satisfied for both sets of data points a,b and a',b':

$$Y = b - \left(\frac{b}{a}\right)X = b' - \left(\frac{b'}{a'}\right)X \qquad (2\text{-}18\text{b})$$

The M-M equation may be rearranged to put it into a form corresponding to equation 2-18a:

$$\frac{V_{max}}{v} - \frac{K_M}{[S]} = 1 \qquad (2\text{-}18\text{c})$$

Crossmultiply to give $vK_M + v[S] = V_{max}[S]$, subtract vK_M from both sides to give $V_{max}[S] - vK_M = v[S]$, then divide both sides by $v[S]$. Given experimental data pairs $v,[S]$, each v (corresponding to b) is plotted on the y-axis, each $[S]$ (corresponding to a) is plotted on the x-axis and a straight line is drawn through the two points. The intersection of the straight lines from two $v,[S]$ data pairs gives unique values for V_{max} and K_M as shown in Figure 2-8A. Additional data pairs define more straight lines; if there were no experimental error in either v or $[S]$ all the lines would pass through the same intersection point.

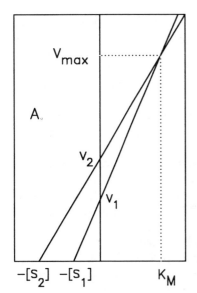
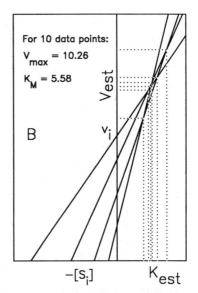

FIGURE 2-8. Direct Linear Plot for analyzing experimental rate data. A. Theoretical plot. B. Plot of data containing experimental error.

In real data the presence of error means that the intersection point of any given line pair is different from that of any other line pair. If initial velocity data are obtained at N values of $[S]$ then plotting each of these $v,[S]$ pairs on the direct linear plot will result in $N(N-1)/2$ intersections. Each intersection is an estimate of V_{max} and K_M, so a set of $N(N-1)/2$ estimates is obtained. The most likely value for V_{max} and K_M is taken as the *median* of the set of estimates. This procedure is statistically sound (Eisenthal and Cornish-Bowden 1974) and has the advantage of minimizing the effect of outliers on the final parameter values. Figure 2-8B is a direct linear plot of four of the $v,[S]$ pairs from the data used in Figure 2-4. The effect of experimental error is obvious; there are six intersection points ($N\{N-1\}/2$ with $N=4$).

It is possible to read off the values of V_{est} and K_{est} for each intersection but this introduces additional error due to plotting inaccuracies. It is easier to calculate each estimate:

$$V_{est,ij} = \frac{s_i - s_j}{s_i/v_i - s_j/v_j}$$

and

$$K_{est,ij} = \frac{v_j - v_i}{v_i/s_i - v_j/s_j}$$

After all the values are calculated they are ranked numerically. The best estimate for V_{max} and K_M is the median value of the list. When there are an even number of items the average of the estimates immediately above and below the center point of the list is used. Most such lists based upon real experimental data show outliers, but their presence does not materially influence the median values. On occasion a particularly poor $v,[S]$ pair will give rise to negative values of estimates; the method of using the median value as the best estimate of V_{max} and K_M is less biased by such data than are the LLS line fitting methods.

Because the number of intersections increases geometrically with increasing N it is impractical to process more than five or six data pairs by hand. However it is easy to write a short program for a personal computer to do this job quite nicely. With an efficient SORT subroutine for ranking the estimates ten to fifteen data pairs can be analyzed within a few minutes. Such a program might also include a section to process the $v,[S]$ data according to the Hanes and the Eadie linear transformations. Comparison of the values of V_{max} and K_M obtained by the three methods is instructive and increases confidence in the reliability of the results.

Computer Analysis.   Until the advent of computers, analyzing rate data by linear transformation and line-fitting was the only practical way to obtain the kinetic parameters. The availability of "number-crunchers" enabled the application of more sophisticated methods which depended upon several iterations of an appropriate algorithm to arrive at values of rates and binding constants which best fit the experimental data in hand. Prof. W. W. Cleland developed FORTRAN programs which addressed almost every conceivable model for enzyme action. The analysis of data using these programs not only gives the "best" estimate of the various kinetic parameters, but also provides estimates of the statistical reliability of the parameters. The results can be shocking to the experimentalist who has labored long and carefully to obtain a good set of $v,[S]$ data. The data shown in Figure 2-4 would be considered good quality, yet the analysis by the HYPER program (Appendix E) shows a standard error of 20% in K_M, but a much smaller standard error of 2.6% in V_{max}. On occasion "reasonable" laboratory data have shown a standard error greater than 100% in K_M; obtaining such a printout motivates one to take more care in making the experimental (re)runs. As a rule of thumb, standard errors of 5% in V_{max} and 25% in K_M are about average; less than that should be sought.

The mathematical rationale behind the operation of these programs is well presented by Cleland (1967, 1979). In the rate equation for the model of interest v equals V_{max} times a nonlinear function of the other kinetic parameters in the system. Starting with an estimated value for these other parameters, a set of simultaneous equations is set up based upon the partial derivative of the rate equation with respect to each of these parameters. This system of equations is solved using matrix methods to arrive at a correction factor for each of the parameters. This correction is made and the process is repeated, either for a set number of iterations or until the correction factor becomes zero. At this point the sum of the square of the quantity (experimental rate minus calculated rate) at each value of [S] is minimized. From this total deviation and the inverse matrix of the last iteration the variance of each of the parameters is calculated. For a fuller exposition of this procedure see Cleland (1967).

While this would seem to be the answer for all initial rate data analysis, that is not so. In some cases the program "blows up," that is, the parameter estimates do not converge after a few iterations. In these cases it is necessary to analyze the experimental data using a response surface of residuals versus possible parameter ranges. Cleland (1979) discusses ways to do this using short computer programs written to address the situation in hand. In such situations the analysis often gives valuable insights into the nature of the enzyme reaction being studied, and the characteristics of the rate model being applied.

Progress Curve Data

No Complications. The Henri (integrated rate) equation (2-11) may be used to obtain V_{max} and K_M from the progress of a single reaction run with enzyme and substrate (Orsi and Tipton 1979). At intervals in time t an aliquot of the reaction mixture is taken and the concentration of product $[P]$ is measured. From these data and the original substrate concentration $[S_0]$ a table of values is constructed according to a slightly rearranged equation 2-11:

$$\frac{[P]}{t} = V_{max} - K_M\left(\frac{1}{t}\right)\ln\left(\frac{[S_0]}{[S_0] - [P]}\right)$$

The values of $[P]/t$ are plotted versus $(1/t)\ln\{[S_0]/([S_0] - [P])\}$ as shown in Figure 2-9. The slope of the straight line fitted to these points is $-K_M$ and the intercept on the y-axis is V_{max}. If the course of the reaction is followed by measuring the decrease in substrate concentration, then since $[S] = [S_0] - [P]$ the plot is of $([S_0] - [S])/t$ versus $(1/t)\ln([S_0]/[S])$.

One useful characteristic of this plot is indicated in Figure 2-9. A line passing through the origin and with a slope numerically equal to $[S_0]$ will intersect the plot at the point corresponding to zero time. This intersection point is the initial velocity, i.e. $\lim\{[P]/t\}$ as $t \rightarrow 0$ and as $[S] \rightarrow [S_0]$. Since the ordinate has the dimension (mole) (L^{-1}) (time^{-1}) and the abscissa has the dimension time^{-1} it is seen that the dimension of the slope is (mole) (L^{-1}) as expected. This property of the Henri plot is most useful in analyzing

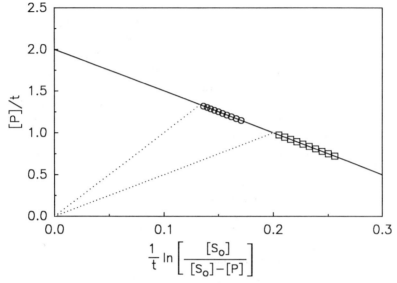

FIGURE 2-9. Henri plot of (theoretical) rate data. $V_{max} = 2.0$, $K_M = 5$. The dotted lines represent two values of $[S_0]$, 10 and 5 (left and right, respectively).

data in those situations where complications arise, either due to denaturation of the enzyme during the course of the reaction (see below) or if product inhibition occurs (see Chapter 3).

Denaturing Enzyme. Often the enzyme is inactivated during the course of an enzymatic reaction due to some extraneous factor such as pH, temperature, or the presence of a destabilizing solute such as an alcohol. These are often the very conditions of practical interest, and measurement of the first-order rate constant for denaturation under these conditions is required. The constant for denaturation in the absence of substrate is relatively easy to find, and the procedure is discussed in Chapter 5. However the presence of substrate usually has a stabilizing effect upon the enzyme, and it is the rate constant in the reacting system which is needed. The Henri equation can be used to find this value.

The reaction is carried out at several values of $[S_0]$. Plots are made of $[P]/t$ versus $(1/t) \ln\{[S_0]/([S_0] - [P])\}$. The intersection of each plot with the respective $[S_0]$ is found; the straight line through these intersection points is the same as the line in the simple Henri plot with a slope of $-K_M$ and a y-axis intercept of V_{max}.

A vertical line is drawn between the experimental line and the undenatured extrapolation (Figure 2-10A). The point on the extrapolation is $[P']/t'$, while

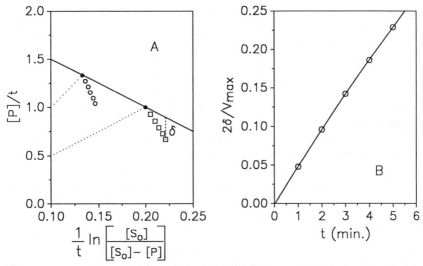

FIGURE 2-10. Analysis of data when the enzyme is denatured during the course of the progress run. A. Henri plot for two runs, V_{max}, K_M, and $[S_0]$ as in Figure 2-9, first-order rate constant for denaturation of 0.05 assumed. B. Plot of $2\delta/V_{max}$ versus time.

the point on the experimental line is $[P]/t$. Note that the x-axis coordinate for both points is the same, so $[P']/t'$ can be calculated from the known values of V_{max} and K_M. From equation 2-11 (for $[P']/t'$) and equation 2-12b (for $[P]/t$) we have:

$$\frac{[P']}{t'} = V_{max} - K_M\left(\frac{1}{t}\right) \ln\left(\frac{[S_0]}{[S_0] - [P]}\right)$$

$$\frac{[P]}{t} = V_{max}\left(1 - \frac{kt}{2} + \frac{k^2t^2}{6} + \cdots\right) - K_M\left(\frac{1}{t}\right) \ln\left(\frac{[S_0]}{[S_0] - [P]}\right)$$

Subtracting the lower from the upper equation gives:

$$\delta = \frac{[P']}{t'} - \frac{[P]}{t} = V_{max}\left(\frac{kt}{2} - \frac{k^2t^2}{6} \cdots\right)$$

Upon slight rearrangement, addition of the next term to the expansion, and multiplication throughout by 2 we have:

$$\frac{2\delta}{V_{max}} = t\left(k - \frac{tk^2}{3} + \frac{t^2k^3}{12} \cdots\right) \qquad (2\text{-}19)$$

Equation 2-19 could be handled in either of two ways. A replot of several points from Figure 2-10A is shown in Figure 2-10B, where $2\delta/V_{max}$ is plotted versus t. The intercept on the x-axis equals 0 and the slope (the tangent to the curve) at the origin equals k. Alternatively, $2\delta/V_{max}$ is a polynomial in t, i.e., $2\delta/V_{max} = A + Bt + Ct^2 + Dt^3 + \cdots$ with $A = 0, B = k$, $C = -k^2/3$, $D = k^3/12$, etc. The set of data pairs for $2\delta/V_{max}, t$ could be fitted to a second or third order equation to get the constants from which k is calculated.

The only actual graphical determinations required are those to find the zero-time intersection points with the original data plots; V_{max} and K_M are obtained from these points. The determination of δ and the subsequent calculations can all be done based upon the table of $[P]/t,(1/t)\ln\{[S_0]/([S_0]-[P])\}$ data pairs constructed for the initial plots. At each value of $[P]/t$ the corresponding $[P']/t'$ and δ are calculated, not obtained graphically as might be inferred from the previous discussion. This whole process could be easily carried out by a rather simple program for a personal computer.

Each of the initial experimental runs yields one such set of δ data. The calculation of k from each run having a different $[S_0]$ gives an indication of the degree to which substrate concentration influences the stability of the enzyme. In theory, if k is a function of $[S_0]$, a plot of k versus $[S_0]$ should, upon extrapolation, intersect the y-axis at a point equal to the denaturation rate measured independently in the absence of substrate. Even in the absence of such a theoretical extrapolation, having an idea of the form of the function would be of great practical use in applying enzymes to processes under denaturing conditions.

3 Inhibitors

Almost all enzymes will combine with some chemical species which reduce their catalytic efficiency. These materials are termed inhibitors. They may be classified into three groups which have different effects upon the kinetics of the enzyme reacting with its usual substrates. These groups are characterized by their affinity for the enzyme molecule, and may be illustrated by referring to inhibitors of the protease trypsin.

1. Low-affinity inhibitors. These are molecules which have an *association* equilibrium constant in the range of 1 to 10^6 M. They often mimic substrates, with the usual reactive bond either being unreactive or in the wrong place because of stereochemistry. An example for trypsin is N-benzoyl-D-arginine-p-nitroanilide, the enantiomer of the commonly-used trypsin substrate often called BAPA. This inhibitor forms a complex with the enzyme with an affinity of 1.25×10^3 M (Erlanger et al. 1961).

2. High-affinity inhibitors. These molecules have an affinity constant in the range of 10^6 to 10^{12} M. They may be either "transition state analogs" which mimic the presumed transition state in the passage from enzyme-substrate complex to enzyme-product complex, or they may be natural molecules which for other reasons bind tightly to the enzyme active site. An example is the Bowman-Birk trypsin inhibitor from soybean, which has an affinity of 3.6×10^9 M (Seidl and Liener 1972).

3. Irreversible inhibitors. These molecules form a covalent bond with a residue in the enzyme active site. This bond is so stable that it is not meaningful to speak of an equilibrium constant. Rather the reaction of the inhibitor with the enzyme is usually pseudo-first order in enzyme (the inhibitor being present in large excess) and proceeds only to inactive enzyme. An example is the reaction of N-tosyl-L-lysine-chloromethylketone (Shaw et al. 1965) with trypsin. The molecule first binds to the active site, followed by formation of a covalent bond with the imidazole ring of the active site histidine.

The differences between these three groups may be related to the "off" rate constant k_{-1} for leaving the enzyme active site. The "on" rate constant k_1 should be about the same for all inhibitors (within an order of magnitude) because it is basically diffusion-controlled, and should be around 10^9 M^{-1} sec^{-1}. Since the affinity constant equals k_1/k_{-1} the large differences noted above must be due to variations in k_{-1} of some 12 or more orders of magnitude. The off rate for irreversible inhibitors may have dimensions of $days^{-1}$ to $months^{-1}$ and will not be discussed any further since their application is beyond the scope of this book. The other two inhibitor groups are quite important in food systems, and the kinetics and methods of data analysis for low-affinity and high-affinity inhibitors are presented.

LOW-AFFINITY INHIBITORS

Initial Velocity Rate Equations

One Substrate Systems. Classical inhibitor kinetic analysis assumes reversible binding of inhibitor to enzyme with the production of a complex with reduced or zero catalytic activity. The concentration of these low-affinity inhibitors is typically in the millimolar range so that the fraction of inhibitor complexed with enzyme is insignificant. The general equilibrium model for reversible inhibition is depicted in Figure 3-1. E, S, I, ES, EI, and EIS represent concentrations of free enzyme, substrate, inhibitor, enzyme-substrate complex, enzyme-inhibitor complex, and ternary complex, respectively (the total enzyme concentration is E_t). The four K values are rapid equilibrium *dissociation* constants; only three constants K_M, K_i, and α are needed, because the closure of the equilibria requires symmetry between the opposite sides. The values k_p and βk_p are rate constants for formation of product P from the ES and EIS complexes, respectively. In the following analysis V_{app} means the maximum rate at large $[S]$, and K_{app} means that $[S]$ which gives half-maximum rate.

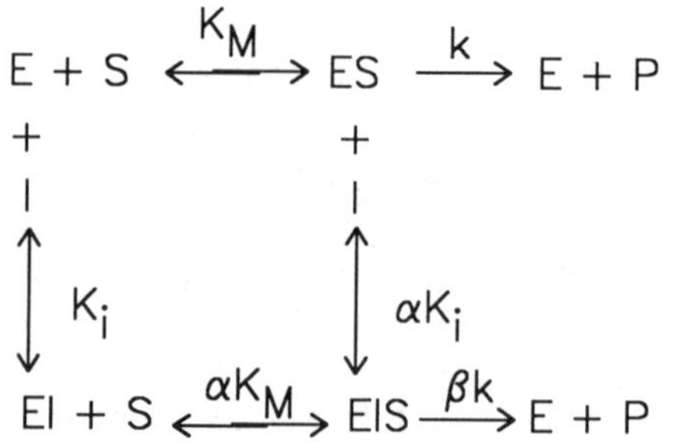

FIGURE 3-1. General equilibrium model for reversible enzyme inhibition.

The equilibrium rate expressions for the various possible inhibitory patterns are easy to obtain from the model of Figure 3-1 (Appendix A). The general rate equation is developed in Appendix F. Each of the rate equations for the specific inhibitory pattern is derived from this general equation by substituting the appropriate values of α and β. The key point is that all of the rate equations for the one enzyme-one substrate system have the same form, related to the M-M equation:

$$v = \frac{V_{app}[S]}{K_{app} + [S]} = \frac{V_{max}\{f_v\}[S]}{K_M\{f_k\} + [S]} \tag{3-1}$$

The inhibitor rate equations differ from the M-M equation in the makeup of the two factors, the velocity factor f_v and the equilibirium constant factor f_k. These factors are a function of the inhibitor concentration $[I]$ and one or more of the equilibrium constants shown in Figure 3-1.

The main types of inhibition can be categorized in terms of the values of α and β. If α equals ∞ there is no binding of I to ES, or S to EI (the dissociation constants are infinitely large). If α equals 1, binding one reactant to the enzyme molecule has no influence upon the affinity for the other reactant. If $1 < \alpha < \infty$ the binding of the second reactant to the complex is less than the affinity of that reactant molecule for the enzyme alone. The rate factor β also has three possible values: if $\beta = 0$, EIS does not break down to give product; if $\beta = 1$, EIS breaks down to yield product at the same rate as ES; and if $0 < \beta < 1$ EIS breaks down to product, but slower than does ES.

Competitive inhibition occurs when $\alpha = \infty$ and $\beta = 0$. The enzyme binds either S or I but not both; I competes with S for the enzyme. The kinetic effect is to increase K_{app}, i.e. it is only on f_k. Sufficiently high concentrations of $[S]$ will overcome the competition, and V_{app} equals V_{max}. The rate equation is:

$$v = \frac{V_{max}[S]}{K_M(1 + [I]/K_i) + [S]} \tag{3-2a}$$

Noncompetitive inhibition is observed when $\alpha = 1$ and $\beta = 0$. All four forms of enzyme are present, but only ES breaks down to product. The kinetic effect is to decrease V_{app} because $f_v < 1$. Sufficiently high concentrations of $[I]$ will reduce V_{app} to zero. The rate equation is:

$$v = \frac{V_{max}\{1/(1 + [I]/K_i)\}[S]}{K_M + [S]} \tag{3-2b}$$

Partially noncompetitive inhibition is the term used to describe the situation when $\alpha = 1$ and $0 < \beta < 1$. As with the previous example all four forms of enzyme are present, but both *ES* and *EIS* break down to form product, albeit *ES* breaks down at a faster rate. Again the kinetic effect is only on f_v, and at high $[I]$, V_{app} will approach βV_{max}. The rate equation is:

$$v = \frac{V_{max}\{(1 + \beta[I]/K_i)/(1 + [I]/K_i)\}[S]}{K_M + [S]} \tag{3-2c}$$

Partially competitive inhibition is the model in which $1 < \alpha < \infty$ and $\beta = 1$. As might be inferred from the "competitive" appellation, the effect is on K_{app}, and is expressed only in f_k. The binding of one reactant molecule to the enzyme decreases the affinity for the second reactant but does not reduce it to zero. The breakdown rate is the same for both *ES* and *EIS*. High values of $[S]$ will raise V_{app} to nearly V_{max}, while at high $[I]$, K_{app} will approach the value αK_M. The rate equation is:

$$v = \frac{V_{max}[S]}{K_M\{(1 + [I]/K_i)/(1 + [I]/\alpha K_i)\} + [S]} \tag{3-2d}$$

Mixed inhibition shows effects on both f_v and f_k, that is, V_{app} is less than V_{max} and K_{app} is greater than K_M. The constant α lies between 1 and ∞, and β is less than 1. When $\beta = 0$ the system is called *Linear Mixed*-type inhibition, and when $0 < \beta < 1$ the system is called *Hyperbolic Mixed*-type inhibition. The rate equation for the linear type is:

$$v = \frac{V_{max}\{1/(1 + [I]/\alpha K_i)\}[S]}{K_M\{(1 + [I]/K_i)/(1 + [I]/\alpha K_i)\} + [S]} \tag{3-2e}$$

For the hyperbolic type the rate equation is the general equation:

$$v = \frac{V_{max}\{(1 + \beta[I]/\alpha K_i)/(1 + [I]/\alpha K_i)\}[S]}{K_M\{(1 + [I]/K_i)/(1 + [I]/\alpha K_i)\} + [S]} \tag{3-2f}$$

These results are summarized in Table 3-1. In general an inhibitor is termed "competitive" if it has the effect of increasing K_{app} and "noncompetitive" if the effect is to decrease V_{app}. In the "mixed" cases, both effects are seen. Figure 3-2 shows the effect of $[I]$ upon the M-M graph and also the Hanes plot for the three types of inhibition. The intercept of the Hanes plot is changed by the factor f_k/f_v, while the slope is changed by the factor $1/f_v$, as seen from the Hanes transformation of the general inhibition equation:

Table 3–1.   Effects of Inhibitors on Rate Parameters

Inhibition Type	f_v	f_k	f_k/f_v
Competitive $\alpha = \infty, \beta = 0$	1	$1 + [I]/K_i$	$1 + [I]/K_i$
Noncompetitive $\alpha = 1, \beta = 0$	$\dfrac{1}{1 + [I]/K_i}$	1	$1 + [I]/K_i$
Partially noncompetitive $\alpha = 1, 0 < \beta < 1$	$\dfrac{1 + \beta[I]/K_i}{1 + [I]/K_i}$	1	$\dfrac{1 + [I]/K_i}{1 + \beta[I]/K_i}$
Partially competitive $1 < \alpha < \infty, \beta = 1$	1	$\dfrac{1 + [I]/K_i}{1 + [I]/\alpha K_i}$	$\dfrac{1 + [I]/K_i}{1 + [I]/\alpha K_i}$
Linear mixed $1 < \alpha < \infty, \beta = 0$	$\dfrac{1}{1 + [I]/\alpha K_i}$	$\dfrac{1 + [I]/K_i}{1 + [I]/\alpha K_i}$	$1 + [I]/K_i$
Hyperbolic mixed $1 < \alpha < \infty, 0 < \beta < 1$	$\dfrac{1 + \beta[I]/\alpha K_i}{1 + [I]/\alpha K_i}$	$\dfrac{1 + [I]/K_i}{1 + [I]/\alpha K_i}$	$\dfrac{1 + [I]/K_i}{1 + \beta[I]\alpha K_i}$
Hyperbolic mixed, rearranged	$\dfrac{aK_i + \beta[I]}{\alpha K_i + [I]}$	$\dfrac{aK_i + \alpha[I]}{\alpha K_i + [I]}$	$\dfrac{aK_i + \alpha[I]}{\alpha K_i + \beta[I]}$

$$\frac{[S]}{v} = \frac{K_M}{V_{max}} \cdot \frac{\{f_k\}}{\{f_v\}} + \frac{1}{V_{max}} \cdot \frac{1}{\{f_v\}} \cdot [S] \qquad (3\text{-}1b)$$

In the discussion on data analysis of inhibited systems below it will be convenient to refer to the quantities being manipulated graphically and/or mathematically in these terms.

Two Substrate Systems.   Inhibition in the two-substrate system can be extremely complicated. Factors such as whether or not the binding of one species (A, B, or I) affects the subsequent binding of a second species, as well as whether or not the quaternary complex $EABI$ breaks down to form product, all lead to a bewildering array of inhibition patterns. For a fuller discussion of many of the possibilities the reader is referred to Segel (1975) or Cleland (1970). The present discussion will be limited in scope but will hopefully address most of the garden-variety two-substrate systems found in connection with food enzymology. The limitations assumed are two:

1. While inhibitor may bind to any of the enzyme species (E, EA, EB, EAB) present in the system indicated by equations 2-5,a-c, this does not affect the equilibrium constants for binding A or B;
2. The complex $EABI$ does not break down to give free enzyme and product.

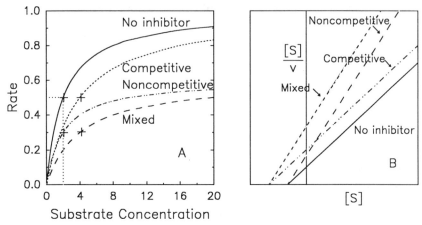

FIGURE 3-2. Form of plots of rate data when plotted according to the; A. Michaelis-Menten (hyperbolic) and; B. Hanes (linear) rate equation.

The kinetic model for the random-addition case is:

$$E(EI) + A \underset{}{\overset{K_A}{\rightleftharpoons}} EA(EAI) + B \underset{}{\overset{\alpha K_B}{\rightleftharpoons}} EAB(EABI)$$

$$E(EI) + B \underset{}{\overset{K_B}{\rightleftharpoons}} EB(EBI) + A \underset{}{\overset{\alpha K_A}{\rightleftharpoons}} EAB(EABI)$$

$$E(EA, EB, EAB) + I \underset{}{\overset{K_i}{\rightleftharpoons}} EI(EAI, EBI, EABI)$$

$$EAB \overset{k_p}{\longrightarrow} E + P$$

These are fully analogous to equations 2-5,a-c, with the addition of an equation expressing the rapid equilibrium binding of I with a dissociation constant K_i. Upon developing the rapid equilibrium rate equation for this model, it is found that the inhibition is noncompetitive, i.e., the presence of inhibitor lowers V_{app} by the factor $1/(1 + [I]/K_i)$, but does not change K_{app}. The rate equation is easily written by comparison with equation 2-6:

$$v = \frac{V_{max}\{1/1 + [I]/K_i)\}[A][B]}{\alpha K_A K_B + \alpha K_B[A] + \alpha K_A[B] + [A][B]} \qquad (3\text{-}3a)$$

A similar result is found when considering the ordered-addition case. The second equation above, analogous to equation 2-5b, disappears, and the species *EB* and *EBI* do not exist, nor does the factor α. Again the inhibition is found to be noncompetitive, and the rate equation by comparison with equation 2-8 is:

$$v = \frac{V_{\max}\{1/(1 + [I]/K_i)\}[A][B]}{K_A K_B + K_B[A] + [A][B]} \tag{3-3b}$$

The methods for analyzing data to find the inhibitor equilibrium constant K_i are fairly straightforward, and will be presented in more detail below.

Product and Substrate Inhibition.   The inhibitors considered thus far are molecular species extraneous to the reaction being catalyzed, added either purposefully or accidentally. It is often observed that the presence of the product of the reaction will inhibit the rate of reaction. Also, it is not uncommon to find that the rate of the reaction under study increases as [S] is increased to some maximum value and then decreases as [S] is made even larger. These two phenomena, product and substrate inhibition, offer some insights into the nature of the enzymatic catalysis, but can also be misleading if not recognized and properly analyzed.

Product inhibition is best understood in terms of reversible binding of product to the active site of the enzyme.

$$E + S \underset{}{\overset{K_M}{\rightleftharpoons}} ES \overset{k_p}{\longrightarrow} EP \underset{}{\overset{K_p}{\rightleftharpoons}} E + P$$

As an example, chymotrypsin binds N-acetyl tyrosine ethyl ester (S), catalyzes the hydrolysis of the ester through an intermediate acyl-enzyme stage (see equation 3-3a,b), and finally releases N-acetyl tyrosine from the active site. The phenolic side chain and the amide bond of the ester substrate are involved in binding the substrate to the enzyme; it would be most curious if they didn't also interact with the active site even though the scissile bond is a carboxylic acid rather than an ester. In fact, there is a significant amount of inhibition of chymotrypsin by this product, and a low level of the acyl-enzyme has been detected when chymotrypsin is incubated with N-acetyl tyrosine.

This phenomenon has been used to study such factors as the thermodynamics of enzyme reactions, but the present discussion is focused upon its perception as inhibition by product of the reaction $S \longrightarrow P$ and how this may affect the analysis of experimental data. The product binding constant can be found in either of two ways.

1. If the product is available independently of the reaction mixture it is treated like any other reversible inhibitor. It is added to the reaction

mixture in an amount $[I]$, and then the initial rate $v(-dS/dt)$ is measured as a function of $[S]$. The data are analyzed according to the methods which will be discussed in the next section.

2. The progress of the reaction may be followed continuously and the data analyzed according to the integrated rate equation. As the reaction proceeds the rate of reaction decreases not only because $[S]$ is decreasing, but also because the increase in $[P]$ leads to increased inhibition. The development of the integrated rate equation for product inhibition is given here, and the method of analysis of data using this equation is shown in the data analysis section below.

If the product is considered to be a competitive inhibitor of the enzyme with an equilibrium dissociation constant K_p, the differential rate equation is:

$$v = \frac{dP}{dt} = \frac{V_{max}([S_0] - [P])}{K_M\{1 + [P]/K_p)\} + ([S_0] - [P])}$$

Rearranging to get the terms in $[P]$ on the left side of the equation and integrating between the limits $(0,t)$ and $(0,P)$ gives, upon rearrangement:

$$[P] = \frac{V_{max}t}{\{1 - K_M/K_p)\}} - K_M\left(\frac{K_p + [S_0]}{K_p - K_M}\right)\ln\left(\frac{[S_0]}{[S_0] - [P]}\right) \quad (3\text{-}4)$$

On a plot of $[P]/t$ versus $(1/t)\ln\{[S_0]/([S_0] - [P])\}$ both the slope and the intercept are different from the true values of V_{max} and K_M. See the data analysis section for the discussion of these plots and their analysis.

Substrate inhibition is seen fairly commonly with enzymes from foodstuffs. Enzyme classes which exhibit this behavior include oxidases, oxygenases, esterases, and carbohydrases. A plot of v versus $[S]$ follows the expected hyperbolic M-M relationship at low $[S]$, but at higher substrate concentration the rate decreases from the maximum. Several different models for this behavior have been proposed. Nonproductive binding of substrate to the active site (the SE complex) is one possibility. Another possibility is that the enzyme may have two binding sites for two parts of the substrate molecule, e.g. invertase may have a "glucose site" and a "fructose site"; if these sites are occupied by the glucose and fructose moieties of two separate sucrose molecules no hydrolysis reaction occurs. This model would imply a termolecular reaction, $E + 2S \longrightarrow SES$. Still another model postulates a four-cornered "box," with E, ES, SE, and SES being at the four corners and in rapid equilibrium via separate binding constants. Product is only formed by the breakdown of ES, and two equilibrium constants plus a symmetry factor suffice to describe the dissociation relationships.

While all these models are interesting and even plausible, kinetic analysis is unable to distinguish among them. Using rapid equilibrium kinetics, they all result in a rate equation which has the same general form:

$$v = \frac{V_{max}[S]}{A + B[S] + C[S]^2}$$
(3-5a)

The Hanes transformation of this equation is:

$$\frac{[S]}{v} = \frac{A}{V_{max}} + \left(\frac{B}{V_{max}}\right)[S] + \left(\frac{C}{V_{max}}\right)[S]^2$$
(3-5b)

From a set of $v, [S]$ data the three coefficients can be obtained by making a LLS fit to a quadratic equation in $[S]$. Having the three equation coefficients we can calculate three rate equation parameters. Since one of these parameters is necessarily V_{max} only two equilibrium constants are obtainable. Several imaginative schemes for modeling the substrate inhibition situation have appeared in the literature, but although they postulate three or four equilibrium constants, in fact the data allow the determination of only two such constants. Constructing a model which includes two constants K_d and K_s in addition to K_M is meaningless if the only two constants mathematically obtainable are K_M and $K_d K_s$. Because of this limitation, the simple model of substrate inhibition is to be preferred. This model implies that inhibition is due to a substrate molecule interacting with the ES complex to form an unproductive SES complex. The model and the significance of the A, B and C parameters of equation 3-5 are:

$$E + S \overset{K_M}{\rightleftharpoons} ES \qquad A = K_M$$

$$ES + S \overset{K_S}{\rightleftharpoons} SES \qquad B = 1$$

$$ES \overset{k_p}{\longrightarrow} E + P \qquad C = \frac{1}{K_S}$$

Dixon and Webb (1979) have proposed a graphical method for obtaining the three parameters. Substituting the values of A, B and C into equation 3-5a and dividing the right side through by $[S]$ gives:

$$v = \frac{V_{max}}{1 + K_M/[S] + [S]/K_S}$$

When $[S]$ is large $K_M/[S]$ becomes negligible with respect to the other two terms in the denominator, and:

$$v = \frac{V_{max}}{1 + [S]/K_S}, \quad \text{or} \quad \frac{1}{v} = \frac{1}{V_{max}} + [S]\left(\frac{1}{V_{max}K_S}\right)$$

A plot is made of $1/v$ versus $[S]$. From the intercept V_{max} is calculated, and then from the slope K_S is found. To obtain K_M the rate data at lower $[S]$ is plotted according to the Hanes transformation. Alternatively a first order kinetic run can be made at very low substrate concentration, yielding V_{max}/K_M, and K_M can be calculated using the value of V_{max} found from the Dixon-Webb plot.

Figure 3-3 shows the data of Bowski et al. (1971) on substrate inhibition of invertase by sucrose. The Michaelis-Menten curve shows the expected rate found by fitting the first six data points with the HYPER program. The three curves showing the inhibition were obtained by using the SUBIN program, the quadratic equation as per equation 3-5b, and the Dixon-Webb procedure. As can be seen, even with the rather different values for V_{max}, K_M and K_S found using these three procedures, the curves give reasonable fits to the data. (The sum of squares of differences indicate that SUBIN gave the best fit, however.)

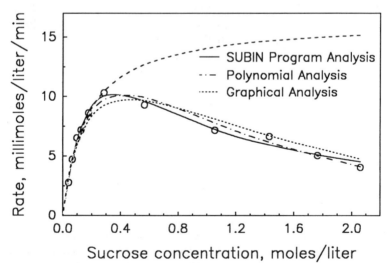

FIGURE 3-3. Substrate inhibition. Data of Bowski et al. (1971). Kinetic parameters for the various methods of analysis are: SUBIN, $V_{max} = 29.7$, $K_M = 0.342$, $K_i = 0.379$; polynomial, $V_{max} = 21.9$, $K_M = 0.244$, $K_i = 1.109$; Graphical, $V_{max} = 17.3$, $K_M = 0.180$, $K_i = 2.161$; HYPER (first six data points), $V_{max} = 16.3$, $K_M = 0.161$.

Data Analysis

An ordinary investigation of the reversible inhibition of an enzyme reaction by an added small molecule usually involves the following steps:

1. Determine initial velocity v at several values of $[S]$ at one concentration of $[I]$. Repeat this procedure at several other values of $[I]$.
2. Plot (or analyze by LLS) each set of $v, [S]$ data at constant $[I]$. This gives several data sets of $V_{app}, K_{app}, [I]$.
3. From suitable plots of V_{app} versus $[I]$ and K_{app} versus $[I]$, determine the values for $V_{max}, K_M, K_i, \alpha$, and β.

These procedures will be described in more detail a little further on. But first I will describe a simpler method for determining K_i in the majority of cases.

First-order kinetic runs. This method depends upon a unique property of the first-order rate constants for enzyme-catalyzed reactions. As pointed out in the data analysis section of Chapter 2, when $[S] \ll K_M$, an enzyme reaction is kinetically first order in $[S]$. This also holds true in the presence of inhibitor. In the earlier section methods for determining the first-order rate constant k' were discussed. If inhibitor is present, the rate constant may be termed k'_i. Then:

$$k' = \frac{V_{max}}{K_M} \text{ and } k'_i = \frac{V_{app}}{K_{app}} = \frac{V_{max}\{f_v\}}{K_M\{f_k\}}$$

The ratio of the two rate constants is:

$$\frac{k'}{k'_i} = \frac{\{f_k\}}{\{f_v\}}.$$

The fourth column in Table 3-1 indicates the significance of $\{f_k\}/\{f_v\}$. Notice that in the three cases where $\beta = 0$ (competitive, noncompetitive and linear mixed inhibition) the significance is the same; $\{f_k\}/\{f_v\} = 1 + [I]/K_i$. The consequences of this fact are both good and bad. In all three cases (which represent the majority of food enzymes) K_i can be determined by this simple procedure. On the other hand, first order kinetic runs cannot distinguish among these three types of inhibition. Knowing the intended use for the value of K_i will determine whether or not the more extensive procedures to be discussed below are necessary.

The procedure is straightforward. The first order rate constant is measured at several values of $[I]$, including $[I] = 0$. Then the ratio k'/k'_i is plotted versus $[I]$. The result should be a straight line with a y-axis intercept of 1. (N.B. A linear graph is also obtained plotting $1/k'_i$ versus $[I]$, but the intercept is $K_M/V_{max} = 1/k'$.) The slope of the line is $1/K_i$. This should be carried out up to a concentration $[I]$ of perhaps $5K_i$ (i.e., so that k'_i is about $k'/6$).

In those cases where $\beta \neq 0$ the plot of k'/k'_i versus $[I]$ is hyperbolic. The intercept on the y-axis is 1, but the curves approach an asymptote at high $[I]$. This value is α in the partially competitive case, $1/\beta$ for partially noncompetitive inhibition, and α/β for hyperbolic mixed inhibition. The mathematical reason for this is more easily grasped by considering the rearranged expression for f_k/f_v for the general model as shown in Table 3-1:

$$\frac{1 + [I]/K_i}{1 + \beta[I]/\alpha K_i} = \frac{\alpha K_i + \alpha[I]}{\alpha K_i + \beta[I]}$$

As $[I]$ gets very large and the terms in K_i become insignificant, this fraction approaches the limiting value α/β. The same argument applies to the partially competitive ($\beta = 1$) and partially noncompetitive ($\alpha = 1$) cases.

These two types of behavior are depicted in Figure 3-4. The hyperbolic plots do not allow discrimination among the three possible inhibitory models, just as the linear plot does not differentiate between competitive, noncompetitive and linear mixed inhibition. A value for K_i cannot be obtained from the hyperbolic plot, but can be found using a slightly different procedure.

If the inhibition is found to be partially competitive or partially noncompetitive ($\beta = 1$ or $\alpha = 1$, respectively) from Hanes plots as described in the next section, then a replot of the first-order rate data will yield a value

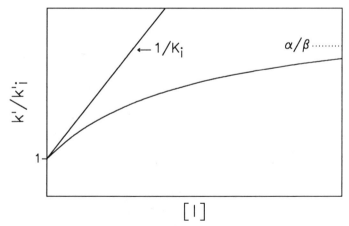

FIGURE 3-4. Inhibition kinetics for first-order enzyme reactions.

for K_i. This is based upon the same concept as the Delta Replots described below, and developed more fully in Appendix F. Taking the difference between the two first-order rate constants:

$$k' - k_i' = \Delta' = \frac{V_{max}}{K_M} - \frac{V_{app}}{K_{app}} = \left(\frac{V_{max}}{K_M}\right)\left(1 - \frac{\{f_v\}}{\{f_k\}}\right) \qquad (3\text{-}6a)$$

$$\Delta' = \left(\frac{V_{max}}{K_M}\right)\frac{[I](\alpha - \beta)}{\alpha K_i + \alpha[I]} \qquad (3\text{-}6b)$$

$$\frac{1}{\Delta'} = \frac{K_M}{V_{max}}\left(\frac{\beta}{\alpha - \beta} + \frac{\alpha K_i}{\alpha - \beta}\cdot\frac{1}{[I]}\right) \qquad (3\text{-}6c)$$

The intercept of the plot of $1/\Delta'$ versus $1/[I]$ equals $1/k'$ times $\beta/(\alpha - \beta)$. Knowing whether α or β equals 1 (noncompetitive- or competitive-type inhibition) allows the calculation of the other parameter. The slope of the plot equals $(1/k')(\alpha/(\alpha - \beta))K_i$, whence K_i is readily calculated. This method of plotting first-order kinetic data is shown in Figure 3-5. Note that in the three cases where $\beta = 0$ a plot of $1/\Delta'$ versus $1/[I]$ is also linear, but passes through the origin.

Making a half dozen measurements of the first-order rate constant at varying concentrations of inhibitor is usually a rather brief experimental task. From the plot of k'/k_i' (or $1/k_i'$) versus $[I]$ the field of possible inhibitory modes can be reduced by half. If you are fortunate enough to obtain a linear plot ($\beta = 0$) the value of K_i is calculated. Then the next step would be to

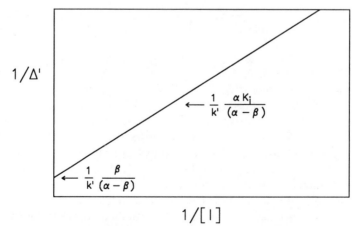

FIGURE 3-5. Delta replots of first-order inhibited enzyme reactions.

perform the more extensive rate measurements appropriate to distinguish between the three cases, if that differentiation is needed. If the plot is hyperbolic then some recalculation is needed, followed by a plot of $1/\Delta'$ versus $1/[I]$. At that point more information (is the inhibition competitive, noncompetitive, or mixed) to decide whether α or β equals 1 is needed; that information would come from the more extensive rate measurements described in the next section.

If the enzyme has two substrates, the first order kinetics are similar, providing that the limitations assumed earlier hold and equations 3-3a and 3-3b are valid. The experimental conditions for obtaining meaningful first-order rate constants for two substrate systems were delineated in the data analysis section of Chapter 2. If these are adhered to, then the ratio $k'/k'_i = 1 + [I]/K_i$, and the same linear analysis as described above can be done. If the graph of $1/k'_i$ versus $[I]$ is not linear, then one or more of the assumptions is not correct, and a more extensive experimental program is necessary. If that appears to be the case and the information is required, the reader is referred to Cleland (1970) and Segel (1975) for guidance.

Initial Rate Data Plots and Replots

While the first-order kinetic plot is a good place to start, the fuller analysis suggested at the beginning of this section is often needed: 1. obtain $v, [S]$ data at various $[I]$; 2. find V_{app} and K_{app} at each $[I]$ and; 3. replot V_{app} and K_{app} as a function of $[I]$ to obtain K_i, α, and β. The rationale for these plots and replots will be discussed in this part.

The Hanes transformation of the generalized one-substrate inhibition rate equation 3-1 is:

$$\frac{[S]}{v} = \frac{K_{app}}{V_{app}} + \frac{1}{V_{app}} \cdot [S] = \frac{K_M\{f_k\}}{V_{max}\{f_v\}} + \frac{1}{V_{max}\{f_v\}} \cdot [S] \qquad (3\text{-}1b)$$

Note that on a Hanes plot three relationships are useful in visual interpretation: 1. the slope equals $1/V_{app}$; 2. the x-axis intercept equals $-K_{app}$; and 3. the y-axis intercept equals K_{app}/V_{app}. These will be applied to the consideration of the plots as shown in Figure 3-6. The expressions for $\{f_v\}$ and $\{f_k\}$ given in Table 3-1 also help in interpreting the patterns of the plots.

Competitive inhibition. Figure 3-6A shows the pattern observed when increasing concentrations $[I]$ produce competitive inhibition. The slopes of the lines are the same, because with $\{f_v\} = 1, 1/V_{app} = 1/V_{max}$ at all $[I]$. The x-axis intercept shifts to larger values, by the factor $\{f_k\}$. This is true for both simple and partially competitive inhibition.

Noncompetitive inhibition. Figure 3-6B shows the pattern for noncompetitive inhibition by $[I]$. Since $\{f_k\} = 1, K_{app} = K_M$ and the x-axis intercept remains constant. With increasing $[I]$ the factor $\{f_v\}$ decreases, and being

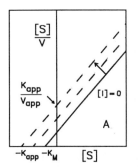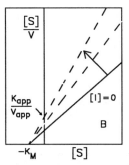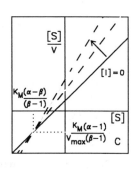

FIGURE 3-6. Hanes plot of inhibition rate data. A. Competitive inhibition. B. Noncompetitive inhibition. C. Mixed inhibition.

in the denominator of the expression for slope, this causes an increase in slope. Again, this pattern holds for both the simple and partial cases.

Mixed inhibition. Figure 3-6C shows the pattern for mixed inhibition by $[I]$. Both the slope and the x-axis intercept increase as $[I]$ increases, because neither $\{f_v\}$ nor $\{f_k\}$ equal 1. Another useful bit of information may be derived from the intersection point in the third quadrant of the extrapolated lines. This point has the y-coordinate of $K_M(\alpha - 1)/V_{max}(\beta - 1)$, and the x-coordinate of $K_M(\alpha - \beta)/(\beta - 1)$. In the case of linear mixed inhibition, with $\beta = 0$, this x-coordinate reduces to $-\alpha K_M$.

Replots versus $[I]$ of one or more of the parameters from the Hanes linear plot give further information about the system. In the simple cases ($\beta = 0$) values can be readily obtained for V_{max}, K_M, K_i and α (in linear mixed inhibition). In those cases where $\beta \neq 0$ the replots are curved, and more manipulation of the data is required to obtain values for K_i, α, and β.

Direct replots are obtained when $\beta = 0$. The plots are linear and are fitted using LLS calculations. Figure 3-7 shows the form of the replots, and the parameters to be obtained from the slope and intercept of the replotted data.

In the case of competitive inhibition K_{app} is plotted versus $[I]$ (Figure 3-7A.). In terms of the Hanes plot parameters, K_{app} = slope/y-axis intercept = $-(x$-axis intercept), so this can be calculated directly from the LLS fit to the $[S]/v, [S]$ data pairs at each constant value of $[I]$. If the rate data is analyzed using a different algorithm (i.e., direct linear plot or the HYPER computer program) K_{app} is directly accessible. The intercept of the replot K_M and the slope equals $1/K_i$. V_{max} is obtained directly from the slope of the Hanes plot.

For noncompetitive inhibition $1/V_{app}$ (the slope of the Hanes plot) is plotted versus $[I]$ (Figure 3-7B). The other methods give V_{app} directly, so the reciprocal must be taken before replotting. The intercept of the replot is $1/V_{max}$ and the slope is $1/K_i$. K_M is calculated from the Hanes plot.

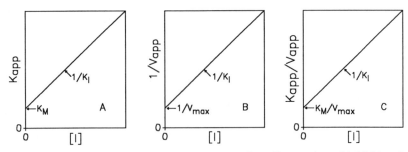

FIGURE 3-7. Direct replots of slopes and intercepts from Hanes plots of inhibition data when $\beta = 0$. Ordinates are: A. the x-axis intercept; B. the slope, and; C. the y-axis intercept from the initial Hanes plots.

If linear mixed inhibition is observed, both the slope ($1/V_{app}$) and the intercept (K_{app}/V_{app}) from the Hanes plot must be replotted against $[I]$ (Figure 3-7 B,C). From the $1/V_{app}$ replot $1/V_{max}$ and $1/\alpha K_i$ are obtained, and from the replot of K_{app}/V_{app} the values of K_M/V_{max} and $1/K_i$ are found. Using these four values, the four parameters of the system, V_{max}, K_M, K_i, and α are calculated.

Delta replots are necessary to find the relevant kinetic parameters when $\beta \neq 0$. The direct plots of Hanes slope or intercept parameters versus $[I]$ are hyperbolic, similar to the plots shown in connection with the first-order kinetic analysis (Figure 3-4). The algebraic justification of the delta plots is detailed in Appendix F, so only the application of them will be discussed at this point.

The data for replotting are obtained from the Hanes plots, as shown in Figure 3-8A. $\Delta V = 1/V_{app} - 1/V_{max}$, or the numerical difference between the slope found in the absence of inhibitor and that found at some concentration $[I]$. $\Delta V = [I](1 - \beta)/V_{max}(\alpha K_i + \beta[I])$; the replot of $1/\Delta V$ versus $1/[I]$ gives the slope and intercept shown in Figure 3-8B. Each of these is multiplied by $1/V_{max}$ (the slope of the Hanes plot at $[I]=0$) to obtain results 1 and 2 (below).

The second delta value, $\Delta K = K_{app} - K_M$, is the numerical difference between the x-axis intercepts in the presence and absence of inhibitor, respectively. This difference is given by $\Delta K = K_M(\alpha - 1)[I]/(\alpha K_i + [I])$. The intercept and slope of the replot of $1/\Delta K$ versus $1/[I]$ are shown in Figure 3-8C. Each of these is multiplied by K_M (the x-axis intercept of the Hanes plot at $[I]=0$) to get results 3 and 4 (below). From the four results the parameters K_i, α, and β are readily calculated. The results are:

$$1. \left(\frac{1}{\Delta V \text{ slope}} \right)\left(\frac{1}{V_{max}} \right) = \frac{\alpha K_i}{1 - \beta}$$

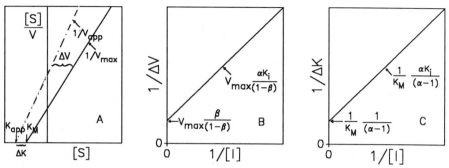

FIGURE 3-8. Delta replots of data from inhibition experiments. A. Hanes plot of inhibition rate data. B. Replot of the difference between slopes. C. Replot of the difference between x-axis intercepts.

$$2. \left(\frac{1}{\Delta V \text{ intercept}}\right)\left(\frac{1}{V_{max}}\right) = \frac{\beta}{1 - \beta}$$

$$3. \left(\frac{1}{\Delta K \text{ slope}}\right)(K_M) = \frac{\alpha K_i}{\alpha - 1}$$

$$4. \left(\frac{1}{\Delta K \text{ intercept}}\right)(K_M) = \frac{1}{\alpha - 1}$$

α is found from 4, and K_i is then calculated from 3. β is obtained from 2, and then K_i can be calculated again from 1, providing a check on internal consistency.

Partial competitive inhibition has $\beta = 1$, so the slope and intercept of a plot of $1/\Delta V$ versus $[I]$ are infinite. The values of K_i and α are calculated from the slope and intercept of the $1/\Delta K$ versus $[I]$ plot.

Partial noncompetitive inhibition has $\alpha = 1$, so the $1/\Delta K$ versus $[I]$ plot is meaningless. From the intercept and slope of the $1/\Delta V$ versus $[I]$ plot, β and then K_i are calculated.

Hyperbolic mixed inhibition has neither α nor β equal to 1, so both replots are required to find all three parameters. All four results are used as stated above.

Two substrate systems. As mentioned earlier, inhibition patterns with those enzymes catalyzing the reaction between two substrates can be extremely complicated. The limitations imposed to make the rate equations both tractable and applicable to most food enzymes resulted in noncompetitive inhibition for both random- and ordered-addition cases. When v, $[A]$, $[B]$ data are collected at different concentrations $[I]$, analysis indicates that $V_{app} = V_{max}\{f_v\}$, $K_{A,app} = K_A$, and $K_{B,app} = K_B$. A replot of $1/V_{app}$

versus $[I]$ is linear, with an intercept of $1/V_{\max}$ and slope of $1/K_i$. This is the same principle as applied to the noncompetitive inhibition of a one-substrate reaction, with the exception that a replot procedure is required to find the value of V_{app} (see the data analysis section of Chapter 2). Since the same result can be obtained with much less experimental effort by using first-order kinetic data, it hardly seems worthwhile to go to these lengths to find K_i in these systems. If the first-order plots are not linear, the replot suggested here will also not be linear, and the reader is referred to Cleland (1970) or Segel (1975) for more information.

Progress Curve Analysis

Measuring K_i. The integrated rate equation in the presence of inhibitor is readily found in the same way as when $[I] = 0$. Referring back to the Henri integrated rate equation of Chapter 2, and substituting $V_{app}(= V_{\max}\{f_v\})$ and $K_{app}(= K_M\{f_k\})$ into equation 2-11, the integrated inhibition equation is:

$$\frac{[P]}{t} = V_{app} - K_{app}\left(\frac{1}{t}\right)\ln\left(\frac{[S_0]}{[S_0] - [P]}\right) \tag{3-7}$$

The same result is found by integrating equation 3-1.

The effect of different inhibition patterns upon the Henri plot is easy to predict if the effects on V_{app} and K_{app} are kept in mind:

Competitive: V_{app} unchanged, K_{app} increases;
Noncompetitive: V_{app} decreases, K_{app} unchanged;
Mixed: V_{app} decreases, K_{app} increases.

The three patterns are shown in Figure 3-9.

Progress curve runs are made at different concentrations $[I]$, and V_{app} and K_{app} are found. These are replotted the same way as if they had been obtained from Hanes plots of initial velocity data (i.e. for competitive inhibition K_{app} is plotted versus $[I]$, for noncompetitive $1/V_{app}$ is plotted versus $[I]$, and for mixed inhibition both $1/V_{app}$ and K_{app}/V_{app} are plotted versus $[I]$). If $\beta \neq 0$ then the appropriate delta replots must be made. The choice between initial velocity runs and progress curve runs depends upon the experimental characteristics of the system being studied. For some enzyme reactions the progress curve is a more convenient way to gather the large amount of data needed.

Product Inhibition. Applying the integrated rate equation is usually the method of choice for analyzing product inhibition. In many cases the product is not readily available for addition as I in initial velocity runs. In the initial paper on the use of integrated rate equations in this way (Foster and Niemann

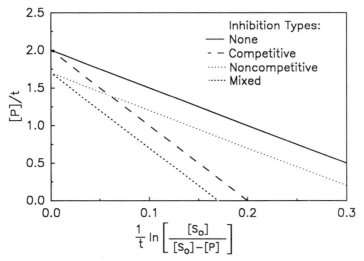

FIGURE 3-9. Henri plots of various inhibition patterns.

1953) the authors concluded they obtained more accurate estimation of K_p using progress curves than by initial rate measurements in the presence of added product.

The integrated rate equation (slightly rearranged) is equation 3-4:

$$\frac{[P]}{t} = \frac{V_{max}}{(1 - K_M/K_P)} - K_M \left(\frac{K_p + [S_0]}{K_p - K_M} \right) \cdot \frac{1}{t} \ln \left(\frac{[S_0]}{[S_0] - [P]} \right)$$

The plot is shown in Figure 3-10A. The y-axis intercept, $V_{max}/(1 - K_M/K_p)$, is constant; the slope, $-K_{app}$, increases with increasing $[S_0]$. (If $K_M = K_p$ the slope is infinite, i.e. the plot is a vertical line; if $K_M > K_p$ the line will have a positive slope and a negative y-axis intercept.) Plotting K_{app} versus $[S_0]$ (Figure 3-10B), according to the identity:

$$K_{app} = \frac{K_M K_p}{(K_p - K_M)} + [S_0] K_M (K_p - K_M)$$

gives an x-axis intercept of $-K_p$ and a slope of $K_M/(K_p - K_M)$ from which K_M can be calculated. With these two values in hand V_{max} is found from the y-axis intercept of the original plot.

Foster and Niemann (1953) used a somewhat different method of analyzing their experimental data. On the plot of progress runs made with different $[S_0]$, they found the intersection points with the lines representing $[S_0]$ as

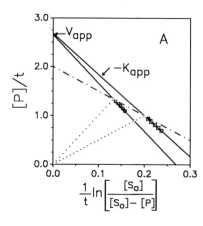
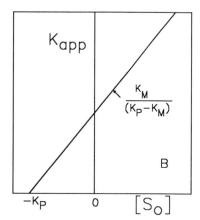

FIGURE 3-10. Analysis of product inhibition using proogress curves. A. The plot of experimental data; B. The replot to find K_p and K_M.

shown in Figure 3-10A (see also Figure 2-9). The line through these points gives the true V_{max} and K_M for the reaction. They then used either of two relationships to calculate K_p:

$$\frac{V_{app}}{V_{max}} = \frac{1}{1 - K_M/K_p}$$

and

$$\frac{K_{app}}{K_M} = \frac{K_p + [S_0]}{K_p - K_M}$$

The plot of $[P]/t$ versus $(1/t)\ln\{[S_0]/([S_0] - [P])\}$ is made first in either method (Figure 3-10A). The subsequent mathematical manipulations are different, but since this takes much less time than gathering the original data it would seem a worthwhile exercise to apply both methods to the data in hand.

HIGH-AFFINITY INHIBITORS

Equations for Effect of Inhibitor on Rate

Low-affinity reversible inhibitors such as have been considered thus far are usually present in the reaction mixture at concentrations of 10^{-2} to 10^{-6} M (i.e. a large molar excess with respect to enzyme since the usual range for $[E_t]$ is 10^{-7} to 10^{-9} M). Thus a negligible fraction of $[I]$ is bound in the form of EI, EIS, etc., and in developing rate equations $[I]$ is taken to be equal to $[I_t]$, the total concentration of inhibitor before enzyme is added. The situation is somewhat different if the inhibitor has a high affinity for the enzyme.

Rapid Equilibrium Rate Equations. If a high-affinity reversible inhibitor were present in the reaction mixture in the 10^{-2} to 10^{-6} M concentration range no appreciable enzymatic reaction would occur, since then $[I]/K_i >> 1$ and essentially 100% of the enzyme would be present as inhibitor complexes (assuming EIS does not break down; otherwise the rate $\beta k_p[E_t]$ would be observed). Studying the high-affinity inhibitor in a concentration range such that $[I]/K_i \approx 1$, then $[I] \approx [E_t]$, and the concentration of free inhibitor is significantly depleted in establishing the equilibrium:

$$E + I \overset{K_i}{\rightleftharpoons} EI$$

$$K_i = \frac{[E][I]}{[EI]} = \frac{([E_t] - [EI])([I_t] - [EI])}{[EI]}$$

This fact necessitates a somewhat different approach to developing the equations for relating reaction rate v to the concentration $[I_t]$ and the equilibrium dissociation constant K_i. The basic approach was proposed by Goldstein (1944) and has been amplified by Bieth (1974).

Operationally enzyme is mixed with inhibitor and the equilibrium given above is established. Then substrate is added and the initial velocity v_i is measured (in the absence of inhibitor the rate is v_0). An implicit assumption in the mathematical development is that EI does not dissociate during the course of the rate measurement to generate additional E which could participate in the enzymatic reaction. This assumption is not always valid. The simple addition of substrate which begins to form a complex ES with E will perturb the inhibition equilibrium. If the addition of substrate solution to begin the rate measurement increases the reaction volume by a significant amount then EI will also tend to dissociate due to the dilution effect. These shortcomings can be overcome by taking two experimental precautions:

1. Minimize the time for rate measurement. A direct-measuring continuous method which gives a value for v within a minute is much preferable to a method which requires a ten-minute incubation. The exact allowable time is related to the "off" rate k_{-1} for the dissociation of EI.
2. Minimize dilution of the reaction. The substrate solution should be as concentrated as practicable so that the volume added to start the enzyme reaction is small.

Assuming valid measurements for v_0 and v_i are obtained, Goldstein defined a ratio α which has several mathematical implications:

$$\alpha = \frac{v_i}{v_0} = \frac{[E]}{[E_t]} = \frac{[E_t] - [EI]}{[E_t]} = 1 - \frac{[EI]}{[E_t]} \tag{3-8a}$$

Note that in the M-M rate equation for v_i, the quantity $[E]$ is equivalent to $[E_t]$ in the M-M equation for v_0. Since $[E] = \alpha[E_t]$ and $[EI] = [E_t](1 - \alpha)$, these are substituted into the expression for K_i:

$$K_i = \frac{([E_t] - [EI])([I_t] - [EI])}{[EI]} = \frac{\alpha[E_t]([I_t] - [E_t] + \alpha[E_t])}{[E_t](1 - \alpha)} \tag{3-8b}$$

$K_i = \alpha\{[I_t] - [E_t](1 - \alpha)\}/(1 - \alpha)$; and, by rearrangement:

$$\frac{[I_t]}{1 - \alpha} = \frac{K_i}{\alpha} + [E_t] \tag{3-8c}$$

A series of runs are made at different $[I_t]$ and the ratio $[I_t]/(1 - \alpha)$ is plotted versus $1/\alpha$ (Figure 3-11A). The slope of the plot is K_i and the intercept on the y-axis is $[E_t]$.

Another useful relationship is obtained by manipulating equation 3-8c as follows. Substitute v_i/v_0 for α, divide through by $v_0[I_t]$ and invert:

$$v_0 - v_i = \frac{[I_t]v_i}{K_i + [E_t]v_i/v_0}$$

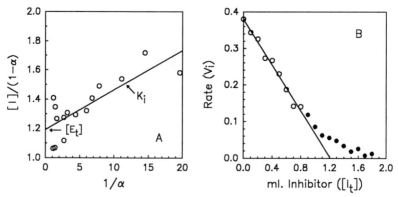

FIGURE 3-11. Analysis of data obtained with a high-afffinity inhibitor (data from Hamerstrand et al. 1981). A. Plot to find inhibitor constant and total amount of enzyme. B. Plot to find total amount of enzyme and/or inhibitor.

Usually $[E_t]$ is greater than K_i so as long as v_i/v_0 is close enough to 1 to maintain the inequality $[E_t]v_i/v_0 >> K_i$, then K_i in the denominator of the equation may be ignored and it simplifies to:

$$v_i = v_0 - \left(\frac{v_0}{[E_t]}\right)[I_t] \tag{3-9}$$

A plot of v_i versus $[I_t]$ (Figure 3-11B) is linear at low degree of inhibition with the intercept v_0 on the y-axis, a slope of $-v_0/[E_t]$ and an x-axis intercept of $[E_t] = [I_t]$.

This plot may be used to quantitate either the enzyme or inhibitor. In the study of Hamerstrand et al. (1981), from which the data in Figure 3-11 were taken, the interest was in finding the amount of trypsin inhibitor present in a heat-treated soy meal, a common method for measuring this anti-nutritional factor. The exact molar amount of trypsin used in the assays is readily found using a titration assay (see the absolute active site measurement section of Chapter 6), so the x-axis dimension "ml inhibitor" can be converted to moles inhibitor per ml of extract. In the opposite sense, the synthetic molecule N-acetyl-leucinaldehyde is a high-affinity inhibitor for papain. Using this material and papain in the same experimental protocol the x-axis dimension is now "micromoles inhibitor" and the amount of papain used in the assay (expressed, say, as mg protein) contains a number of μmoles of active enzyme equivalent to the x-axis intercept.

All situations are not so straightforward, however. If two or more inhibitors with significantly different K_i's are present in the extract being studied, the plot of Figure 3-11A is curved and interpretation is difficult. Furthermore, if K_i is large enough so that the inequality assumed in deriving equation 3-9 does not hold, then the plot in Figure 3-11B will not provide any linear portion which can be extrapolated to the x-axis with any confidence. Several examples of natural enzyme inhibitors found in foods which show behaviors much like these are discussed by Richardson (1981).

There are a number of high-affinity inhibitors for proteases which would work in this method of titrating enzyme. For serine proteases there are many protein inhibitors (Laskowski and Kato 1980) and synthetic fragments of some of these (Nishino and Izumiya 1982). For sulfhydryl proteases we have the egg protein cystatin (Anastasi et al. 1983) as well as N-acetyl-leucinaldehyde (Westerik and Wolfenden 1972). For acidic proteins there is the depsipeptide pepstatin (Barrett and Dingle 1972). And for metalloproteases such as thermolysin there is the natural inhibitor phosphoramidon (Komiyama et al. 1975; Weaver et al. 1977) and other phosphonate transition state analogs (Wolfenden 1977; Nishino and Powers 1979). Similar analogs are applicable to the titration of carboxypeptidase A (Jacobsen and Bartlett 1981).

Pre-Steady State Rate Equations.   The method discussed above determines only the equilibrium dissociation constant $K_i = k_{-1}/k_1$. This is all that is required for most food application (e.g., in measuring the amount of anti-nutritional factors (proteinase or amylase inhibitors) in various foodstuffs). Sometimes it may be of interest to know the rate at which the inhibitor and enzyme react (i.e., k_1). An approach which finds both the inhibitor constant K_i and the "on" rate k_1 has been developed by Mangel and his colleagues (Leytus et al. 1983a, 1984). The experimental procedure is novel in that the high-affinity inhibitor and substrate are mixed in the spectrophotometer cuvette and the reaction is started by adding the enzyme sample. There is an immediate kinetic competition between the inhibitor and the substrate; as substrate interacts with an enzyme molecule product is formed, but as inhibitor reacts with enzyme it is removed from the competition. Thus the rate of product formation rapidly goes to zero. However, an analysis of the rate of product formation before this occurs allows one to determine k_1 and K_i. The mathematics of the procedure and the analysis are presented in section 2 of Appendix G. While the instrumentation required is somewhat more complex than a simple colorimeter it is much less sophisticated than the stopped-flow apparatus previously used for investigations of this sort. A recording spectrophotometer which can accurately plot the change in light absorbance in the time period from 1 to 60 seconds will do quite well. If serious research on a proteinase inhibitor is being done this approach is highly recommended.

4   Effect of pH on Activity

It is well known that changes in the pH of the reaction milieu may affect enzyme activity. In fact, the concept of pH was developed by Sorensen and his coworkers at the Carlsberg Laboratories in the early 1900's in order to rationalize the effects of varying concentrations of acid on the activity of pepsin. The changes are due to protonation or deprotonation of ionizing groups in the enzyme, in the substrate, or in the enzyme-substrate complex. Such an ionization is characterized by the negative log of its dissociation constant, pK, which corresponds to the pH at which the ionizing group is half in the protonated state and half in the unprotonated form.

The alterations in enzyme rate as a function of pH may be due to any of three factors.

1. The protonation status of amino acid side chains in the active site in the enzyme-substrate complex may change, resulting in a change in the ability of *ES* to break down to *P* (i.e., a change in V_{max}).
2. Changes in the ionic charge of the substrate molecule or of the active site may alter the propensity of the two molecules to combine to form *ES* (i.e., a change in K_M).
3. Shifting the pH away from neutral may weaken the forces stabilizing protein conformation, leading to an enhanced rate of enzyme denaturation (irreversible loss of activity) at the temperature of the assay.

Studies of enzyme activity as a function of pH are performed for one of two purposes, either to learn more about the nature of the enzymatic reaction itself, or to define conditions for the most efficient practical application of the enzyme. In either case it is important to adequately assess all three of the above factors. Most published reports on the pH-dependence of enzyme activity, whether in refereed scientific publications or in commercial "technical data sheets," are at best inadequate in this respect.

Most books which deal with enzymes touch on the role of pH to some extent. Two thorough recent reviews are those by Tipton and Dixon (1979) and by Cleland (1982). We will limit the discussion in this chapter to a brief theoretical explanation of the involvement of pH in the M-M equation,

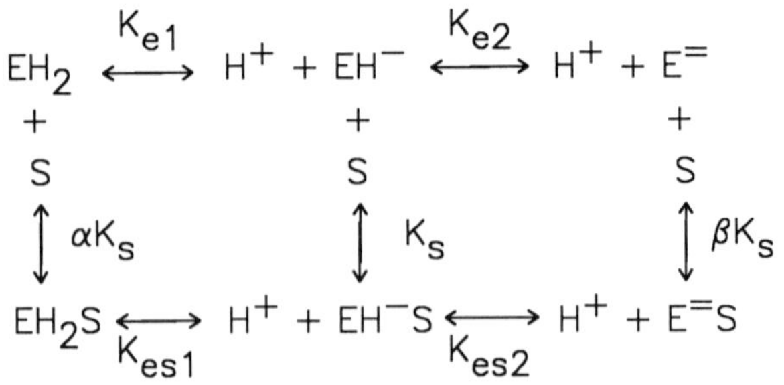

$$EH_2 \xrightleftharpoons[]{K_{e1}} H^+ + EH^- \xrightleftharpoons[]{K_{e2}} H^+ + E^=$$

$$+ \qquad\qquad\qquad + \qquad\qquad\qquad +$$

$$S \qquad\qquad\qquad S \qquad\qquad\qquad S$$

$$\updownarrow \alpha K_s \qquad\qquad \updownarrow K_s \qquad\qquad \updownarrow \beta K_s$$

$$EH_2S \underset{K_{es1}}{\xrightleftharpoons{}} H^+ + EH^-S \underset{K_{es2}}{\xrightleftharpoons{}} H^+ + E^=S$$

FIGURE 4-1. General equilibrium model for pH effects.

methods for finding the relevant pK's in the most common cases, the involvement of pH with enzyme stability, and some practical information on buffers and effects of pH on experimental measurements of enzyme rate.

EFFECTS ON ENZYME ACTIVITY

Enzyme Ionization and Michaelis-Menten Parameters

Figure 4-1 shows the general model for the ionization of enzyme E and enzyme-substrate complex ES. The model makes the following assumptions:

1. Two ionizing groups on the enzyme surface are relevant to activity. In fact, of course, many amino acid side chains protonate/deprotonate, but it is extremely rare for more than two groups to directly influence activity. In many cases only one group affects the actual rate; this will be discussed below.

2. Only the ESH^- complex breaks down to give product. Sometimes one of the other forms (ESH_2 or $ES^=$) does form product at a reduced rate, but this is an uncommon situation, and the more complicated analysis of this situation is not warranted in this chapter.

3. The substrate does not change its ionization state within the pH range of interest. While substrate ionization is an important consideration, it is covered in the next section. Alternatively, the ionization state of the substrate may be irrelevant with respect to binding and break down to product.

4. Enzyme stability is not affected by the pH range being considered. The means of differentiating denaturation and direct effects on rate are considered later.

Protonation is a very fast reaction with respect to the other kinetic steps so the rapid equilibrium rate expression is used. The central complex is EHS^- and the various dissociation constants and expressions for the concentration of the other five forms of enzyme are:

$$K_m = \frac{[EH^-][S]}{[EHS^-]} \qquad [EH^-] = [EHS^-]\left(\frac{K_m}{[S]}\right) \qquad \text{(4-1a)}$$

$$K_{e1} = \frac{[EH^-][H^+]}{[EH_2]} \qquad [EH_2] = \frac{[EH^-][H^+]}{K_{e1}} \qquad \text{(4-1b)}$$

$$= [EHS^-]\left(\frac{K_m}{[S]}\right)\left(\frac{[H^+]}{K_{e1}}\right)$$

$$K_{e2} = \frac{[E^=][H^+]}{[EH^-]} \qquad [E^=] = [EH^-]\left(\frac{K_{e2}}{[H^+]}\right) \qquad \text{(4-1c)}$$

$$= [EHS^-]\left(\frac{K_m}{[S]}\right)\left(\frac{K_{e2}}{[H^+]}\right)$$

$$K_{es1} = \frac{[EHS^-][H^+]}{[EH_2S]} \qquad [EH_2S] = [EHS^-]\left(\frac{[H^+]}{K_{es1}}\right) \qquad \text{(4-1d)}$$

$$K_{es2} = \frac{[ES^=][H^+]}{[EHS^-]} \qquad [ES^=] = [EHS^-]\left(\frac{K_{es2}}{[H^+]}\right) \qquad \text{(4-1e)}$$

Note that the expressions for αK_m and βK_m are not needed in order to fully specify all the relationships, although $K_{es1} = \alpha K_{e1}$ and $K_{es2} = K_{e2}/\beta$. If, for example, $\alpha = \infty$ the low-pH form EH_2 would not bind substrate at all and if EHS^- were protonated (K_{es1}) it would rapidly and completely dissociate.

The next step is to set up the ratio expression for rate divided by total enzyme:

$$\frac{v}{[E_t]} = \frac{k_p[EHS^-]}{[EH_2] + [EH^-] + [E^=] + [EH_2S] + [EHS^-] + [ES^=]}$$

$$\text{(4-2a)}$$

Since $V_{max} = k_p[E_t]$, substitute and cancel the central complex terms, to get:

$$v = V_{max}\Big/\left\{\left(\frac{K_m}{[S]}\right)\left(\frac{[H^+]}{K_{e1}}\right) + \frac{K_m}{[S]} + \left(\frac{K_m}{[S]}\right)\left(\frac{K_{e2}}{[H^+]}\right) + \frac{[H^+]}{K_{es1}} + 1 + \frac{K_{es2}}{[H^+]}\right\}$$

$$\text{(4-2b)}$$

Gather the denominator terms into those in $K_m/[S]$ and the others, and then multiply through by $[S]$, to arrive at:

$$v = \frac{V_{max}[S]}{K_m(1 + [H^+]/K_{e1} + K_{e2}/[H^+]) + [S](1 + [H^+]/K_{es1} + K_{es2}/[H^+])}$$

(4-2c)

The parenthetical factor for K_m is the Michaelis function f_{E^-} for the free enzyme and the parenthetical factor for $[S]$ is the Michaelis function f_{ES^-} for the enzyme-substrate complex. For a diprotic species such as we are considering here the reciprocal of the Michaelis function gives the fraction of the total amount of material which is in the intermediate ionization state (i.e., $[EH^-]/[E_t] = 1/f_{E^-}$ if there were no substrate present). Put another way, the total amount of the material equals the Michaelis function times the ionic species (i.e., $[E_t] = [EH^-]f_{E^-}$). For the three ES complex species, $[EHS^-]/([EH_2S] + [EHS^-] + [ES^=]) = 1/f_{ES^-}$. The Michaelis functions for the three ionization states are, generically:

$$f^0 = 1 + \frac{K_1}{[H^+]} + \frac{K_1 K_2}{[H^+]^2}$$

(4-3a)

$$f^- = 1 + \frac{[H^+]}{K_1} + \frac{K_2}{[H^+]}$$

(4-3b)

$$f^= = 1 + \frac{[H^+]}{K_2} + \frac{[H^+]^2}{K_1 K_2}$$

(4-3c)

A plot of $1/f^-$ versus pH gives the familiar bell-shaped curve (Figure 4-2). When the two pK's are well-separated (3.5 pH units is the usual guideline) the value of $1/f^-$ comes very close to 1 at the midpoint between them (the "pH-optimum"). When they are separated by only one pH unit the peak height of the curve is only 61% of the potential maximum. We will return to this point when we are considering the analysis of experimental pH-activity data.

Equation 4-2c may be rewritten as follows:

$$v = \frac{V_{max}(1/f_{ES^-})[S]}{K_M(f_E^-/f_{ES^-}) + [S]}$$

(4-4)

For an analysis made at any given pH, the apparent kinetic constants have the following values: $V_{app} = V_{max}/f_{ES^-}$; $K_{app} = K_M f_{E^-}/f_{ES^-}$; and the first

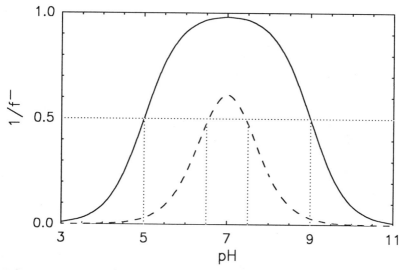

FIGURE 4-2. Concentration of singly-ionized species. Solid curve; pKs of 5 and 9. Dashed curve, pKs of 6.5 and 7.5.

order rate constant $k' = V_{app}/K_{app} = (V_{max}/K_M)/f_{E^-}$. A plot of V_{app} versus pH describes a bell-shaped curve similar to Figure 4-2, with the pK's being pK_{es1} and pK_{es2}. A plot of the first order rate constant versus pH also follows a bell-shaped curve, but now the pK values are those for the free enzyme (i.e., pK_{e1} and pK_{e2}). The plot of K_{app} versus pH may be rather more complicated; the interpretation of the possible bends in this plot will be given in the next section on data analysis.

If there is only one ionization affecting enzyme rate, the model is similar to that for a monoprotic ionization such as for acetic acid. The Michaelis functions are:

$$f^0 = 1 + \frac{K_1}{[H^+]} \tag{4-5}$$

$$f^- = 1 + \frac{[H^+]}{K_1}$$

If the active form of the enzyme is the deprotonated species as is the case with serine proteases such as chymotrypsin, the plot of V_{app} versus pH is a sigmoid curve defined by $1/f^-$, zero at low pH and approaching true V_{max} at high pH, with the mid-point of the curve corresponding to pK_{es1}. The plot of the first-order rate constant versus pH has the same shape, but the midpoint of the curve is pK_{e1}.

It is possible for the reverse situation to occur, in which the high-pH form of the enzyme is inactive and the protonated species is the active form. In that case the sigmoid curve would be given by $1/f^0$, approaching full activity at low pH and falling to zero activity at high pH. An example of such an enzyme is wheat germ acid phosphatase. The value of V_{max} is constant from pH 3 to 7, then decreases at higher pH. The loss of activity is due to deprotonation in the *ES* complex of a group with pK of about 8. This ionization is not observed in the free enzyme; the plot of the first order rate constant versus pH is level between pH 3 and 9.

Substrate Ionization

Many food enzyme substrates change ionization state in the usual pH range of interest (i.e., from about 3 to 9). A few examples are phosphate esters such as phytic acid and the esters used for phosphatase assays; linoleic acid as a substrate for lipoxygenase; and nitro-catechol, oxidized by polyphenol oxidase. The simple measurement of rate across a pH range at a single substrate concentration may yield a "bell-shaped" curve; the reason for the decrease in rate at low or high pH may be that the substrate is protonated (or ionized) and in that state does not interact with the enzyme. There are numerous examples in the literature where this factor has not been considered, and curves purporting to demonstrate the presence of ionizing groups in the enzyme active site serve as the basis (wrongly) for a long "Discussion" section.

The Michaelis function for substrate molecules are exactly the same as those given above. If the substrate is monoprotic, then the applicable functions are those in equation 4-5. The ionization constant K_1 can be readily determined by titrating a sample of substrate in solution. If the ionized form is the true substrate (e.g., linoleate for lipoxygenase) then f^- is the function f_S for correcting rate results. If the unionized form is the substrate (e.g., catechol for polyphenol oxidase) then f^0 is f_S. If the substrate is diprotic, then one of the Michaelis functions from equation 4-3 apply. The singly-ionized phosphate esters are the substrate form which reacts with, and is hydrolyzed by, wheat germ acid phosphatase. In this case f_S is given by f^- (equation 4-3b) (Hickey et al. 1976; Van Etten 1982).

The general equilibrium model of Figure 4-1 can be expanded to include substrate in three ionization states. When the rapid equilibrium rate expression is developed, an equation analogous to 4-4 is obtained.

$$v = \frac{V_{max}\left(1/f_{ES}\right)[S]}{K_M\left(f_E f_S/f_{ES}\right) + [S]} \tag{4-6}$$

Substrate ionization does not affect V_{max}. This is intuitively reasonable, since by definition V_{max} is the rate in the presence of enough substrate (strictly speaking, an infinite concentration) to maintain all the enzyme present in the productive ES complex. If a fraction of the substrate is in a non-reacting form, this does not affect the infinite concentration which will give V_{max}.

Substrate ionization increases K_{app} by the factor f_S. Since $f_S \geq 1$ the binding of substrate is either unaffected or else diminished by ionization. The value of K_{app} is affected by all three Michaelis functions, which complicate the interpretation of plots of K_{app} versus pH. Dixon (1953) gives rules for interpreting the inflections in the plot of pK_{app} versus pH which sort out the various ionization possibilities.

The first order rate constant k' (V_{app}/K_{app}) is proportional to $1/f_E f_S$. If the plot of k' versus pH has an inflection this may be due either to ionization of the enzyme or of substrate. Multiplying k' by f_S corrects for the latter factor; if the corrected plot still has an inflection it is due to enzyme ionization.

The correction for substrate ionization is straightforward. At each pH, f_S for the substrate is calculated; the value for K_{app} is divided by f_S and the value for k' is multiplied by f_S. These corrected values are then used for the diagnostic plots described in the next section, for determining the relevant ionization constants of the enzyme and enzyme-substrate complex.

Data Analysis

Figure 4-3 shows a plot for a hypothetical enzyme reaction as a function of pH. The system has several features which will illustrate the analysis of pH-kinetic data, as well as the inadequacy of the usual practice of measuring a rate at one substrate concentration across a pH range. The substrate is monoprotic with a pK of 7, and the unionized form is the real substrate ($f_S = 1 + K_1/[H^+]$). The enzyme has two relevant ionizing groups, with $pK_{e1} = 5$ and $pK_{e2} = 8.5$. Binding substrate affects only the lower ionization, so $pK_{es1} = 6$ and $pK_{es2} = 8.5$.

The first point to be made is that with $[S] = 3K_M$ as in the figure, a relatively small increase in K_{app} due to either substrate ionization or ES ionization will sharply reduce v. The reduction in v is less drastic if $[S]$ is a larger multiple of K_M. A second point is that if this "pH-optimum" plot is made using several different $[S]$, the pH of maximum activity will vary. In other words, "optimum pH" defined as the pH of maximum velocity at a non-saturating substrate concentration is dependent on $[S]$. A third point is that if a substrate with a much different pK is used the "pH-optimum" shifts quite markedly. It is confusing to read publications from different laboratories investigating the same enzyme which report quite different pH optima; the use of different ionized substrates is often the reason for these discrepancies.

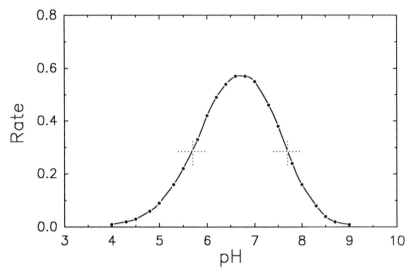

FIGURE 4-3. Rate as a function of pH for a hypothetical enzyme; $V_{max} = 1$, $K_M = 1$, $[S_o] = 3$, $pK_s = 7$, $pK_{e1} = 5$, $pK_{e2} = 8.5$, $pK_{es1} = 6$, $pK_{es2} = 8.5$.

The only meaningful report of the effect of pH on enzyme activity is in terms of changes in the M-M constants as a function of pH. Such a study may be conveniently done in two stages, similar to the approach suggested in Chapter 3 for investigating reversible inhibition. These stages are: 1. obtain the first-order rate constant k' as a function of pH; and 2. determine V_{app} and K_{app} as a function of pH. From analysis of these data (and a knowledge of ionization constants for the substrate, if needed) all the ionizing steps affecting enzymatic rate may be specified.

The most useful way to plot the data is by Dixon plots of $\log(V_{app})$, $-\log(K_{app})$, and $\log(k')$ versus pH. The Dixon plots for the enzyme of Figure 4-3 are shown in Figure 4-4. These logarithmic plots have certain characteristics which are helpful in drawing and interpreting them.

1. The graphs consist of straight-line segments joined by short curved portions.
2. The straight-line segments have integer slopes (i.e., 0, +1, -1, -2, etc.).
3. The extrapolated segments intersect at pH values corresponding to an ionization, either of enzyme, of substrate, or of enzyme-substrate complex.
4. Each pK produces a change of 1 unit in the slope. Two pK's at the same pH produce a slope change of 2 units (or 0 units if the change is in the opposite sense).

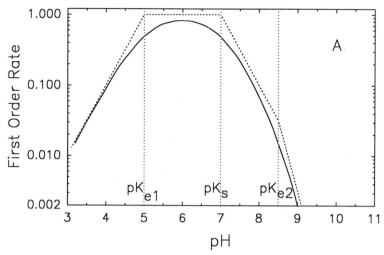

FIGURE 4-4A. Dixon plot for the first order rate.

5. A pK of a group situated in the *ES* complex produces an increase in the slope of the plot (i.e., from -1 to 0 or from 0 to $+1$); a pK of either the free enzyme *E* or the substrate *S* produces a decrease in the slope of the plot (i.e., from $+1$ to 0 or from 0 to -1).

6. The graph of (ideal) data points is displaced from the intersection point by a vertical distance of 0.3 (log 2). If two pK's occur at the same pH the displacement is 0.47 (log 3).

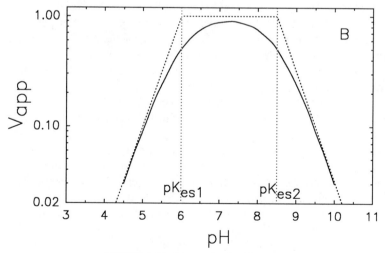

FIGURE 4-4B. Dixon plot for V_{app}.

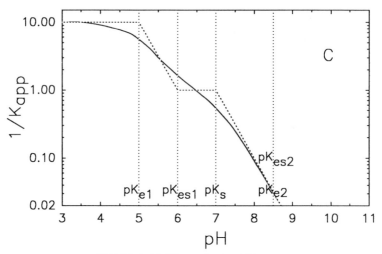

FIGURE 4-4C. Dixon plot for $1/K_{app}$.

The plot of $\log(k')$ versus pH (Figure 4-4A) has four straight line sections. The rate constant increases as pH increases from 3. The first intersection point at pH 5 represents an ionization in free enzyme, pK_{e1}; the slope changes from $+1$ to 0. There is no change in slope at pK_{es1} but the substrate ionization pK_s produces another change in slope to -1. Finally the second enzyme ionization pK_{e2} produces yet another change in slope to -2. From this curve one might conclude that the pH optimum of the enzyme is 6.

The plot of $\log(V_{app})$ versus pH (Figure 4-4B) has a somewhat different appearance. Starting with the segment with slope $+1$ at lower pH, the first bend in the graph is produced by the ionization of the ES complex, pK_{es1}, giving a slope of 0. The next slope change is from 0 to -1 at the pH corresponding to pK_{es2}. Neither pK_{e1} nor pK_s affect this plot. From it the pH optimum might be calculated to be 7.25.

The final plot of $pK_{app}\{-\log(K_{app})\}$ versus pH (Figure 4-4C) is more complicated. The interpretation of these plots is simplified by rule 5 listed above. Starting from low pH, where the graph has a slope of 0, the first ionization is in free enzyme and so causes a downward bend in the plot. The second ionization pK_{es1} causes an upward bend, and the third ionization due to free substrate causes another downward bend. We have specified that the last two ionizations, pK_{e2} and pK_{es2}, occur at the same pH. Since they change the slope in the opposite senses, there is no change in the plot as drawn here. Between pH 5 and 7 the plotted points appear to delineate a straight line with slope -0.5. By inspection of the points alone it would be

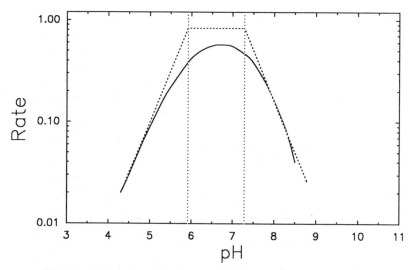

FIGURE 4-5. Semi-logarithmic plot of the rate data of Figure 4-3.

difficult to conclude that this region includes three pK's. The lesson is that in dealing with real-life experimental data, especially in the plot of pK_{app} versus pH, draw all estimated lines through the points with integer slopes, and strive to have the vertical distance between the plot and the intersection equal 0.3 units.

Graphical analysis of experimental data is done as follows. First, plot the log of the parameter (V_{app}, K_{app}, or k') versus pH. Draw a line with integer slope through the data at either the right or left side of the plot. Determine the point where the best smooth line through the experimental data is 0.3 units above or below the first straight line; this is the intersection with the next line segment, and from inspection determine what the integer slope of this line segment will be. Draw the second line segment, find where the experimental data is 0.3 units above or below the line, and draw the next line segment of integer slope through this intersection point. Repeat the process as often as needed to cover the whole range of data.

Figure 4-5 shows this process applied to the rate data used to construct the pH-activity curve of Figure 4-3. The results demonstrate the fallacy of using simple rates at one [S] versus pH for supposedly determining ionization constants in an enzyme. From this plot we might conclude that the ionizing groups in the enzyme active site would have pK's of 5.9 and 7.3. In fact neither of these values represent pK's in the present system. Data of this sort are of little use in learning anything substantive about the relevant ionizing groups in an enzyme system.

If the plot of the log of the parameter versus pH indicates that one or two ionizing groups are involved, that is, it represents a bell-shaped curve such as Figure 4-4B or a sigmoid curve, then more precise estimates of the pK's in the system may be obtained by using one of the computer programs listed in Appendix E. The BELL program fits the two-pK system where the singly-ionized species is the active form. The HABELL program is for the system in which one ionization is involved and the active species is the high-pH form, while the HBBELL program assumes the active species is the low-pH form. Before these programs can be applied the data must be corrected for substrate ionization as outlined earlier, if this is a factor.

Mention must be made of the common practice of determining pK's as the pH at which the rate is half of the maximum rate. In Figure 4-3, for example, these points are at pH 5.7 and 7.7. When the two pK values of a diprotic curve are less than 3.5 pH units apart the half-height pH's deviate from the true pK's. Using H_0, H_1, and H_2 for $[H^+]$ at the peak and the two half-height points, and K_1 and K_2 for the two true ionization constants, then: $H_0 = \sqrt{K_1 K_2}$ and $K_1 = H_1 + H_2 - 4H_0$. Knowing the three $[H^+]$ concentrations from the experimental curve K_1 and then K_2 can be calculated. In the data of Figure 4-3, this correction yields pK's of 5.88 and 7.52. These numbers bear no more relationship to any of the known pK's than do the two original half-height estimates of 5.7 and 7.7, although they are closer to the two "pK's" estimated from the Dixon plot of the rate data. This seeming internal consistency, however, only confirms the statement made at the beginning of Chapter 2: GIGO (Garbage In, Garbage Out). An analysis of the pH dependency of an enzyme rate measured at only one substrate concentration provides little useful information.

The reasons for determining meaningful pK's for the enzyme are two-fold: 1. to predict the activity at any pH of practical interest; and 2. to have some identification of the amino acid side chains which participate in the reaction. The prediction of activity is performed using equations 4-4 and 4-6. The tentative identification of participating amino acid side chains requires additional information.

Table 4-1 shows pK ranges and enthalpy of ionization for the ionizable groups found in proteins. If the side chains are in a normal environment (i.e., exposed to the aqueous solution and able to freely take up or give off a proton) then an experimentally determined pK should correspond to one or two of the indicated groups. If there is a question, the enthalpy of ionization (see Chapter 5) may provide additional direction. Sometimes the side chains are not freely accessible to water, and their pK is different from that indicated in Table 4-1. An example is the aspartyl carboxyl group in the active site of the serine proteases; it is "buried" in the tertiary structure of the enzyme, and is responsible for the pK_e of about 6 seen for the

Table 4–1. pK Values and Heats of Ionization[a] of Ionizing Groups in Proteins

Group	pK(25°C)	ΔH_i (Kcal/mol)
Carboxyl		0 ± 1.5
C-terminal	3.0-3.2	
Aspartyl	3.0-4.7	
Glutamyl	4.0-4.8	
Ammonium		11 ± 1
N-terminal	7.6-8.4	
Lysine ϵ-amino	9.4-10.6	
Complex amines		
Imidazolium	5.6-7.0	7.5 ± 0.5
Guanidinium	11.6-12.6	12 ± 1
Sulfhydryl[b]		
Cysteine-SH	8.4	8.6
Hydroxyl		
Tyrosine phenolic	9.8-10.4	8.6

[a] Source: Cohn, E.J. and J. T. Edsall (1943).
[b] Source: Sober, J.A. (1970). Page J-98.

activity of these enzymes. It is not warranted to categorically state that an experimental pK corresponds to a given ionizable group as listed in this table; nonetheless, if the only data available are properly run pH activity curves, then a pK of 4.5 strongly suggests that a carboxyl group is involved in the catalytic mechanism.

EFFECTS ON ENZYME STABILITY

It is common to find that enzymes denature at room temperature at high and/or low pH. This fact is responsible for part of the so-called "pH optimum" often seen in literature. Figure 4-6 demonstrates the results of two kinds of experiments which discriminate between diminution of activity due to ionization and loss of activity due to denaturation (irreversible loss of enzyme activity). The solid curve represents results which might be obtained by performing the assay over a wide pH range. The dotted curve, on the other hand, might be obtained if enzyme in solution is adjusted to the indicated pH, held for a period of time, then assayed at a neutral (and constant) pH.

The interpretation of these two curves is straightforward. At pH 4 the enzyme is stable, as shown by the dotted curve. Therefore the decrease in activity in moving from pH 7 down to pH 4 is in fact due to protonation of some group rendering the enzyme unable to catalyze the reaction. At high pH the enzyme denatures when incubated for a period of time. The loss of

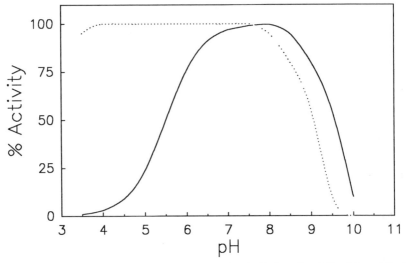

FIGURE 4-6. Effect of pH on enzyme activity (*solid line*) and stability (*dotted line*).

activity when the enzyme is assayed at, say, pH 10 is due to denaturation during the incubation period for the assay. A method of confirming this conclusion is to arrange the experiment so that the enzyme is exposed to the pH for the shortest possible time, and measure the reaction velocity as quickly as possible. As an example, most bacterial serine proteases are reported to have a "pH optimum" between 7 and 9, with the activity falling off at higher pH. Using a sensitive spectrophotometric arrangement so that the rate could be accurately measured within the first 10 seconds after addition of enzyme, it was found that the rate remained at a constant maximum plateau up to pH 13.

The influence of pH upon the rate of denaturation of enzymes is connected to thermal denaturation of enzymes. When we speak of pH regions of stability, we mean that at ambient temperatures the rate constant for denaturation is so small that we do not detect any loss of activity over the length of time we are willing to spend monitoring the enzyme solution. Put another way, at 25°C the half-life of the enzyme may be 3 weeks at pH 7, 2 days at pH 6, 5 hours at pH 5, 30 minutes at pH 4, 3 minutes at pH 3, and 20 seconds at pH 2. In normal parlance we would consider the enzyme stable above pH 5 and unstable at lower pH's. If the temperature were raised by 10°C the half-lives might be decreased by a factor of 50, to 10 hours at pH 7, 1 hour at pH 6, 6 minutes at pH 5, etc. (this would be an unstable enzyme, but certainly not uncommonly so), and the "range of pH stability" would be above 6.5.

In other words "stability" and "instability" of an enzyme are relative adjectives, taking their significance from our intended use of the enzyme. If the work is going on in a laboratory and the enzyme is kept refrigerated except for short periods of time then the characteristics given in the previous paragraph would represent adequate stability. If the work entails application to a food processing line with temperatures around 50°C for several hours, then the enzyme would be woefully unstable and inapplicable. For precise definition, then, "stable" should be replaced by a numerical expression which gives the pH and temperature dependence of the first-order rate constant for denaturation. (For easy comprehension this might be accompanied by a few half-life examples such as given above.) In Chapter 5 the method for determining the rate constant under given conditions will be explained, as well as the manner of expressing the temperature dependence of the rate. The way of finding the pH dependence is explained next.

The denaturation of an enzyme may be modeled as follows:

$$E_{act} + n[H^+] \xrightarrow{k} E_{inact} \qquad \frac{-d[E_{act}]}{dt} = k[E_{act}][H^+]^n \qquad (4\text{-}7a)$$

In words, the denaturation rate constant k is first-order in enzyme and n-order in H^+. The pseudo-first-order rate constant for enzyme denaturation is measured at several pH values and has the significance:

$$k = k_0[H^+]^n$$

$$\log(k) = \log(k_0) + n\log[H^+] = \log(k_0) - n(\text{pH}) \qquad (4\text{-}7b)$$

A simple plot of the log of the measured first order denaturation rate constants versus pH should be a straight line with a slope of $-n$ and an intercept of $\log(k_0)$. This value of k_0 is the rate constant for denaturation at pH 0 (i.e., $[H^+]_{act} = 1\ M$). Like the intercept A in the Arrhenius equation (see Chapter 5) it represents an unrealistic condition in practical terms, but is needed to predict the observed k under use conditions.

If n were 0 the enzyme would have the same denaturation rate at all pH values, a situation not often seen. More commonly in the acid range n is found to be 1 or 2, and in the alkaline range -1 to -3. The interpretation of the significance of these numbers is speculative and not often convincing in the present state of our knowledge about protein structure stability. In practice, as indicated earlier, knowing the value of n and the pH range over which it applies allows the prediction of k for any desired conditions.

BUFFERS FOR ENZYME ASSAYS

Since enzyme activity is sensitive to the pH of the reaction, buffers are often used to keep that pH relatively constant during the course of a rate measurement or an assay. The choice of the weak acid or base to use as the buffer ion can cause unexpected complications if some care is not exercised. There are at least three factors to consider.

1. Many weak acids or bases complex metal ions. Citrate, for example, binds Ca^{++}, and glycine will complex Cu^{++}. If the enzyme requires one of these metal ions for activity the use of that buffer ion could result in a lowering of observed rate.
2. Inorganic anions are sometimes observed to exert either inhibitory or activating effects on enzymes. The mode of activation is not well understood, but it is not surprising that phosphate anion might inhibit a phosphatase (product inhibition).
3. The weak amine bases usually have a rather large enthalpy of ionization (i.e., the pK changes as temperature changes). Thus a particular composition of glycine/glycine hydrochloride might buffer at pH 8 at 25°C, but at 35°C the pH would be 7.7.

In general it is a good idea to run the assay with two or three different buffers at the pH and temperature of interest, to find out if any inhibitory or activating effects are present. While it is not always possible to completely avoid such complications, if they are explicitly recognized they should not lead to any nasty surprises when the research results are put into use in practical situations.

Anionic Buffers

The most commonly used buffers, and indeed almost the only ones available for many years, are mixtures of the salt and acid of a weak acid. These may be either inorganic acids such as phosphoric or boric, or organic carboxylic acids such as acetic or citric. These buffers have several advantages: they are inexpensive, quite stable in aqueous solution, and have low enthalpies of ionization. On the negative side they often complex metal ions (an advantage if the metal ion is inhibitory) and may inhibit or activate the enzyme in an unpredictable manner. If the possible disadvantages can be ruled out in the case of the enzyme under study, anionic buffers are probably the best choice.

The most useful buffer ions are listed in Table 4-2. Since a buffer is usually considered useful over a pH range of ±0.9 pH units from the pK, the acids (and conjugate salts) listed in Table 4-2 cover the pH range 1.22 to 13.22, quite wide enough for almost any enzyme research. If a study is to cover the pH range 7 to 10, for instance, a mixed buffer of phosphate plus borate might be used. If the desired pH range is narrower, a single anion might be sufficient (i.e., phosphate for pH 6.5 to 7.5, or acetate for the range 4 to 5.5).

Universal buffer. If the study being planned covers quite a broad pH range the buffer used should be a mixture of weak acids such that a good buffer capacity is obtained anywhere in the range, but the anions to which the enzyme is exposed do not vary. Teorell and Stenhagen (1938) constructed such a wide-range buffer which has fairly uniform buffering capacity from pH 2 to pH 12. One liter of stock solution is made from the amounts of citric acid, sodium phosphate, boric acid and NaOH given below. For use, 20 ml of stock solution is mixed with x ml 0.1N HCl and diluted to 100 ml. The composition of the stock solution and some representative pH mixtures are given.

Universal buffer stock solution	pH	x
3.54 g boric acid, H_3BO_3	2	73.3
7.00 g citric acid monohydrate	4	50.5
4.60 g monobasic sodium phosphate,	6	39.4
$NaH_2PO_4 \cdot H_2O$	8	28.0
310 ml 1 N NaOH	10	17.9
Make to 1 L with distilled water.	12	0.4

Cationic Buffers

Because a large number of enzyme systems require metal ions for proper functioning, the metal-binding properties of anionic buffers presented a problem. About 20 years ago Good et al. (1966) prepared and studied a large number of buffers based upon weak amine bases of various kinds. Table 4-2 lists several of the more commonly used examples of this category. The accessible pH range is narrower than with the anionic buffers, covering about 5 to 9, but this range is suitable for the bulk of studies which are concerned with enzymes and their physiological functioning.

As mentioned earlier, these compounds have a fairly large enthalpy of ionization, and a ten degree increase in temperature will cause a decrease in the pH of a given buffer solution of 0.1 to 0.3 pH units. If the pH of the buffer is checked with a glass electrode and pH-meter at the temperature of use (a good practice in any event) this should not cause any problem. As with anionic buffers, it is desirable to check the activity of an enzyme at a given pH with two or three of these buffers. For unknown reasons some of them are inhibitory towards certain enzymes. For example, at pH 7 the bacterial metallo-protease thermolysin has only 40% as much activity in TRIS buffer as in PIPES or HEPES buffer (Stauffer 1971).

Table 4–2.   Some Buffers for Enzyme Assays

Anionic Buffers

Acid	pK_1	pK_2	pK_3
Phosphoric	2.12	7.21	12.32
o-Phthalic	2.89	5.41	-
Citric	3.06	4.74	5.40
Succinic	4.19	5.48	-
Acetic	4.76	-	-
Carbonic	6.37	10.25	-
Boric	9.24	-	-

Cationic Buffers

Chemical name	Acronym	$pK(20°C)$
2-(N-Morpholino)- ethanesulfonic acid	MES	6.15
Piperazine-N,N'- Bis(2-ethanesulfonic acid)	PIPES	6.8
N-2-Hydroxyethylpiperazine- N'-2-ethanesulfonic acid	HEPES	7.55
Tris(hydroxymethyl)- aminomethane	Tris	8.3

A few points about the experimental use of buffers should be made.

1. Do not rely on mixing stated proportions of stock solutions to achieve the desired pH. Always check a newly-made buffer with a properly-calibrated pH-meter and glass electrode.
2. Enzymes may be sensitive to the ionic strength of the assay solution, but this is usually a factor only when $\mu < 0.01$. Buffers are usually greater than 0.05M, so this should not be a problem. If ionic strength effects are suspected, make the buffer 0.1M in NaCl to swamp them out.
3. Most buffer salts are chemically stable in solution at room temperature. However, the solutions are normally not microbiologically sterile. Concentrated stock solutions may be stored under refrigeration, and diluted working buffer should not be held for more than a day or two.
4. It is recommended that the distilled water used to make up buffers for standardizing glass electrodes should be maintained free of carbon dioxide. In enzyme assays the various solutions are usually open to the air, so it is more convenient to have the buffer in equilibrium with ambient CO_2 and correct for the carbonate buffering involved by checking the buffer with the pH-meter.

5   Effect of Temperature

Increasing the temperature of an enzyme reaction has two disproportionate effects; it increases the rate of the reaction proper, and it increases the rate of inactivation of the enzyme. As a rule of thumb, an increase of 10°C will cause the reaction rate to double, while the inactivation rate will increase 64-fold. The experimental perception of this unequal effect depends upon time. At room temperature the denaturation rate constant is so small that even if it is multiplied by 64 the extent of inactivation is unnoticeable during the length of time taken for the assay. The doubling of the amount of product formed during this time, however, is readily apparent.

Figure 5-1 illustrates this idea. The enzyme is incubated with substrate at various temperatures for a period of time. The curvature in the graph of product concentration versus time (Fig. 5-1A) is due to progressive denaturation of enzyme at the higher temperatures. The perception of the relative temperature effects on activity and inactivation will depend upon which time scale is used. The plots in Figure 5-1B are typical of the so-called "temperature optimum" curves which appear in "technical data sheets." As is apparent the "optimum" depends upon the length of time required for the assay. If the initial velocity of the reaction could be measured quickly before any inactivation occurred, the rate versus temperature plot would increase exponentially without limit.

This discussion of the effect of temperature upon enzymes is in two parts, dealing with the energetics of the reaction being catalyzed, and with the effect upon the rate of enzyme denaturation. The topics are interdependent, but separating them makes the explanations clearer.

ACTIVATION ENERGY AND ENZYMATIC CATALYSIS

The Van't Hoff Equation

When a system is in rapid equilibrium, the free energy difference between the reactants in the beginning state and those in the final state is:

$$\Delta F° = \Delta H° - T\Delta S° = -2.3RT \log K \qquad (5\text{-}1)$$

79

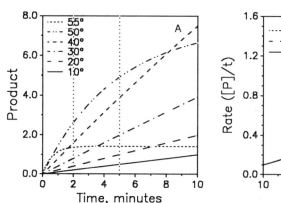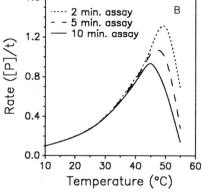

FIGURE 5-1. Product concentration versus time at various temperatures (*left*) and effect of assay time on apparent "temperature optimum" (*right*).

K here may be any dissociation equilibrium constant, K_M, K for ionization of a charged group, etc; R is the gas constant (1.987 cal deg^{-1} mole^{-1}) and T is temperature in °K (°C + 273.16). From equation 5-1 $\Delta F°$ may be calculated directly from the equilibrium constant.

Equation 5-1 may be rearranged to:

$$- \log K = \left(\frac{\Delta H°}{2.3R} \right)\left(\frac{1}{T} \right) - \frac{\Delta S°}{2.3R}$$
(5-2a)

Plotting $- \log K$ versus $1/T$ gives a straight line with a slope of $\Delta H°/2.3R$ and an intercept of $-\Delta S°/2.3R$. The differential form with respect to $1/T$ is the Van't Hoff equation:

$$\frac{d \ln K}{d(1/T)} = \frac{-\Delta H°}{R}$$
(5-2b)

If the equilibrium constant is the ionization pK determined experimentally at several temperatures, then 2.3R times the slope of the plot of pK versus $1/T$ is the enthalpy of that ionization, and may be used to tentatively identify the nature of the group from the data presented in Table 4-1. If the constant is K_m and the system is truly in a rapid equilibrium, $\Delta F°$ is the free energy difference between ES and $E + S$; if the system is in fact a steady-state system then $\Delta F°$ is an apparent value which is actually a complicated function of the various rate constants which make up the M-M parameter K_s. Since inhibitor constants K_i are true equilibrium constants the free energy parameters are actual thermodynamic values.

The Arrhenius Equation

Energy of Activation. The frequency with which molecules in a homogeneous solution collide depends upon their concentration (i.e., collision frequency $= Z[E][S]$ where Z is a constant normalizing to unit time and unit volume). Formation of a reaction product $[ES]$ depends upon the probability P that the colliding molecules have the proper orientation (substrate must contact that part of the enzyme which comprises the active site) and also upon the presence of sufficient energy to enable the reaction to occur. This minimum energy is called the *activation energy*, E_a. From the Maxwell-Boltzman law the fraction of collisions which entail at least this much energy is proportional to $\exp(-E_a/RT)$. The rate at which ES forms by collision of E and S is:

$$\text{rate of productive collisions} = PZ[E][S]\exp\left(\frac{-E_a}{RT}\right) \qquad (5\text{-}3a)$$

But this rate is also given by:

$$\text{rate} = k_1[E][S] \qquad (5\text{-}3b)$$

so by comparison, we arrive at the Arrhenius equation:

$$k_1 = PZ\exp\left(\frac{-E_a}{RT}\right) \qquad (5\text{-}3c)$$

This may be linearized by taking logs, substituting the more usual A for the constants PZ, and noting that k is the specific reaction rate constant for any rate.

$$\ln k = \ln A - \left(\frac{E_a}{R}\right)\left(\frac{1}{T}\right) \qquad (5\text{-}4a)$$

The plot of $\ln k$ versus $1/T$ is (usually) linear with a slope of $-E_a/R$ (Figure 5-2). The intercept A reflects both frequency and probability factors, and is usually of little practical interest. Knowing the activation energy for the reaction, however, is useful in predicting the rate of the reaction at a temperature T_2 when the rate is known at T_1. The integrated form of equation 5-4a is:

$$\ln\left(\frac{k_2}{k_1}\right) = \left(\frac{E_a}{R}\right)\left(\frac{T_2 - T_1}{T_2 T_1}\right) \qquad (5\text{-}4b)$$

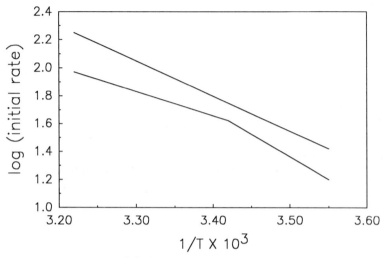

FIGURE 5-2. Arrhenius plot.

While the Arrhenius plot is generally linear over the temperature range of interest, occasionally plots are found which bend in the middle. This is often due to a change in the rate-limiting step of a sequential reaction. For example, the hydrolysis of certain esters by carboxypeptidase A at very low temperatures occurs with intermediate acyl-enzyme formation. At $T < 273°K$ the rate limiting step appears to be deacylation, while at higher temperatures the acylation step is rate limiting. The Arrhenius plot of log(overall rate) versus $1/T$ shows an inflection around that temperature. This behavior can provide insight into the mechanism of enzyme action if the researcher is astute enough to interpret it correctly. If the plot breaks sharply, giving a positive slope on the left side, this is usually due to denaturation of the enzyme.

Transition State Theory.   While collision theory allows us to systematize temperature effects on rate data it does not provide an explanation of the need for a minimum energy for fruitful reaction. Eyring in 1935 published his transition state theory which explained some observations, but more importantly stimulated other hypotheses and approaches to studying enzymatic reactions. The theory proposes that in a reaction the reactants pass through a high energy, unstable "transition state" before proceeding on to the product. The activation energy is needed to reach this transition state and is used to distort bonds and orientations so that they can then be reordered and/or reoriented into products. If reactants do not acquire this much energy, they simply return to the original state. The reaction can be monomolecular;

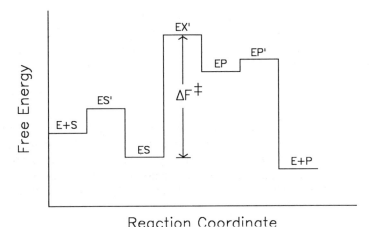

Reaction Coordinate

FIGURE 5-3. Free energy plot for the simplest enzyme reaction.

the reaction from enzyme-substrate complex ES to enzyme-product complex EP is first order, even though there is not a covalent bond between enzyme and substrate/product.

Figure 5-3 depicts the energy levels at the different stages of the enzyme reaction characterized by the following equilibria:

$$E + S \overset{ES'}{\rightleftharpoons} ES \overset{EX'}{\rightleftharpoons} EP \overset{EP'}{\rightleftharpoons} E + P$$

The three primed species represent the transition state between the reactants and products. It is an assumption that if reactants achieve sufficient energy to reach the transition state the reaction will proceed to product. In other words, if ES' is reached from $E + S$ the "momentum" will carry the reaction on to form ES, while if ES' is formed from ES the "momentum" will result in dissociation to $E + S$. (In transition state theory terms, this is the statement that the partition coefficient is 1.)

The energy for converting ES to EP can be analyzed as follows. If we assume that ES molecules have varying energies according to the Maxwell-Boltzman distribution, and some molecules are climbing towards the threshold energy level while others are falling back to the ground state, then the ground and transition states are in a pseudo-equilibrium described by the constant:

$$K^{\ddagger} = \frac{[EX']}{[ES]} \text{ and } [EX'] = K^{\ddagger}[ES] \qquad (5\text{-}5a)$$

This equilibrium constant bears the same relationship to $\Delta F^{\ddagger}$, $\Delta H^{\ddagger}$, and $\Delta S^{\ddagger}$ as given in equation 5-1 for the true equilibrium constant.

The rate of formation of EP is equal to a rate constant τ times the concentration of the transition state, or:

$$\text{rate} = \tau[EX'] = \tau K^{\ddagger}[ES] \tag{5-5b}$$

This rate constant is the vibrational frequency of the reacting bond in the transition state, and is given by the expression $\tau = k_B T/h$ where k_B is the Boltzman constant (1.38×10^{-16} erg deg^{-1}) and h is Planck's constant (6.624×10^{-27} erg-sec). The rate of formation of EP is also given by: rate $= k_2[ES]$ and since (from equation 5-5b) rate $= (k_B T/h)K^{\ddagger}[ES]$ then:

$$k_2 = \left(\frac{k_B T}{h}\right)K^{\ddagger} = \left(\frac{k_B T}{h}\right)\exp\left(\frac{-\Delta F^{\ddagger}}{RT}\right) \text{ and} \tag{5-5c}$$

$$\Delta F^{\ddagger} = -RT\ln\left(\frac{k_2 h}{k_B T}\right) \tag{5-5d}$$

If we substitute $\Delta H^{\ddagger} - T\Delta S^{\ddagger}$ for $\Delta F^{\ddagger}$ in equation 5-5c we get:

$$k_2 = \left(\frac{k_B T}{h}\right)\exp\left(\frac{-\Delta H^{\ddagger}}{RT} + \frac{\Delta S^{\ddagger}}{R}\right) \tag{5-6a}$$

$$\ln k_2 = \ln\left(\frac{k_B}{h}\right) + \ln T - \frac{\Delta H^{\ddagger}}{RT} + \frac{\Delta S^{\ddagger}}{R} \tag{5-6b}$$

Differentiating with respect to $1/T$ we get:

$$\frac{d(\ln k_2)}{d(1/T)} = \frac{-(\Delta H^{\ddagger} + RT)}{R} \tag{5-6c}$$

A plot of $\ln k_2$ versus $1/T$ has a slope of $-(\Delta H^{\ddagger} + RT)/R$. Comparison with the Arrhenius equation, 5-4a, shows that the activation energy $E_a = \Delta H^{\ddagger} + RT$. The correction is not negligible. For most enzyme reactions the activation energy is in the range of 6 to 15 Kcal/mole. The correction factor RT equals about 600 cal/mole at room temperature, or 4% to 10% of the activation energy.

From the plot shown in Figure 5-2 the enthalpy of activation for the reaction in question can be calculated. If the absolute value of k_2 is known $\Delta F^{\ddagger}$ can be calculated from equation 5-5d, and $\Delta S^{\ddagger}$ from $(\Delta H^{\ddagger} - \Delta F^{\ddagger})/T$. This is possible if V_{max} were measured, and if the true molar concentration of enzyme in the reactions were known, since $k_2 = V_{max}/[E_t]$. In the absence of such absolute values for rate constants, the only useful parameter which can be obtained is the enthalpy of activation for the step.

In certain cases the individual steps in an enzyme catalyzed reaction can be dissected thermodynamically. For the most part this has been done with proteolytic enzymes such as trypsin for which absolute molar concentrations can be obtained, making possible the determinations of total free energy, entropy, and enthalpy of activation for each catalytic step. However, for most practical enzyme applications this much detailed information is not necessary; knowing the activation energy for V_{max} and K_M allows the calculation of reaction rate at some temperature other than that at which the fundamental measurements have been made. If a fuller exploration of enzyme thermodynamics is of interest, the review by Laidler and Peterman (1979) is a good place to start reading.

THERMAL ENZYME DENATURATION

Measurement of Inactivation Rate

The spontaneous, irreversible denaturation of enzymes at elevated temperatures is a process which is first-order with respect to enzyme.

$$\frac{-d[E]}{dt} = k[E]; \quad \frac{-d[E]}{[E]} = k \ dt \ ; \quad \ln[E]_t = \ln[E]_0 - kt \qquad (5\text{-}7)$$

Operationally, an enzyme solution is raised to the test temperature as rapidly as possible. (This might be done by diluting a small amount of concentrated enzyme stock solution into a large amount of buffer which is at a temperature slightly higher than the test temperature.) At various times an aliquot of the solution is cooled quickly to stop the denaturation reaction, the activity in the aliquot is assayed, and ln of activity is plotted versus time. The result is a straight line of slope $-k$. If the results are plotted on semi-logarithmic paper (log[activity] vs. t) then the slope equals $-2.303k$ (Figure 5-4).

The entire process is quick, uncomplicated, and provides a very useful parameter, the first-order denaturation rate constant k. Using k measured at several different temperatures (at constant pH) the activation energy (enthalpy) of denaturation is calculated. If the absolute molar concentration of the enzyme is known (i.e., if a titration assay is available), then the free energy and entropy of denaturation can also be found. Finally, the pH dependence of k is found at one temperature, chosen so that the denaturation rates are measurable at both ends of the pH range studied.

Having this data one may write an expression which will predict the rate of denaturation of the enzyme at any temperature and any pH within the range of valid extrapolation. This extrapolation should not be too far beyond the

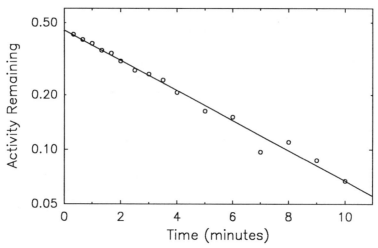

FIGURE 5-4. First-order rate of loss of enzyme activity (Stauffer and Treptow 1973).

bounds of the experimental data, because changes in the mechanism (rate-limiting steps) of inactivation can change the parameters for energy and H^+ influences. The denaturation rate constant dependency upon pH is given in equation 4-7b:

$$\log(k) = \log(k_0) - n(\text{pH})$$

Let us specify that this determination is made at 25°C, so $T = 298.16°\text{K}$. By the integrated form of the Arrhenius equation, 5-4b, the rate at pH = 0 at any other temperature T is given by:

$$\log(k_0)_T = \left(\frac{2.303E_a}{R}\right)\left(\frac{T - 298.16}{298.16 \times T}\right) + \log(k_0)_{25} \qquad (5\text{-}8a)$$

Then the rate constant for denaturation at any pH and temperature T (within the bounds of valid extrapolation) is given by:

$$\log(k)_{\text{pH},T} = \left(\frac{2.303E_a}{R}\right)\left(\frac{T - 298.16}{298.16 \times T}\right) + \log(k_0)_{25} - n\text{pH} \qquad (5\text{-}8b)$$

This equation can be quite useful for predicting the stability of enzymes in practical applications. For example, the stability of an enzyme at ambient conditions may be needed in planning a process. To measure the denaturation rate in a convenient time-frame (minutes to an hour) measurements may be made in the pH range of 3 to 5 and temperatures of 40°C to 60°C. With the parameters E_a, k_0 and n the value of k at, say, 25°C and pH 6 is readily calculated. The half-life, or the length of time required to lose 50% of the activity, is given by $\ln 2/k = .693/k$. This time may be a matter of days or even weeks which is inconvenient to measure experimentally but may be quite significant in a practical sense. Other uses for equation 5-8b will no doubt occur to the researcher involved in useful applications of enzymes.

Other solvent characteristics besides pH and temperature can influence the rate of enzyme inactivation. Chaotropes, detergents, and organic solvents also have an influence. Remarkably, in a nearly dry ($< 0.5\%$ water) n-butanol solvent, pancreatic lipase retains its activity for days at 100°C (Zaks and Klibanov 1984). Whenever you are investigating the effect of these other experimental parameters on enzyme stability, the only meaningful characterizing value is the first-order rate constant for activity loss.

Apparent Non-First-Order Inactivation

On occasion the plot of log[activity] versus time will appear to be nonlinear. While the data may seem to fit the second-order rate law, such an interpretation of the reaction should be tested most stringently before it is accepted. I know of no instances in which enzyme denaturation has been shown to be a true non-first-order process. An excellent test of the hypothesis is as follows. The inactivation rate may be written as $-d[E]/dt = k[E]^n$ where n is the molecularity of the reaction. The rate of loss of activity at very short times (i.e., the slope of the plot of $-$[activity] versus t) is measured at several different initial enzyme concentrations $[E_0]$. Then log(rate) is plotted versus $\log[E_0]$; the slope of this plot is n. A reaction which is in fact second order will have $n = 2$. This is a minimum requirement for establishing that inactivation of a particular enzyme is not a first-order process.

There are two factors which may lead to apparent non-first-order inactivation kinetics: stabilization by substrate; and the presence of two or more enzymes which catalyze the assay reaction.

Substrate stabilization is particularly likely to appear in studies on proteolytic enzymes. As enzyme denatures it becomes ordinary protein (i.e., substrate for the remaining active enzyme). In a study on subtilisin (Stauffer and Treptow 1973) it was found that in the basic pH range where the enzyme is active the plot of log[activity] versus time was linear for only about the first half-life of the denaturation reaction. The log(rate) versus $\log[E_0]$ plot, however, confirmed that the process was indeed first order. At acidic pH's, where the enzyme does not bind to substrate, the plot of log[activity] versus time was linear for at least three half-lives.

With other enzymes the presence of unsuspected endogenous substrate in the enzyme preparation would contribute to stabilization. The substrate would be removed by conversion to product at the same time the inactivation was occurring, leading to an increase in the rate constant. This would

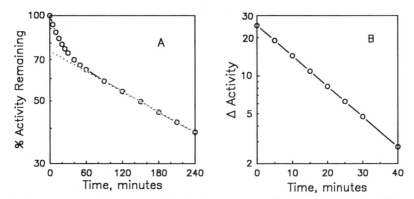

FIGURE 5-5. Loss of enzyme activity in the presence of two species having different stabilities (Erlanger et al. 1964). A. Plot of loss of total activity. B. Plot of difference at early times.

produce a downward curvature in the log[activity] versus time plot. The experimental data might fit as well to a zero-order line as to the first-order plot. Again the log(rate) versus log$[E_0]$ plot will clarify the situation.

The situation involving more than one enzyme is more common. It can appear unexpectedly in studies of what is thought to be a pure enzyme. For instance Erlanger et al. (1964) studied the thermal denaturation of thrice-crystallized α-chymotrypsin in the absence of Ca^{++}. They found that 25% of the enzyme denatured quickly, with a half-life of 12.5 minutes, while the remaining 75% had a half-life of 252 minutes. When the data of activity versus time is analyzed, it gives a rather good fit to a second-order line. The actual analysis which revealed the true situation is shown in Figure 5-5.

This is the plot of two concurrent first-order reactions, one often seen in dealing with enzymes and proteins (e.g., in chemical derivatization of amino acid side chains). The experimental data is plotted on semi-logarithmic paper. The early part of the plot is curved, followed by a straight line portion. The straight line is extrapolated back to zero time; the intercept represents the fraction of the overall response which is slow-reacting, and the slope is $-2.303k_{slow}$. The vertical distance between this straight line and the actual data represents the amount of fast-reacting enzyme present. This distance is replotted versus t to give the first-order line representing the fast reactant, with a slope of $-2.303k_{fast}$. Theoretically any number of concurrent first-order reactions could be analyzed in this fashion, but practically the limit is about three. Above that number the decisions as to what constitutes a straight-line portion of the plot become quite problematical.

Two factors contributed to the unexpectedness of this result. First, as mentioned earlier, the enzyme had been crystallized three times, which would certainly be expected to result in high purity. Second, the denaturation conditions were 37°C, pH 7.6; mild conditions indeed for an enzyme such as chymotrypsin. The difference was that Ca^{++} was carefully excluded from the medium. This single change in experimental conditions made the heterogeneity of the enzyme preparation detectable. The moral of the story is: (almost) Never trust a non-first-order rate of enzyme inactivation!

6 Assay Principles

An enzyme assay is performed to answer one of two questions: How much enzyme is present? or How much substrate is present? The response of the assay is a measurement of product concentration (sometimes detected as a decrease in substrate concentration) which, expressed as a function of the time during which the reaction proceeds, is the rate v which is related to the two quantities in question by the Michaelis-Menten equation:

$$\frac{d[P]}{dt} = v = k_p \left(\frac{[E_t][S]}{K_M + [S]} \right)$$

The conditions of the assay are fixed (in general) so that k_p and K_M are constant and/or known. If the question being asked is how much enzyme is present, then $[S]$ is fixed; if the question relates to substrate concentration, then the amount of enzyme used is fixed. Examples of the uses of the two kinds of assays are: standardizing an enzyme preparation for a baking adjunct; and measuring glucose concentration with glucose oxidase.

A third use of enzyme assays is to measure the amount of an inhibitor. In its most common guise (e.g., determining the amount of trypsin inhibitor in legume meal) this is really a variant on the question: How much enzyme is present (after incubation with the high-affinity trypsin inhibitor)? The assay response v is assumed to be proportional to $[E]$.

In general the values of k_p and K_M are not accurately known. The usual practice is to run a "standard curve" to get an implicit proportionality factor which involves k_p, K_M and either $[E_t]$ or $[S]$. The concentration of enzyme is then expressed in some sort of "units," either arbitrary ones defined by the researcher, or the EC-defined "katal," the amount of enzyme activity which catalyzes the reaction at the rate of one mole per second. Under usual assay conditions a convenient unit is the nkatal, or 10^{-9} moles per second. The standard curve for substrate measurement is constructed with known molar concentrations of $[S]$, so the assay results are readily expressed in moles per liter.

In choosing an assay the first set of questions to be settled is: What are we really trying to measure, and how accurately do we need to measure it? The second set of questions is: How might the proportionality factor change during the course of our proposed assay, and how can we prevent this or correct for the change? Subsequent questions deal with more practical factors: speed of performance, cost of reagents, convenience of preparation, sophistication (instrumental and personnel) required, possible interference by materials in the enzyme sample, etc. When good answers to the first two sets of questions are established, the possibilities inherent in the answers to the subsequent questions can be sifted and the best assay to suit your particular needs can be chosen.

INITIAL RATE MEASUREMENTS

The Michaelis-Menten expression for relating enzyme rate to $[E]$ and $[S]$ implicitly assumes that v is the tangent at time zero to the plot of $[P]$ (or $-[S]$) versus t. In practice this tangent cannot be unequivocally determined. Initial rate measurement assays try to establish conditions so that the rate which is experimentally accessible is the same as the true initial rate within an acceptable margin of error.

The initial rate assay is the most generally useful type of enzyme assay. A sound initial rate assay is required in order to study reversible inhibition and substrate inhibition, and for making meaningful kinetic studies of two-substrate systems. Monitoring of enzyme concentrations during purification is best done with initial rate assays since these are usually designed to be rapid and accurate. The fixed time assay, discussed later in this chapter, is really a form of initial rate assay in which two points ($0,0$ and $[P], t$) are used to calculate the rate $d[P]/dt$ under the assumption that the non-linearity described next is avoided.

Some finite period of time is required to record enough data sets $[P], t$ to determine the slope of the $[P]$ versus t plot. One factor, then is how many points are required to define the slope with an acceptable experimental error, and how rapidly can those points be recorded. At one extreme, two points will define a slope with zero statistical variance. If one point is taken as $0,0$ then the possible error in the slope (in v) is a function of the experimental errors in $[P]$ and in t. The experimental error in these data is proportional to $1/[P]^4$ and $1/t^4$, so small values will contribute to a large experimental error in v. This method of establishing the initial rate is related to the fixed time measurements to be considered below, although in those measurements the time is usually an order of magnitude greater than is being considered here.

In practice one tries to gather a large number of $[P], t$ data points for determining the slope, most often by continuously recording some function of [P] or [S]. Then the slope of the [P] versus t plot is obtained by drawing a straight line through the recorder trace or, in a more sophisticated laboratory, sampling the instrumental output at small time intervals with a computer and making a Linear Least Squares fit to the points.

The recorder trace deviates from strict linearity after some period of reaction time. This is primarily due to the fact that as [S] decreases the proportionality factor (the M-M equation) decreases. Two methods for minimizing this deviation are used.

1. Start the assay with [S_0] large enough that the fraction ([S_0] − [P])/(K_M + [S_0] − [P]) does not sensibly deviate from its initial value. If, for example, [S_0] = 5 K_M then when 5% of [S_0] has been converted to [P] the fraction is 4.75/5.75 or .82609 compared to the initial 5/6, or .83333. The deviation is 0.87%. If 20% of [S_0] is converted the fraction is 4.00/5.00 = .80, a 4.0% deviation.

2. Increase detection sensitivity so that the instrumental experimental error in measuring [P] is small, and the amount of [P] required to achieve this low experimental error is a small fraction of [S_0]. This is particularly necessary if some factor (solubility or large instrumental response by S) limits the usable concentration of [S_0]. If [S_0] = K_M, then at 5% reaction the proportionality fraction is .95/1.95 or .48718, a deviation of 2.56%. If [P] is accurately measurable at 1% reaction the deviation is 0.5%.

Two other factors may also contribute to a decrease in the slope of the [P] versus t plot; product inhibition and enzyme denaturation. The fall-off in slope is more rapid in these cases than that due simply to substrate depletion.

Many ways have been proposed for finding the true initial velocity v_0 from curvilinear [P], t plots. Some require more mathematical manipulation of data than others, but a simple yet accurate method is shown in Figure 6-1, based upon the computer-generated data given in Table 6-1. The measure of [P] at several times might be of any nature: absorbance from the recording of a spectrophotometric assay, ml titrant added in a pH-stat assay, mV reading in an O_2-electrode assay, etc. The value of [P]/t is calculated at several times, and the best straight line is fitted to the [P]/t, t data pairs to find the intercept ([P]/t at $t = 0$). The extent of the reaction has been exaggerated to show the principal involved. It is unlikely that one would run an initial rate assay to the extent of 30% to 50% conversion of substrate. The results are surprisingly good. For the uncomplicated, product inhibited, and denatured cases the percentage deviations from the "true rate" are 0.36%, 0.02% and 3.54% respectively. If only the first four data points are used to make the extrapolation the deviations are even less: 0.08%, 0.02% and 0.94%.

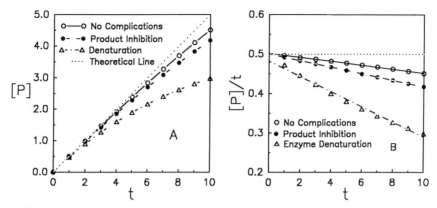

FIGURE 6-1. Obtaining initial rate from time-dependence of product concentration. A. [P] as a function of time. B. Extrapolation of [P]/t to time zero.

The improvement in predicted v_0 when fewer points are used is an expression of the second principle given above. When accuracy and sensitivity in measuring [P] are improved so that the data points can be gathered at shorter times and/or at lower percent reaction the accuracy of estimating v_0 is also improved. However, unless you are quite sure that the [P], t plot is linear, merely fitting a straight line through the early data in this plot is not the method of choice. The slope of the LLS line through the first four [P], t

Table 6–1. Nonlinearity of Product Versus Time Plots[a]

Time	No Complications [P]	[P]/t	Product Inhibition[b] [P]	[P]/t	Denaturation[c] [P]	[P]/t
1	0.4958	0.4958	0.4917	0.4917	0.4723	0.4723
2	0.9828	0.4914	0.9668	0.4834	0.8927	0.4464
3	1.4607	0.4869	1.4253	0.4751	1.2674	0.4225
4	1.9288	0.4822	1.8673	0.4668	1.6014	0.4004
5	2.3868	0.4774	2.2928	0.4586	1.8995	0.3799
6	2.8342	0.4724	2.7020	0.4503	2.1656	0.3609
7	3.2703	0.4672	3.0950	0.4421	2.4034	0.3433
8	3.6947	0.4618	3.4720	0.4340	2.6160	0.3270
9	4.1068	0.4563	3.8335	0.4259	2.8063	0.3118
10	4.5061	0.4506	4.1794	0.4179	2.9768	0.2977
Pred. v_0	0.5018		0.4999		0.4823	

[a] $V_{max} = 0.75, K_M = 5, [S_0] = 10$, true $v_0 = 0.5000$
[b] $K_p = 10$
[c] $k_{denat} = 0.1$

data points of each of the examples gives an apparent initial velocity which is in error by 4.5%, 8.3%, and 24.8% respectively.

It is important to remember that this is calculated data with no experimental error. In real life each value for $[P]$ would have some uncertainty associated with it. The percentage uncertainty is greater at smaller $[P]$, so the reliability of each point must be taken into account in deciding where to establish the cut-off point of the data set. Further, at short times t may include a major experimental error. Using an ordinary spectrophotometric arrangement, for instance, zero time will have an uncertainty of one to two seconds. If you are trying to get $[P]/t$ at intervals of five seconds, the first several data points may have significant experimental timing errors. Special precautions can be taken to minimize this difficulty, but don't overlook timing as a possible source of experimental error.

To summarize, assays for measuring initial velocity of an enzyme reaction should be constructed with the following parameters in mind.

1. Maximize $[S_0]$ so that the fraction $[S]/(K_M + [S])$ changes negligibly during the course of the assay.
2. Maximize sensitivity so that the extent of reaction necessary to give accurate measurements of $[P]$ is minimized.
3. If the plot of $[P]$ versus t is still curvilinear, estimate v_0 from an extrapolation of a plot of $[P]/t$ versus t to zero time.

During the initial work establishing the validity of the assay, it is good practice to check the apparent linearity of the $[P]$ versus t plot using the more sensitive $[P]/t$ versus t plot. If the slope of the former equals the intercept of the latter (within an acceptable margin of error) the imputation of linearity for the assay is probably valid.

PROGRESS CURVE MEASUREMENTS

In some systems it is difficult to measure $[P]$ at low extent of reaction. Sometimes the detection methods (chemical reaction, absorbance, etc.) are simply not sensitive enough to allow accurate measurement of $[P]$ corresponding to 5% of $[S_0]$. In other cases $[P]$ is determined as the decrease in $[S]$; for example, following the reduction of pyruvate with NADH by the decrease in absorbance at 340 nm, and a significant amount of reaction might be necessary before the difference reading between two large absorbances is trustworthy.

In the study of certain situations such as product inhibition or enzyme denaturation during the reaction, the use of progress curve assays is the method of choice for obtaining accurate results with minimum effort. Several miscellaneous forms of enzyme assays—first-order assays, fixed-product end point assays, and the average $[S]$ assay—are all variants of the progress curve measurement.

The Henri equation as given in Chapter 2 (equation 2-11) is:

$$[P] = V_{max}t - K_M \ln\left(\frac{[S_0]}{[S_0] - [P]}\right) \quad (N. B. \quad [S] = [S_0] - [P]) \quad (6\text{-}1a)$$

This is rearranged to forms useful for our subsequent discussions:

$$\left(\frac{1}{t}\right)\ln\left(\frac{[S_0]}{[S]}\right) = \frac{V_{max}}{K_M} - \left(\frac{1}{K_M}\right)\left(\frac{[P]}{t}\right) \quad (6\text{-}1b)$$

$$\frac{t}{[P]} = \frac{1}{V_{max}} + \left(\frac{K_M}{V_{max}}\right)\left(\ln\left\{\frac{[S_0]}{[S]}\right\}\right)\left(\frac{1}{[S_0] - [S]}\right) \quad (6\text{-}1c)$$

$$[E_t]t = \frac{[P]}{k_p} + \left(\frac{K_M}{k_p}\right)\ln\left(\frac{[S_0]}{[S_0] - [P]}\right) \quad (6\text{-}1d)$$

These three rearrangements correspond to the three subsequent discussions on 1. first-order kinetic assays; 2. average $[S]$ assays; and 3. fixed end point assays.

First-Order Kinetic Assays

When $[S_0] \ll K_M$ the progress of the enzymatic reaction is first order in substrate, as has been pointed out earlier (the data analysis section of Chapter 2). Considering the kinetics in light of the integrated rate equation 6-1b, this means that at complete reaction $[P] \ll K_M$, so the second term on the right hand side of 6-1b is negligible compared to V_{max}/K_M. The plot of $\ln[S]$ versus t has a slope of $-V_{max}/K_M = -k'$. This is a general conclusion. The plot may be slightly curved at early times if $[S_0]$ is not sufficiently smaller than K_M, so more weight should be put upon the later points if the curvature is noticeable.

The use of first-order kinetics has been discussed in connection with inhibitor studies (Chapter 3) and pH studies (Chapter 4). In terms of assays, first-order kinetics are useful in situations where $[S_0]$ is limited due to experimental factors (e.g., limited solubility, signal interference). The first-order rate constant k' is strictly proportional to $[E_t]$ when k_p and K_M are constant. If the absolute value of k_p/K_M can be measured, perhaps by correlating k' with $[E_t]$ determined in an absolute active site assay, then k' will yield the molar concentration of enzyme in subsequent assays under the same conditions.

The value of $[P_\infty]$ equals $[S_0]$ so if this is determined either by allowing the reaction to go to completion or from the Swinbourne plot, first-order kinetic assays may be used to measure the amount of substrate in a material or solution of interest.

Average [S] Assays

When an assay reaction is allowed to proceed for some time, [S] no longer equals [S_0]. The apparent rate, [P]/t, is not equal to the initial rate, nor is it equal to the true rate at time t. Lee and Wilson (1971) showed how to relate the average rate $v_{avg} = [P]/t$ to substrate concentration (Figure 6-2). The overall slope [P]/t is identical to the slope of the [P], t curve at some concentration of substrate lying between [S_0] and [S]. Lee and Wilson showed that this concentration of substrate is very nearly the average; $[S_{avg}] = ([S_0] + [S])/2$. In making runs at different values of [S_0] for analysis by the Hanes plot or using the HYPER program, using v_{avg}, [S_{avg}] data pairs leads to only a 1% error in the estimation of K_M even if as much as 30% of the substrate has reacted; if v_{avg}, [S_0] data pairs are used the error is 18%.

The principle is clarified by comparing the reciprocal transformation of the M-M equation, $1/v = 1/V_{max} + (K_M/V_{max})(1/[S])$, to equation 6-1c. To the extent that 1/[S_{avg}] corresponds to $\ln\{[S_0]/[S]\}/([S_0] - [S])$ the relationship holds. By substituting various values for [S_0] and [S] the error introduced by different extents of reaction can be evaluated.

If a fixed endpoint assay is being developed, gather data for determining V_{max} and K_M using this method. From $v_{avg} = [P]/t$ and $[S_{avg}] = [S_0] - [P]/2$

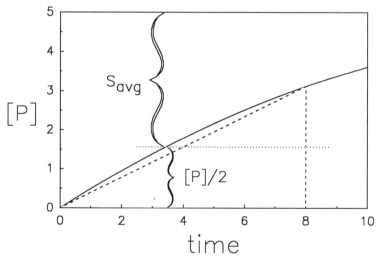

FIGURE 6-2. Significance of S_{avg} for an enzymatic reaction with $V_{max} = 1$, $K_M = 5$ and [S_0] = 5.

obtain values for those two parameters and then the results of subsequent measurements of $[P]/t$ can be related to enzyme concentration by substitution into the M-M equation. The value $[S_{avg}]$ should be used in this substitution, since then the error introduced by having a different extent of reaction in various runs is minimized. In other words, the working equation for measuring $[E_t]$ by this method is:

$$[E_t] = \left(\frac{[P]}{t}\right)\left(\frac{K_M + [S_{avg}]}{k_p[S_{avg}]}\right) \tag{6-2}$$

Fixed Endpoint Assays

Sometimes assays are structured to determine the length of time required to attain a certain extent of reaction (i.e., a particular value for $[P]$). By looking at equation 6-1d, it is apparent that $[E_t]t =$ constant, when $[S_0]$ and $[P]$ are fixed. In other words, if 1 "unit" of enzyme attains the end point in 10 minutes, then 2 "units" will require 5 minutes, 5 "units" will need 2 minutes, etc. If the endpoint happens to correspond to, say, 1 mM product, then while 2 "units" will generate 1 mM product in 5 minutes, it will not necessarily generate 2 mM product in 10 minutes.

Fixed endpoint assays are most often applicable when an enzyme is being used to effect a certain amount of modification of a substrate. Some typical endpoints might be the clotting of milk casein brought about by the proteolytic activity of rennet, conversion of starch to malto-dextrins of given D.E. by α-amylase, limited modification of pectin gel strength by pectin esterase, a limited hydrolysis of fat in cheese by lipase, and a slight weakening of gluten in bread dough by fungal proteases. In all these instances the formation of too much product in the specified time is as undesirable as too little reaction. Often the reactions are complex and not well understood, and the event of technological importance is not linearly related to a chemical test for the appearance of product. The clotting time of milk casein is not strictly predictable by following the rate of generation of free amino groups during the early stages of the rennet proteolysis, nor is the rheological effect of proteases on wheat gluten predicted by following proteolysis by chemical tests.

ABSOLUTE ACTIVE-SITE MEASUREMENT

Absolute enzyme active-site assays measure the moles of enzyme present without regard to the kinetic rate constants (as long as they are not zero, i.e., the enzyme is active). Sometimes these assays are termed all-or-none assays. A reaction is performed in which there is a one-to-one correspondence between moles of product formed and moles of active enzyme catalytic centers present. These may be of two sorts: irreversible, in which the reaction goes to completion; and kinetic, involving a rapid reaction of all or nearly all the active centers, followed by a slower turnover reaction. The kinetics of the latter type of assay are detailed in Appendix G.

Irreversible assays usually involve some sort of chemical inhibitor which reacts specifically at the enzyme active site. Many of these are called active-site-directed reagents in the literature. They are often planned and synthesized to resemble the substrate of the enzyme so they are specifically bound at the enzyme active site, but they also contain a chemically-reactive grouping which then forms a stable covalent bond with the enzyme. Many of the organophosphate insecticides inhibit enzymes by forming a bond with a residue in the active site. These can also be used in assaying esterase or protease enzymes.

The inhibiting molecule must be measurable in some fashion. The simplest way is to use a radioactive atom in the synthesis of the inhibitor. After incubation with the enzyme and complete inactivation the excess inhibitor is removed (gel-filtration, precipitation of the protein, or any similar technique) and the amount of protein-bound radioactivity is measured. The calculation of moles of enzyme active site is then straightforward. The measurement may be spectrophotometric. N-phosphoryl-L-Leu-L-Trp is a high-affinity inhibitor of metalloproteases such as thermolysin (Komiyama et al. 1975). When it binds to the active site of the enzyme a difference spectrum is generated (due to the interaction with the tryptophan residue) and from the magnitude of this the moles of inhibitor bound can be calculated.

The inhibiting molecule does not have to form a completely irreversible bond to the enzyme, if the reverse step is so slow that the enzyme-inhibitor complex is not dissociated during a subsequent velocity assay for unreacted enzyme. This situation was discussed in Chapter 3 with respect to high-affinity inhibitors. The use of these inhibitors for absolute measurement of enzyme active sites was outlined there. Varying amounts of inhibitor, to give up to about 50% inhibition of enzyme, are mixed with the enzyme, and after an incubation period the amount of free enzyme in the mixture is assayed using a suitable initial rate assay. Plotting the rate versus moles of inhibitor used according to equation 3-9 gives a line which, upon extrapolation to the x-axis, gives the number of moles of enzyme present in the initial sample.

In addition to the natural anti-nutritional enzyme inhibitors such as the anti-trypsin or anti-chymotrypsin inhibitors found in legumes, one might also use the natural inhibitors of α-amylase (Silano 1987) to titrate those enzymes. Some other natural inhibitors include pepstatin (pepsin), cystatin (sulfhydryl proteases), and phosphoramidon (metalloproteases). The growing number of transition state analogs (Lienhard 1973) are synthetic molecules more easily quantitated than the natural inhibitors, and with a wider range of potential applications.

Kinetic active site assays depend upon a two-step enzymatic reaction and the ability to so order the experimental conditions that the extent of the first reaction is measurable. The first example of such an assay was the observation by Hartley and Kilby (1954) that when a large amount of chymotrypsin was added to a p-nitrophenyl ester of an amino acid substrate there was an initial rapid release of p-nitrophenol followed by a slower turnover reaction forming initial product. This was the first evidence that an intermediate acyl-enzyme was an integral part of the mechanism of action of trypsin and related proteases. The extrapolation of the slow turnover reaction to zero time gave an estimate of the number of moles of trypsin taken for the experiment. To date most of the kinetic active site assays developed have been for application to proteolytic enzymes, primarily serine proteases and sulfhydryl proteases.

The kinetics of these "burst" assays are presented in Appendix G. The usual experimental condition requirements are: 1. $[S_0] \gg K_M$, and 2. $k_2 \gg k_3$. Given a substrate which fulfills these constraints, the course of product formation appears somewhat as shown in Figure 6-3. The straight line portion of the curve represents a steady-state turnover reaction, with a slope equal to v_0 in the M-M equation. The intercept of the straight line with the y-axis equals $[E_t]$. The curved portion is the pre-steady state transient phase; from the pseudo-first order rate constant associated with this phase, information about the rate constants k_2 and k_3 and the true rapid equilibrium dissociation constant K_m is obtained.

In setting up a kinetic titration assay it is desirable to have a substrate which will give a product P_1 with a large signal (i.e., spectrophotometric

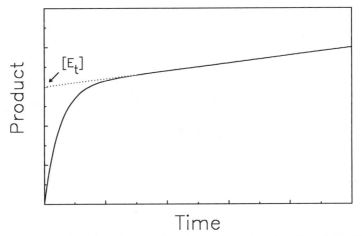

FIGURE 6-3. Formation of product as a function of time when there is an initial "burst" followed by a slower turnover reaction.

absorbance), and also one in which the rate of acylation (first reaction) is much greater than the rate of deacylation (second reaction). Often k_2 and k_3 will have different pH dependencies and it is possible to achieve the desired inequality by choosing a reaction pH somewhat removed from the optimum pH for normal enzyme reaction. In the titration of chymotrypsin with N-*trans*-cinnamoylimidazole (CI), for example, the reaction is performed at pH 5 rather than the more usual pH 7 to 8, in order to bring k_3 to a low value (Schonbaum et al. 1961). The amount of enzyme required for a determination will, of course, depend upon the signal size. In the titration with CI about 5 mg of chymotrypsin (2×10^{-7} moles) per 3 ml reaction volume is needed. This gives an absorbance change of about 0.3, a conveniently measurable change. The fluorogenic trypsin substrate fluorescein mono-p-guanidinobenzoate hydrochloride will accurately titrate 2×10^{-12} moles of trypsin in 3 ml reaction volume (Melhado et al. 1982), demonstrating the increased sensitivity available with fluorescence spectrophotometry.

Active site assays are most often used for "titrating" a stock solution of an enzyme which is then used in calibrating some other assay: an initial rate assay for monitoring a purification, or for determining the amount of a high-affinity inhibitor in a foodstuff (soy meal, for example). The initial titration then allows the results of the subsequent assays to be expressed in absolute molar concentrations, rather than in some ill-defined "units/ml." This is a definite advantage in making results more meaningful than in just the immediate time and place. Because the concentration of enzyme being titrated must often be relatively high, on the order of mg/ml, relatively crude preparations cannot be easily assayed in this fashion. Also, the enzymes for which titration assays have been developed are relatively few, being mostly proteases of the serine and sulfhydryl groups and a few esterases. Active site assays for carbohydrases, isomerases, and oxidases would greatly aid studies of enzymes in these classes.

FIXED TIME ASSAYS

Of the various kinds of assays used by enzymologists, fixed time assays are by far the weakest, theoretically speaking, and they are the most popular and widely used. This popularity arises from their practical advantages: fixed time assays are generally the easiest to perform; they make it possible to run many assays concurrently; and they are usually the type of assay with which the enzymologist is most familiar. A well-designed, properly-validated, and meticulously-performed fixed time assay can be just as accurate as an initial rate assay or a progress curve result. The problem is that far too often fixed time assays are not well-designed and certainly not properly validated. This section will address some of the factors which must be considered in accomplishing those goals.

The key point in validating a fixed time assay is linearity, with respect to time of reaction, and with respect to the amount of enzyme present. In the strictest sense linearity is not achieved. The substrate concentration begins decreasing as soon as the reaction is initiated, and the rate of formation of product begins to slow, regardless of how much $[S]$ is greater than K_M. The assay is usually designed to restrict this slowdown to less than some acceptable margin of error. The situation is that shown in Figure 6-1 above. The "rate," $[P]/t$, has a negative dependence on t in any situation (with the possible exception of substrate inhibition, where the lessening of inhibition as substrate concentration decreases may offset the decrease in true reaction velocity). The "limit of the assay" corresponds to that time in equation 6-1b at which the difference between $[P]/t$ and the true zero time rate exceeds the acceptable margin of error.

The fixed time assay corresponds to the Henri integrated rate equation:

$$\frac{[P]}{t} = k_p[E_t] - K_M\left(\frac{1}{t}\right)\ln\left(\frac{[S_0]}{[S_0 - [P]]}\right) \qquad (6\text{-}3)$$

In a given assay the first term on the right hand side (V_{max}) is constant, as is K_M. The time linearity of the rate (i.e., constancy of $[P]/t$) depends on two offsetting factors. As t increases $1/t$ decreases; simultaneously the parenthetical term increases, as does $\ln([S_0]/[S])$. At 1% reaction this is $\ln(1/.99) = 0.01005$, at 5% reaction it is $\ln(1/.95) = 0.05129$, and at 10% reaction $\ln(1/.90) = 0.10536$. To keep the "correction term" constant the corresponding times would need to be 1.005, 5.129, and 10.536 (arbitrary units). But the three corresponding values of $[P]/t$ (in percent per t) are 0.995, 0.975, and 0.949. The compensation effect reduces the linearity error (at 10% reaction the deviation is 5.1% instead of 10.5%) but does not completely offset it.

This argument leads to a useful generality:

> In a fixed time assay, the percentage deviation from time linearity is equal to approximately one-half the percentage of substrate which is converted to product during the assay.

This statement holds regardless of the extent to which $[S_0]$ exceeds K_M. One goal in designing a fixed time assay, then, is to choose a sensitive method for measuring $[P]$ so that as small a percentage as possible of substrate is converted to product.

The other linearity desired in a fixed time assay is enzyme linearity (i.e., $[P]/t$ is directly proportional to $[E_t]$). Surprisingly, this relationship is less sensitive to the extent of substrate reaction than is time linearity. In equation 6-3 the change due to substrate depletion is in the second term on the right hand side which is a correction factor to the direct relationship. We can rewrite equation 6-1a in the following form, where Y_1 is the fraction of substrate converted to product during an assay (note that $[P] = Y[S_0]$):

$$k_p[E_t]t = Y_1[S_0] + K_M \ln\left(\frac{1}{1 - Y_1}\right) \tag{6-4a}$$

A second run is made using X times as much enzyme, and observing a conversion of a fraction Y_2 of substrate:

$$k_p X[E_t]t = Y_2[S_0] + K_M \ln\left(\frac{1}{1 - Y_2}\right) \tag{6-4b}$$

Substituting $Z = [S_0]/K_M$, and dividing 6-4b by 6-4a, we get:

$$X = \frac{Y_2 Z + \ln\{1/(1 - Y_2)\}}{Y_1 Z + \ln\{1/(1 - Y_1)\}} \tag{6-4c}$$

In order for the plot of $[P]$ versus $[E_t]$ to be strictly linear, the two logarithmic terms would of necessity be negligible, so Y_2/Y_1 would exactly equal X, showing that if X were 2, doubling the amount of enzyme taken would double the percentage conversion of substrate to product.

In fact the logarithmic terms are not usually negligible; we have to find a value of X' (where $Y_2 = X'Y_1$) which will satisfy equation 6-4d, and the discrepancy between X and X' is the measure of the deviation from enzyme linearity of the assay.

$$XY_1 Z + X \ln\left(\frac{1}{1 - Y_1}\right) = X'Y_1 Z + \ln\left(\frac{1}{1 - X'Y_1}\right) \tag{6-4d}$$

There is no analytical solution for this equation. However, with the help of a simple program on a personal computer it is easy to find an approximate value of X' which will make the equality hold at various combinations of X, Y_1 and Z. From such an exercise, some rules of thumb can be stated.

1. As $Z([S_0]/K_M)$ increases the error decreases. Increasing Z five-fold decreases the error by three- to four-fold. The decrease factor is greater at smaller values of X and Y_1.
2. As X increases the error increases approximately twice as fast. Thus a five-fold increase in X produces roughly a ten-fold increase in error.
3. As Y_1 (percent substrate reacted) increases the error increases slightly faster. An increase in Y_1 from 2% to 10%, will produce about a six-fold increase in the error.

These numerical results confirm earlier statements about design parameters for minimizing errors in fixed time assays : 1. Set $[S_0]/K_M$ as large as practicable; 2. At wider ranges of enzyme concentrations expect larger errors; 3. Keep the percentage conversion of substrate as small as practicable. Yet having repeated these dicta, it is still worth noting that with $Z = 5$, $X = 10$, and $Y_1 = .05$, the deviation at the top of the plot is only 5.2% (i.e., instead of measuring 10 "units" of enzyme, the assay would report the presence of only 9.48 "units"). Comparing this result with the earlier discussion on deviations from time linearity supports the statement that enzyme linearity is less error-prone.

When the substrate for the reaction is a small molecule of known structure and molecular weight it is rather easy to evaluate how well a given assay fulfills these design conditions. Often this is not the case. The substrate may be polymeric (proteins, starch) or insoluble (triglycerides) and the evaluation of a meaningful K_M or Y_1 is difficult if not impossible. Some guidelines for working with substrates of this sort will be given in the section on perturbing factors, below.

COUPLED ENZYME ASSAYS

The kinetics relevant to designing a valid coupled enzyme assay were discussed in Chapter 2 and in Appendix B. The points made there may be briefly restated here.

1. The initial enzyme reaction (the one of actual interest) should be of apparent zero order (i.e., the rate should be constant during the assay).
2. The kinetic parameter of interest for each auxiliary enzyme is its first-order rate constant under the pH and temperature conditions of the assay. The ratio of auxiliary enzyme activity to initial enzyme activity is irrelevant.
3. The lag period before the measured reaction becomes linear is a function of the first-order rate constant of the auxiliary enzyme(s). The relationship has been given earlier. The choice of a suitable lag period depends upon the characteristics of the initial reaction, the instrumentation to be used, and perhaps the cost of the auxiliary enzyme.

The concentrations of the first product and the second product as a function of time are represented in Figure 6-4. Note that the relationship between the activities of the two enzymes governs only the steady state concentration of the first product. The slope of the line of second product versus time is the same as the slope of the line of disappearance of substrate versus time, with a vertical offset equal to the steady state concentration; both slopes are a function strictly of the reaction rate of the first enzyme.

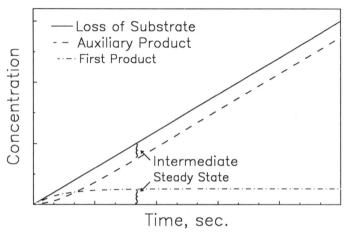

FIGURE 6-4. Coupled-enzyme reaction. Concentration of substrate, initial product and auxiliary product as a function of time.

Coupled enzyme assays are usually employed to render the product from the initial reaction easily measurable. As an example, phytase releases phosphate ion from phytic acid. The chemical determination of phosphate is somewhat tedious, time-consuming, and does not allow continuous monitoring of the reaction. In a possible three-step auxiliary scheme, the phosphate reacts with glycogen (phosphorylase a) to form glucose-1-phosphate. G-1-P is isomerized to glucose-6-phosphate (phosphoglucomutase) and the G-6-P is oxidized with $NADP^+$ (G-6-P dehydrogenase) with the production of NADPH. This latter product may be monitored spectrophotometrically at 340 nm, or amperometrically with a suitable immobilized enzyme electrode. Thus a long, insensitive, fixed point assay is converted into a short, sensitive, continuous assay.

The major pitfall to be aware of is the action of inhibitors. Usually the coupled enzyme assay is developed in a "clean" system. Then the initial enzyme activity is sought in a crude system such as a tissue homogenate or extract. Unsuspected inhibitors of the auxiliary enzyme present in the extract will reduce the expected first-order rate constant. This will appear as a longer lag period, but the deviation from linearity may not be obvious and a lower slope might be accepted. For instance, the assay may have been set up with a lag period of 5 seconds, and the rate measured between 5 and 30 seconds. Inhibition of the auxiliary enzyme may lengthen the lag period to 25 seconds, but the instrument trace from 10 to 30 seconds may appear nearly linear; using the best straight line through this region of the trace will obviously result in a false estimate of the rate of the initial reaction. If a complication of this sort is suspected the first-order rate constant of the auxiliary enzyme acting on its substrate alone (i.e., with no initial reaction occurring) should be measured in the presence and absence of extract.

Sometimes the inhibitor is added on purpose, to measure its effect on the initial enzyme. If such an inhibition study is being planned the effect of the inhibitor on the auxiliary enzyme must also be noted. If the inhibitor is less effective vis-a-vis the auxiliary enzyme, the study may still be carried out by adding extra second enzyme to keep the lag period down. The steady-state, linear rate of second product formation will then reflect the effect of the inhibitor on the initial enzyme. If the inhibition of the auxiliary enzyme is large enough (at concentrations of inhibitor of interest for the study) that it cannot be overcome by "brute force," another means of following the first reaction must be sought.

Many times coupled enzyme systems are used to measure the amount of the initial reaction substrate by driving the overall reaction to completion. As an example, glucose concentration may be measured by oxidizing it with glucose oxidase, then reacting the H_2O_2 with pyrogallol (peroxidase catalyzed). If both reactions go to completion, the amount of purpurogallin formed is directly proportional to the amount of glucose initially present. In such a system the steady-state kinetics developed in Appendix B do not apply. Instead, we have two consecutive first-order reactions:

$$G \xrightarrow{k_1} H \xrightarrow{k_2} P \qquad (6\text{-}5a)$$

This model is often discussed in elementary kinetics texts, so the mathematics will not be elaborated here. The resulting expression for P (as the fraction of total product which corresponds to the initial amount of G) at any time t is:

$$P = \frac{k_2(1 - e^{-k_1 t}) - k_1(1 - e^{-k_2 t})}{k_2 - k_1} \qquad (6\text{-}5b)$$

The reaction of G to H will be $> 99\%$ complete in seven half-lives, so if k_1 is 0.1 min^{-1} this point is reached in 50 minutes. The second reaction lags somewhat, because the concentration of H builds up at first then declines again as the first reaction goes to completion. The second rate constant k_2 must be somewhat greater than k_1 in order to reach a complete end point within a reasonable time. As an example, taking k_1 as 0.1 min^{-1}, and if the end point of the reaction is read at 60 minutes (where 99.75% of G has reacted), then with k_2 equal to 0.13 min^{-1}, P equals 99.1% of the original value of G, and with k_2 equal to 0.20 min^{-1}, P is 99.5% of the original G.

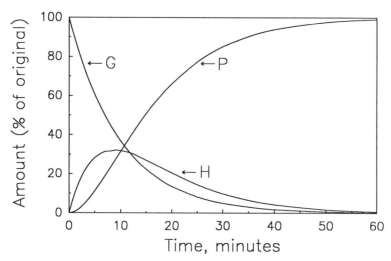

FIGURE 6-5. Two sequential first-order reactions.

Thus with $k_2/k_1 \geq 1.3$ the analysis is $> 99\%$ complete after 9 half-lives of the first reaction. The concentrations of G, H and P as a function of time are shown in Figure 6-5. While the decrease in G is a simple logarithmic decay curve, the increase in P is somewhat sigmoidal (the early lag). The size of the lag is equal to H, where $H = G_0(1 - e^{-k_1 t}) - P$.

Note that merely increasing the amount of auxiliary enzyme does not appreciably speed up the rate of reaching the end point. A larger k_2/k_1 ratio just decreases the amount of time by which the production of P lags the disappearance of G (decreases the concentration of H). If a decrease in end point time from 60 minutes to 30 minutes is needed, it is necessary to double the amounts of both enzymes, keeping the ratio the same. With $k_1 = 0.2$ min^{-1} and $k_2 = 0.26$ min^{-1} the reaction is $> 99\%$ complete in 30 minutes. To further decrease the reaction time to 15 minutes both enzymes must be doubled again.

PERTURBING FACTORS

Non-Simple Substrates

All the kinetic discussions thus far make some implicit assumptions about the nature of the substrate.

1. The substrate is a well-defined chemical species of known molecular weight so that the molar concentration in assay medium is readily ascertained.
2. One molecule of substrate is transformed to one molecule of product via one turnover reaction catalyzed by the enzyme.

3. The substrate is soluble so the kinetic analysis may be made using molarity as a measure of the substrate's reactivity.
4. The substrate is stable under the conditions of the assay, so all formation of product is due to enzymatic catalysis.

Unfortunately many of the substrates for enzymes of interest to food scientists violate one or more of these assumptions. Here are some examples:

1. Hammarsten soluble casein is a commonly used substrate for protease assays. This is a mixture of molecular species.
2. A starch molecule has multiple bonds susceptible to hydrolysis catalyzed by amylase. One turnover reaction forms product plus more substrate.
3. Triglycerides are insoluble in water. The hydrolysis by lipase has been shown to depend upon the area of the lipid/water interface as the equivalent of $[S]$.
4. p-Nitrophenyl esters are convenient substrates for numerous hydrolytic enzymes. Their hydrolysis is also catalyzed by OH^- and a buffered solution of substrate slowly turns yellow. The measured enzyme-catalyzed rate of hydrolysis must be corrected for the OH^--catalyzed rate; $[S_0]$ is not usually equal to the original substrate concentration.

The detailed effects of each of these perturbations will be discussed in connection with specific enzyme assays in Part II. Sometimes more than one assumption is violated in the same assay. Insoluble substrates with a dye molecule attached have become popular for certain "depolymerases." The extent of reaction is assumed to be proportional to the concentration of solubilized dye. Examples are diazotized collagen for proteases and stabilized starch granules with dye attached for α-amylase. The physical events at the molecular level which lead to the solubilization of one dye molecule are very poorly understood in these cases. Another ill-defined system is the Anson assay for proteases, where "product" is short peptides which are soluble in the precipitating reagent, trichloroacetic acid. The number of peptide bonds in the protein substrate molecule which must be hydrolyzed to form a molecule of soluble peptide product is not accurately known.

This is not to say that such substrates should be avoided in designing enzyme assays. Rather, these complications should be recognized and kinetic analysis should not be pushed into areas of invalidity. In the Anson method for protease assay an apparent K_M (expressed as mg casein per ml) may be determined for the purpose of defining limits of linearity of the assay. However, this number should not be expected to give any meaningful information about the enzyme-substrate system at the level of molecular interaction. In using insoluble substrates the degree of subdivision of the substrate influences the rate of reaction. For reproducible substrate preparation some means of measuring the surface area per ml of suspension is important. With unstable substrates the extent of non-enzymatic decomposition should be minimized. With p-nitrophenyl esters, for example, the substrate might be prepared as a concentrated, stable solution in a non-aqueous solvent and an aliquot added to the buffered enzyme solution to initiate the assay. The main point is to explicitly recognize the potential perturbations and design the assay so that they have a minimal effect upon the accuracy of the desired information—How much enzyme is present.

Unintended Inhibition

Enzyme reactions can be retarded by inhibitors. Often these inhibitors are not added intentionally, for the purpose of studying their interaction with the enzyme, but appear unintentionally. If their action is not recognized they can lead to erroneous conclusions about the system and/or inaccurate assay reports. We will mention three kinds of such unintentional inhibitors: products, substrate, and endogenous in the enzyme preparation.

Product inhibition has the effect of causing the plot of $[P]$ versus t to be more strongly concave that would be seen in its absence (Figure 6-1, Table 6-1). Since in a specific real system one does not have an "uninhibited" curve for comparison, the presence of product inhibition is not always easy to recognize. If a fixed time assay is being used, the plot of $[P]$ versus $[E]$ will tend to deviate downwards from a straight line somewhat more than expected. If $[S_0]$, $[P]$, and K_M can be measured and expressed as molarity, then from equation 6-4d and the attendant discussion the expected deviation at 10 "units" of enzyme can be calculated, given the $[P]$ versus $[E]$ results at 1 and 2 "units." If the deviation is much more than calculated, product inhibition should be suspected. If an initial rate velocity assay is being used, of course, product inhibition is not a factor.

The best diagnosis for product inhibition is to make progress curve runs at several different values of $[S_0]$ (Chapter 3). If product inhibition is absent, the experimental points of $[P]/t$ versus $(1/t)\ln([S_0]/[S])$ should all lie along the same line. If they form a set of different lines with the same y-axis intercept competitive inhibition by product is the probable cause (I assume denaturation of enzyme during the reaction has been ruled out). If this is the case, the inhibition constant K_p can be determined as discussed in Chapter 3.

Substrate inhibition is easy to recognize from a plot of rate versus $[S]$ (Chapter 3). This phenomenon makes it difficult to design a good fixed time assay, because the recommendation to have $[S_0] >> K_M$ is, in this case, deleterious to maximum formation of product. If the substrate concentration for the assay is chosen to be slightly greater than the maximum of the v, $[S]$ plot (Figure 3-3) then the plot of $[P]$ versus t will initially curve upward then start to curve downward. Depending upon the magnitude of these excursions from a straight line, this could give an approximate compensation for the usual decrease in rate with time, and extend the apparent time linearity of the assay. Thus, with the proper background work, this phenomenon could be used to advantage in designing a routine fixed-time assay for such a system.

If for various reasons this approach is not desirable, an alternative is to design the assay as a first-order kinetic assay. Since $[S_0] = [P_\infty]$, in routine work $[P_\infty]$ is known, and after reacting for time t, $\ln\{[P_\infty]/([P_\infty] - [P])\}/t = k' = [E_t]k_p/K_M$, i.e., k' is directly proportional to the amount of enzyme present. This requires slightly more arithmetical manipulation of data than simply measuring $[P]$ at time t, but this should not be a problem with an inexpensive scientific calculator. This method of assay design also has the advantage that enzyme linearity is exact; there is no deviation due to substrate depletion as is the case with ordinary fixed-time assays.

Endogenous inhibitors may be present in crude extracts containing the enzyme being assayed. These are usually low-affinity reversible inhibitors which are removed at some subsequent purification step (if enzyme purification is being performed). This appears as a recovery of activity in that step of more than 100%, a not-unusual phenomenon. If the enzyme study is being carried out on the crude extract (e.g., surveying several sources for enzyme content) endogenous inhibition may be more of a problem. This is manifested by a marked non-linearity in the plot of $[P]$ versus ml extract (Figure 6-6). This may be differentiated from product inhibition by using an initial rate assay; the plot of v_0 versus ml extract is linear in the presence of product inhibition, but curved in the presence of endogenous inhibitor.

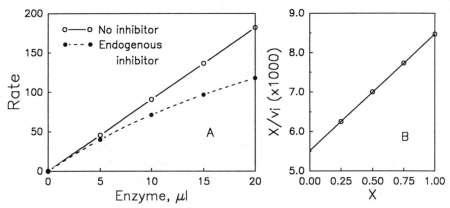

FIGURE 6-6. Endogenous inhibitor in the enzyme preparation. A. Effect on assay linearity. B. Plot to obtain the uninhibited rate.

A value for the uninhibited enzyme rate may be extracted from this data by the following method. Let $[e]$ equal the amount of enzyme in the largest aliquot of extract used in making the plot of Figure 6-6. The velocity of the uninhibited reaction with this amount of enzyme is (N.B. the subscript 0 designates lack of inhibition, not necessarily an initial velocity as used elsewhere.):

$$v_0 = \frac{[e]k_p[S]}{K_M + [S]} \tag{6-6a}$$

Rearranging, we have:

$$\frac{[e]}{v_0} = \frac{K_M + [S]}{k_p[S]} \tag{6-6b}$$

The amount of enzyme in each of the aliquots will be $X[e]$, where $0 < X < 1$. The amount of endogenous inhibitor in the largest extract aliquot is $[I]$, and in the other aliquots is $X[I]$. We will assume the inhibition is competitive, although the final equation is the same for noncompetitive inhibition. The velocity of the inhibited (actual) reaction is:

$$v_i = \frac{X[e]k_p[S]}{K_M(1 + X[I]/K_i) + [S]} \tag{6-6c}$$

$$\frac{X[e]}{v_i} = \frac{K_M + [S]}{k_p[S]} + \frac{(XK_M[I]/K_i)}{k_p[S]}$$

$$= \frac{[e]}{v_0} + \frac{X(K_M[I]/K_i)}{k_p[S]} \tag{6-6d}$$

Dividing through by $[e]$ gives the final form:

$$\frac{X}{v_i} = \frac{1}{v_0} + \frac{X(K_M[I]/K_i)}{[e]k_p[S]} \tag{6-6e}$$

A plot of X/v_i versus X has an intercept of $1/v_0$; $1/$intercept equals the uninhibited rate of reaction catalyzed by the amount of enzyme $[e]$.

This treatment could have several uses. If a survey is being made of the amount of enzyme found in several related sources, an extraction of v_0 should be done so that the comparisons are not confounded with possible inhibitors present in the various sources. In monitoring the recovery of enzyme during a purification scheme it would be preferable to establish the amount of uninhibited activity present at each step to have a more accurate picture of the recovery (this obviates the 100% + recoveries alluded to above). Finally, if K_M and V_{max} are being determined on a crude enzyme preparation for the purposes of designing an assay, v_0 as defined by equation 6-6e should be obtained at each value of $[S_0]$ for the subsequent analysis.

Another unsuspected source of possible inhibition is metal ions. Trace heavy metals (lead, mercury, copper) are usually absent from the water and buffer salts used in the laboratory. However certain ions (calcium, magnesium, zinc) are often added to stabilize the enzyme during purification and handling. If the substrate complexes with these metals it may alter the mode of reaction with the enzyme. Phosphate and sulfate substrates are particularly likely to show this effect. The presence of large amounts of calcium, for instance, will complex with phytic acid (inositol hexaphosphate) and change its character as a substrate for phytase. Little work has been published about inhibitory metal-substrate interactions, but it is a possibility which should not be overlooked.

Finally, in crude extracts of plant materials some of the non-enzyme high molecular weight material present may bind substrate and thus decrease the rate of reaction in the routine assay. For example, some polysaccharides present in plant extracts will interact with the protein substrate in a trypsin assay; the extracts appear to have trypsin inhibitor activity (Ikeda and Kusano 1983). In fact, the polysaccharides are not true enzyme inhibitors. Many of the other "enzyme inhibitor activities" reported to reside in crude extracts of various foodstuffs will probably turn out to be similar sorts of "substrate inhibitor activities" if and when they are critically studied.

7 Measurement Methods

The most fundamental operation in any enzyme assay is the measurement of a concentration, either of product or substrate. Through the application of a chain of other measurements and/or assumptions this is expressed as a function of time, either as a rate (a differential term) or directly as in a progress curve measurement. Operationally, we make a reading: of a meter, as on a spectrophotometer; of a burette, as in a pH-stat assay; or of a printout from a radioisotope counting machine. In some cases the conceptual connection between this reading and concentration is quite direct; an absorbance may be related to molarity of product via an absorption coefficient which is determined in a separate experiment. In other cases the concatenation of causes and effects between product concentration and operational readout is ill-defined, such as the change in turbidity of a starch suspension due to amylolytic action. In general, the longer the conceptual chain between product and readout, the smaller the variety of information which can be extracted from the assay system. The irreducible minimum of useful information would seem to be some value which would give a number for $[E_t]$ as read from a "standardization curve," but in some assays the reliability of even this much information appears to be questionable.

Measurement methods may be conveniently divided into two groups: direct and indirect. A direct measurement gives some reading which is related to the concentration of product or substrate while the enzymatic reaction is proceeding, without the need for intermediate manipulations. Detection is by means of some instrument whose response changes as a solution property changes because of the reaction (e.g., the change in absorbance at 400 nm registered by a spectrophotometer as a p-nitrophenyl ester is being hydrolyzed). Direct methods are generally the quickest, give continuous information about the reaction (especially if the instrumental readout is recorded on a stripchart or by computer time-sampling), are most convenient for initial rate assays, and are almost a necessity for "burst" titrations of enzymes.

Indirect methods require a fixed incubation period followed by other manipulations before a reading related to product or substrate concentration is obtained. An example is the assay for amylase in which the increase in concentration of reducing saccharides is measured by a chromogenic reaction, followed by quantitation in a spectrophotometer and reference to a standard curve. By their nature indirect methods are discontinuous, although by taking samples for analysis from a reaction mixture at frequent intervals an approximation of a continuous curve may be obtained. Indirect methods lend themselves well to performing a large number of assays concurrently, with the manipulations being carried out batchwise after the enzyme runs are finished. Several books which describe particular assay methods are available (Barman 1969; Bergmeyer 1974; Guilbault 1976, 1984).

DIRECT METHODS

Optical Methods

Spectrophotometry.   The use of a spectrophotometer with ultraviolet and visible wavelength capabilities is almost universal in enzyme laboratories today. The UV-VIS spectrophotometer is to the enzymologist as the gas chromatograph is to the synthetic organic chemist. While this is not the place for a treatise on the fine points of spectrophotometry, some comments on the use of spectrophotometers for enzyme assays should be made.

Since enzyme reaction rates are moderately temperature dependent, it is advisable to have some form of temperature control. This may be as simple as keeping the room temperature fairly constant. A better method is to have a thermostatted jacket for the cuvette compartment with water from a constant temperature bath circulating through it. This is a necessity if the temperature dependence of the reaction is being investigated.

A double-beam spectrophotometer is desirable. The "blank" cuvette holds the solution of substrate in buffer while the "reaction" cuvette includes enzyme. This arrangement allows for continuous and accurate subtraction of absorbance due to non-enzymatic reaction, such as the OH^--catalyzed hydrolysis of p-nitrophenyl esters. The double-beam configuration is also useful when the absorbance being measured is due to substrate. In this case the blank cuvette is placed in the "reaction" beam and the enzyme cuvette is placed in the "blank" beam. The recorded absorbance change is then positive with time, although it is actually due to the decrease in the "blank" as substrate concentration decreases.

The bandwidth of the spectral beam does not usually affect the accuracy of the assay. However, be aware of the effect of bandwidth on absorption coefficients. As an example, NADH has a molar absorption coefficient at 339 nm of 6.22×10^3 M^{-1} cm^{-1}. This value was determined using an instrument with a 2 nm bandwidth. If your spectrophotometer has a 20 nm bandwidth and is set at 339 nm the absorption is actually averaged over the range 329 nm to 349 nm. The absorbance of a 0.1 mM solution in a

1 cm pathlength cuvette will be less than 0.622 so an absorbance change of 0.622 will represent *more* than 0.1 mmoles/L of NADH. It is advisable to check actual absorbancies on your instrument for solutions of products of interest (e.g., NADH, p-nitrophenol, p-nitroaniline) at a few accurately known concentrations.

The accuracy of a given measurement depends upon the accuracy with which the spectrophotometer responds to small changes in the intensity of the light beam impinging on the detector (the stability of the light source may also be a factor). When the instrument is "zeroed," the detector responds to a light beam of 100% intensity. When an absorbing species is placed in the light beam the intensity is reduced, for example to 80% (20% of the light is absorbed). Then absorbance $= \log (100/80) = .097$. If the stability of the instrument (light source, detector) is $\pm 1\%$, then the ratio could be 99/81, or absorbance $= .087$, a 10% error. As a rule of thumb, the most accurate absorbance range is from 0.3 to 1. Modern spectrophotometers enable accurate measurement at low absorbancies by improved electronics and split-beam configurations which reduce instabilities to much less than 1%.

As the light-absorbing substrate is depleted the absorbance decreases. The addition of small increments of a concentrated solution of the substrate may be made to keep the absorbance constant. The rate of addition, i.e., μl per minute times the substrate concentration, defines the rate of the enzyme reaction in moles per minute. This constant-absorbance scheme was termed a UV-absorptiostat by its inventors (Pantel and Weisz 1979). They used a cuvette with a 10 ml capacity and a 2 cm path length, with magnetic stirring and a control system operated from the spectrophotometer output which added the substrate in small increments. The concept is quite similar to the pH-stat (see below) and has particular application to oxidation-reduction enzymes.

Spectrofluorometry. Although less common than spectrophotometric measurements, fluorometric measurements have the advantage of vastly increased sensitivity. It is not uncommon to find that an assay using a fluorometric analog of a standard chromogenic substrate is three orders of magnitude more sensitive. Rotman (1961) used a fluorescent substrate and a micro-fluorometer to study the activity of single molecules of β-D-galactosidase. This enhanced sensitivity can be applied to both direct assays (e.g., the fluorescein substrates for trypsin) or to indirect assays (converting NAD^+ to a fluorescent derivative).

A spectrofluorometer irradiates the sample with monochromatic light of a wavelength which is absorbed by the compound of interest (usually product). The compound emits light at a higher wavelength which is filtered through optical filters or a monochromator so that the lower wavelength light is removed and only the emitted light impinges on the detector. The intensity of emitted light is directly proportional to the concentration of the product. The measurement is direct (i.e., some value above a zero background), rather than a difference measurement as in a spectrophotometer. This allows greater amplification of the current from the detector and more sensitivity in the measurement.

The comment about a thermostatted cuvette compartment applies in this case as well as to spectrophotometers. There are few absolute standards for fluorometers which would correspond to an absorbance standard, and it is common to use a fluorescer such as Rhodamine B embedded in plastic to check the response of the instrument each day. A major problem (at least in early instruments) was stability of the light source. As the lamp aged, or if there were minor fluctuations in line voltage, the change in light intensity translated into some rather large measurement errors. This can be rectified by improving the stability of the power source, which is done in newer machines.

Several factors can cause non-linearity in the dependence of emitted light on product concentration. If the absorbance of the reaction medium at the excitation wavelength is high the light energy interacting with the species of interest is decreased. It is recommended that the solution absorbance be no greater than 0.2. Since the fluorescent molecules are present at very low concentrations, adsorption to the cuvette wall may significantly decrease the bulk concentration and hence the emission energy. One method of obviating this error is to siliconize the internal walls of the cuvette by treating them with a dilute solution of chlorotrimethylsilane in toluene. Fluorescent impurities may enter the reaction cuvette from unsuspected sources: rubber stoppers, stopcock grease, skin oils, filter paper, and unwashed dialysis tubing are some of the more common ones.

The fluorescence can be decreased by the presence of quenchers, molecules which decrease the fraction of absorbed light energy which is re-emitted at the detection wavelength. Some common ones are halide ions, polycyclic ring systems (e.g., adenine, naphthalene), and cyclic amines. These materials should be avoided if possible; don't use Tris hydrochloride as the buffer salt, for instance. A major quencher of many fluors is molecular oxygen. If the fluorogenic reaction is run in an evacuated sealed tube and the emission yield is significantly higher than in an open cuvette, this mode of operation might be considered if the gain in sensitivity warrants the extra pains.

For good discussions of both theory and application of fluorescence to enzyme assays, consult the volume by Udenfriend (1962) and the chapter by Brewer et al. (1974).

Nephelometry. A suspension appears turbid because incident light is scattered by suspended particles which are much larger than the wavelength of the light. In a nephelometer the detector is placed perpendicular to the direction of the incident light falling on the reaction cuvette and so responds to that fraction of the light which is scattered at 90°. The scattered light is proportional to the concentration of particles (number per cubic centimeter) and the square of the average particle volume. If the particles consist of substrate (e.g., droplets of triglyceride) then as the enzyme (lipase) acts on this substrate and reduces the particle volume it "disappears" (i.e., becomes nonscattering). The connection between progress of the reaction and the decrease in light scattering has not been evaluated mathematically, and is usually empirically correlated with the rate of enzymatic reaction.

This technique has been used in certain practical applications such as in clinical work for detecting lipase, and for the assay of α-amylase using a suspension of starch granules. Within well-defined bounds this is a useful assay. It is often done as a fixed end point assay (Chapter 6) arranged to measure the time required to achieve a certain decrease in light scattering. However any change such as altering the concentration of the substrate suspension may introduce major complications. Since the measurement is actually not of $[P]$ nor $-[S]$ but rather of some complicated physical interaction of these values with light, the extrapolation of the principles inherent in equation 6-1d to this system is not strictly valid.

The proportion of light scattered by a particle is a function of the wavelength, and thus, at least in theory, the accuracy of the system would be improved by using monochromatic light. However, given all the other factors in the system which are poorly understood and/or controlled, the use of white light for the assay is probably a minor source of error.

Optical Rotation. The application of optical rotation for following enzyme reactions was limited in the past because of instrument limitations. Filling a polarimeter tube with the reaction mixture and then monitoring the change in optical rotation in a manual polarimeter is a tedious process which made the acquisition of a reasonable amount of rate data a time-consuming task. Nonetheless this is still an excellent direct signal for the reaction of many carbohydrolases such as invertase, glucosidase, and galactosidase. The advent of optical rotatory dispersion/circular dichroism (ORD/CD) instruments for rapidly and continuously measuring changes in optical rotation have made this technique as simple as measuring absorbance changes in a spectrophotometer.

ORD/CD instruments are expensive and less commonly found in biochemical laboratories than spectrophotometers. They are usually used for studies of protein and polymer configurations. But if you are working with carbohydrases or other enzymes which produce a significant optical rotatory change during catalysis, arranging for the use of the instrument is well worthwhile. An example of such an enzyme is the hydrolytic action of a protease upon the L enantiomer of a racemic N-acyl amino acid ester, generating a rotation of the polarized light incident to the reaction cuvette.

In general, all the precautions listed for spectrophotometers also apply to ORD/CD studies. The sensitivity of the measurement is roughly the same as for absorbance measurements, although this may vary by an order of magnitude depending upon the respective optical rotations of the substrate and product.

Colorimetric with Indicators.   Many enzyme reactions generate (or consume) $[H^+]$. To keep the pH essentially constant a buffer is included, or else base (or acid) titrant is added (see pH-stat, next section). The rate of change in $[H^+]$ may also be followed by including an indicator dye, allowing the pH to slowly drift, and measuring the change in absorbance due to the dye. By titrating the dye with an appropriate acid (or base) one calibrates the slope of absorbance versus moles of acid, and so converts a given absorbance change into a known change in $[H^+]$. The solution is usually lightly buffered so that the total pH change is less than 0.1 unit. This technique is widely used in stopped-flow studies of enzyme kinetics in the millisecond time range.

Khalifah (1971) explored the theory of this method and showed that the ionization of the indicator dye pK_{in} should be as near as possible to the ionization of the buffer ion pK_{buf}. If they are the same then the calibration slope Q is independent of pH; the more the ratio pK_{in}/pK_{buf} deviates from 1, the more Q depends upon pH, complicating the conversion of absorbance changes to $[H^+]$ changes. Rowlett and Silverman (1982) published a table of buffers covering the pK range 6.1 to 9.2 along with the recommended indicator dyes to use with each buffer. This listing, which includes the absorbance maximum and difference in specific absorbance between the acid and base form of each dye, is given in Table 7-1. By selecting the appropriate buffer/dye pair, one may cover the pH range 5 to 10, which ought to serve most enzyme investigations.

Potentiometric Methods

These methods depend upon the change in electric potential between two electrodes as some property of the solution changes. The most familiar example is pH electrodes, where the potential difference between the glass electrode and the calomel reference electrode is logarithmically related to the concentration of $[H^+]$ in the solution. Other electrodes have been made which respond to concentrations of other ions (Ca^{++}) or gases (O_2) in the solution, and the potential is read in the same way as for pH.

Table 7–1.   Buffers and Indicator Dyes for Monitoring [H$^+$]

Buffer	pK$_{buf}$	Indicator Dye	pK$_{in}$	Abs. Max. nm	$\Delta\epsilon$ M^{-1} cm^{-1}
MES[a]	6.1	Chlorophenol	6.3	574	17.7×10^3
3,5 lutidine	6.2	red			
3,4 lutidine	6.6	Bromocresol	6.8	588	30.0×10^3
2,4 lutidine	6.8	purple			
1-methyl-	7.2	p-Nitrophenol	7.1	400	18.3×10^3
imidazole					
HEPES[a]	7.5				
Triethanolamine	7.8	Phenol red	7.5	557	55.9×10^3
4-me-imidazole	7.8				
1,2-dimethyl-	8.2	m-Cresol	8.3	578	38.1×10^3
imidazole		purple			
CHES[a]	9.2	Thymol blue	8.9	590	24.3×10^3

[a] MES—2-(N-morpholino)-ethanesulfonic acid; HEPES—N-2-hydroxyethylpiperazine-N′-2-
ethanesulfonic acid; CHES—2-(N-cyclohexamino)-ethanesulfonic acid

pH, pH-stat. Many reactions occur in which the products are more or less acidic than the substrate. The most obvious is the hydrolysis of an ester, with the generation of a carboxylic acid from a neutral substrate. This product acid ionizes, and causes a drop in pH. If the reaction is unbuffered the drop can be several tenths of a pH unit in a few minutes. While the direct measurement of pH is a measure of rate, the change in pH is likely to change the catalytic rate.

The pH-stat overcomes this problem by intermittently adding small amounts of base (or acid) to the reaction mixture to keep the pH constant. The signal is the ml of titrant added, and this is usually continuously recorded to give a curve of base (viz. acid) consumption versus time. In such a system buffers are not usually used, although some neutral salt may be added to keep the ionic strength of the reaction medium constant. The assay is usually quick, although not as rapid as a spectrophotometric assay, and the reaction volume is usually relatively large, 5 to 10 ml. Since the reaction is vigorously stirred so that the additions of titrant are quickly mixed in, this system works well with substrate suspensions as in the lipase/triglyceride reaction.

Esters are not the only substrate class to which the pH-stat may be applied. The hydrolysis of a peptide bond liberates a carboxylic acid of pK 3.0 to 3.2 and an amino group of pK 7.6 to 8.4. If the reaction occurs at pH 5.5 there will be no net change in pH. If it occurs at pH 8 each cleavage event will release $\frac{1}{2}$ H$^+$. If the reaction is maintained at pH 3.1 each peptide bond cleavage will consume $\frac{1}{2}$ H$^+$. Thus the action of a proteolytic enzyme on a polypeptide substrate could be easily followed in a pH-stat.

The signal would be more directly related to the absolute rate of enzyme action than is the case with most of the protease assays presently in use. It is interesting that a pH-stat is routinely used to maintain a constant pH during the hydrolysis of a protein with trypsin or chymotrypsin for amino acid sequence studies, but to my knowledge no one has applied this technique to an assay of proteolytic activity.

Because the pH-stat is nothing more than an automated version of a manual burette, the sensitivity of the signal might be questioned. Actually it is about the same as a spectrophotometric assay. A typical low-end rate of titrant addition might be 10 μl/min. If the normality is 0.01N, this equals 0.1 μmole per minute. In a 10 ml reaction volume this gives $d[P]/dt$ of 0.01 mM/min. A reaction generating NADH at this rate would have an A_{340} change of 0.062/min, while if the product is p-nitrophenoxide ion from hydrolysis of an ester the change in A_{400} would be 0.018/min. The sensitivity of pH-stat assays is quite adequate for most purposes.

Dissolved Gas and Specific Electrodes. Dissolved gases may react at a semi-permeable membrane to generate a potential. This principle is used to measure the concentration of dissolved gases such as O_2 or NH_3. The most commonly applied gas electrode in enzymology is the oxygen electrode in connection with reactions in which molecular oxygen is consumed. Examples are the various sugar oxidases, most frequently glucose oxidase. As the reaction proceeds the oxygen concentration drops, the potential at the electrode membrane decreases, and instrument meter readings reflect this decrease. The meter circuitry is designed so that the reading goes from 100% (saturated solution of O_2) to 0%. In distilled water saturated with air at 25°C and 1 atm pressure the oxygen concentration $[O_2] = 0.258$ mM.

The oxygen electrode could, in principle, be used to monitor the reaction of any oxidase or oxygenase. A decrease of 5% per minute represents a rate of product formation $d[P]/dt = -d[S]/dt = 0.0129$ mM/min. This is about the same as that discussed above for the pH-stat, so the sensitivity of the oxygen electrode would also be adequate for assaying the oxygenase enzymes.

The ammonia electrode is not often referred to, although it could be used to monitor the formation of NH_3 from urea by the action of urease. Additional enzymes which could potentially be measured with this electrode are: proteases and peptidases hydrolyzing amino acid amides; deaminases, found in fish muscle and in garlic and onion; and L-amino acid dehydrogenases. Again, the response sensitivity of this electrode is of the same order of magnitude as that of spectrophotometric assays for other products from the reaction of these enzymes. The ammonia electrode deserves more attention from enzymologists than it has received to date.

Another electrode of interest responds to the concentration of hydrogen peroxide. This has primarily been applied in conjunction with an immobilized enzyme which forms H_2O_2 as a reaction product, such as glucose oxidase. The hydrogen peroxide concentration is relatively high in the semipermeable membrane near the site of the reaction, and generates a potential. The membrane concentration of H_2O_2 equals the rate of formation via the glucose oxidase reaction minus the rate of diffusion out of the membrane. A steady state is reached which is correlated with the concentration of the substrate glucose in the solution. Similar electrodes are available for other oxidases which also form hydrogen peroxide as one product. It would be interesting to explore the use of the simple hydrogen peroxide electrode for monitoring the decrease in substrate during the catalase reaction.

Mention should be made of immobilized-enzyme electrodes (Guilbault and Kauffmann 1987). An enzyme (e.g., urease) is immobilized in a thin permeable membrane on the tip of a gas electrode (e.g., NH_3 electrode). Dipping this into a solution of urea results in formation of ammonia at a rate proportional to urea concentration. The electrode response defines the rate and so, from a standard curve, the concentration of urea is determined.

Specific Ion Electrodes.   While specific ion electrodes have found extensive use in analytical and inorganic chemistry laboratories they have not been as widely applied by enzymologists. Since few ions (with the exception of NH_4^+) are substrates or products of enzyme reactions this lack of interest is understandable. However, some substrates or products will chelate metal ions, and it might be possible to follow the course of a reaction by the change in chelation of an added ion during the course of the reaction. Some possible combinations might be phytate-phosphate-magnesium, triglyceride-free fatty acid-copper, and glucose-gluconate-iron.

Physical Properties

Viscosity.   Solutions of large water-soluble polymeric molecules tend to be viscous, with the magnitude of the viscosity depending upon several factors, one of them being the average molecular weight of the molecules. If the solute in question is also a substrate for some sort of depolymerase (e.g., starch/amylase or carboxymethylcellulose/cellulase), the action of the enzyme will decrease the average molecular weight of the substrate and the viscosity of the solution will decrease.

The technique has been used since at least 1923. Northrop and Hussey (1923) incubated a solution of gelatin mixed with trypsin in Ostwald viscometers and periodically measured the outflow time of the solution in the viscometer. A plot of outflow times (proportional to viscosity) versus incubation time gave a curve which looks much like a typical [S] versus t plot, at first linear in the downward trend, then curving towards the horizontal.

Extrapolation determined the outflow time at the beginning of the incubation, and Northrop (1933) used the time to reach a certain decrease in outflow time (usually 5% decrease) as an inverse measure of the amount of enzyme present (i.e., a fixed end point assay).

Numerous instrumental methods for the measurement of viscosity have been applied to assaying enzymes. Some of them are: the force exerted upon a paddle or cylinder placed in the container holding the reacting mixture, when either the container (Visco-Amylograph) or cylinder (Brookfield, Haake) is rotated; the time required for a plunger to drop through the reaction mixture (Falling Number); the pressure drop across a capillary tube when the reaction mixture is pumped through it at a fixed rate; and automated adaptations of the Ostwald viscosimeter. A continuous recording of viscosity change, as is possible with the Haake Roto-Viscometer, is probably the most useful for a detailed investigation of the enzyme reaction, but several of the other methods are more convenient and quicker when the only information required is the amount of enzyme present in a sample.

A theory to enable quantitation of the enzyme reaction by the decrease in viscosity has been developed by Manning (1981). The mathematics of this theory is given in detail in Appendix H. A useful simplification is that applied by Hulme (1971) to cellulase assay. The main factor to keep in mind is that for any of these assays to be fundamentally sound the hydrolysis pattern of the substrate by enzyme must be random so that the ratio of the number average molecular weight M_N to weight average molecular weight M_W does not change. Only in this case can the change in solution viscosity be related to actual number of bonds hydrolyzed (i.e., increase in [P]). This circumstance holds for the action of cellulases upon carboxymethylcellulose and possibly for the action of proteases upon gelatin. In most other instances of depolymerizing hydrolytic enzyme action (α-amylase is the best known instance) random attack does not occur.

Viscosimetric assays can be very sensitive. By consideration of the change in molecular weight to give a 5% change in specific viscosity we can estimate the number of bonds hydrolyzed. If the substrate solution viscosity is three times that of the solvent (a normal ratio) then only 2% of the substrate molecules need to undergo one bond cleavage to have the 5% decrease. If the solution is 5 g/L of carboxymethylcellulose with a number average molecular weight of 100,000 the substrate concentration is $50\mu M$ and 2% cleavage corresponds to the formation of 1 μmole of new reducing ends per liter. Measuring this amount of hydrolysis by chemical means would be difficult.

The major experimental concern in using viscosity is reproducibility of substrate. This is most apparent when non-homogeneous polymers such as starch, carboxymethylcellulose, pectin, or gelatin are being used. It is almost impossible to have complete identity from lot to lot of such materials, so it is necessary to prepare new "standard curves" with known enzyme each time a new lot of substrate is used. Also, the temperature dependence of viscosity for many polymers is hysteretic (i.e., it depends upon whether you are warming up or cooling down the substrate to the temperature of the assay bath). It is necessary to control this factor. Starches are mixtures of amylose and amylopectin, and the composition is not precisely fixed from lot to lot, even those coming from the same plant source. This inhomogeneity can affect viscosity and the progress of an α-amylase assay. Overall, the best practice is to prepare a large supply of substrate at one time, spend some time validating the assay conditions with that lot of substrate, and be prepared to repeat the entire validation procedure when a new lot of substrate is required, including several comparison runs between the old and the new lots.

Colligative Properties. Hydrolytic reactions increase the number of solute molecules present in a given volume of reaction mixture; one molecule of sucrose forms two molecules of sugars upon reaction with invertase, one molecule of an ester forms a molecule of acid plus a molecule of alcohol, etc. These reactions thus affect the colligative properties of the solution (i.e., those properties which depend only upon the total number of solute molecules present). Two such properties, freezing point depression and vapor pressure lowering, have been used to follow the reactions of certain enzymes.

Cryoscopy (freezing point depression) requires taking samples from the reaction mixture at various times and measuring the freezing point. This is plotted versus incubation time, and the rate of change in freezing point is a measure of the increase in concentration of solute molecules. The sensitivity of the method is not particularly high. A 1M solution of a non-electrolyte will depress the freezing point of water by 1.86°C. If the freezing point of the reaction mixture is decreasing at the rate of 0.001°C/min, this is a change of 0.54 mM/min. Methods using a pH-stat, oxygen electrode, or spectrophotometry typically have lower end sensitivities around 0.01 mM/min, or 50-fold greater than cryoscopy. On the other hand, the use of the cryoscope obviates indirect chemical methods for quantitating the product, and there is no requirement for a chromophore in either the substrate or product. If a cryoscope is present in the laboratory, and you want to measure hydrolytic reactions with an expected rate of 1 mM/min or more, this is a quick, simple, theoretically sound method for accomplishing that purpose.

Osmometry (vapor pressure lowering) is based upon the same colligative properties as cryoscopy. Instruments for measuring the difference in vapor pressure between a sample and the solvent (water) are generally available. Again, samples are taken from the reaction mixture at various times. It is necessary to add something to halt the enzyme reaction, because the measurement of vapor pressure lowering is usually done at room temperature and some time elapses before an equilibrium reading related to the difference between sample and solvent is obtained. (In the cryoscope the reaction is stopped by the rapid lowering of temperature.) The molar boiling point constant for water is 0.51°C, so this technique is inherently only 27% as sensitive as cryoscopy. Given the other operational difficulties, it would seem that cryoscopy is preferable to osmometry for enzyme assays.

Elasticity. Certain biopolymers contribute to elastic properties when included in a matrix. An example is the elasticity due to gluten protein in bread dough. The theoretical understanding of this phenomenon is rudimentary at best, so the finding that the inclusion of proteases in the dough while mixing causes a progressive loss of elasticity is technologically important but enzymologically vague. We don't understand the causative chain between hydrolysis of peptide bonds in gluten molecules and the change in the reading on the Do-Corder or other rheological instruments for measuring elasticity. It is mentioned here, because it is of the same type as the milk-clotting assay for rennet; poorly understood on the molecular level but significant at the practical level, and, because of specificity factors, other assays of proteolytic activity do not accurately correlate with observed changes in the real-life food system. If the instrument signals can be related to the effects desired in the practical situation (e.g., in the bakery) it might make sense, at some future time, to develop elasticity measurements into an assay for proteases, seeking to quantify the effects to be expected in actual use. We might someday see proteolytic enzymes for these applications assayed and specified in terms of "Elasticity Response Units."

Elasticity is an important phenomenon in many foods. Collagen and connective tissue in meat; pectin gels and matrix in fruits; cellulose and hemicellulose structures in vegetables; these are all substrates which, upon modification by an enzyme, change the processing properties and/or eating texture of the food. In each case the conceptual link between the enzyme molecular reaction and the reading on the measuring instrument is as tenuous as is the case with gluten and bread dough. Again, much more research is necessary to quantify the connection between enzyme action and practical goal, but it would seem to be effort well spent.

INDIRECT METHODS

Chemical Reactions

After incubating substrate with enzyme for a fixed time the reaction is stopped, usually by raising the temperature, adjusting pH, or adding some chemical which inactivates the enzyme. Then a series of physical and/or chemical operations are performed which convert either substrate or product into a quantifiable species. The development of indirect signal, fixed time enzyme assays during this century has fostered a great deal of imaginative chemistry for analytical purposes.

Colorimetry.   The largest group of these chemical reactions are ones which produce a chromophore which can be quantitated with a spectrophotometer, and whose concentration bears some relationship (direct, linear is preferred but not always obtained) to the concentration of substrate or product. A few examples are: Folin-Lowry reaction, for soluble peptides formed during proteolysis; Fiske-Subbarow reaction, for phosphate released during phytase reaction; reaction with 3,5-dinitrosalicylic acid, for reducing sugars. The variety of chemical reactions applied are at least as varied as the chemical natures of the products of enzyme reactions. In some cases the reaction product may be fluorescent (e.g., o-phthalaldehyde with amino groups from protease or peptidase reactions), and the sensitivity is correspondingly increased.

The reaction is calibrated by running a standard curve (i.e., carrying out the chromogenic reaction with known amounts of the product: peptides, ortho-phosphate, reducing sugar). This procedure compensates for slight variations found in the laboratory performing the test. These variations may be, for instance, errors in the delivery pipettes used, deviations in the temperature of the boiling water bath from 100°C, timing differences, spectrophotometer errors (bandwidth, wavelength calibration), etc. The material used for constructing the standard curve should be of the purest grade obtainable, and should be stored in a cool desiccator to prevent moisture pickup, which could lead to errors of several percent. It is advisable to choose the crystalline form of the material which is the most stable (i.e., dextrose monohydrate rather than anhydrous dextrose). Many of the reagent solutions used in these tests are rather unstable, and should either be stored in the refrigerator or made fresh each day. In sum, taking a number of elementary precautions should give a stable, reproducible standard curve so that assay results can be compared even when done several months apart.

A second factor is the experimental design for running the standard curve; how many values of X (the product) should be used, and how should they be spaced? I strongly recommend reading Draper and Smith (1981, pp. 51 ff.) or a similar book on statistical experiment design. There are two questions to consider: Is the relationship between Y (absorbance) and X truly linear and What is the confidence level in the prediction of X made from this fit? Without going into the whole topic of testing goodness of fit, I will simply present Draper and Smith's conclusions about the optimum experimental design. Assuming that the range of X will run from 0 to 10, and 14 reaction tubes will be run for the standard curve, most biochemists would set X at roughly equal intervals, say, 0.7, 1.4, 2.1 . . . 10. This is the poorest possible design. Considering several different alternatives, from two runs each at 1.4, 2.8 . . . 10 to the other extreme of seven runs at 1 and seven runs at 10, Draper and Smith concluded that two designs are about equally commendable. The first involves five runs at 1, two runs at 4, two runs at 7, and five runs at 10. The second uses six runs at 1 and at 10, and two runs at 5.5. These designs seem outlandish, but statistically they are the strongest for accomplishing three aims: 1. test to see if the regression line is truly linear, or whether perhaps a quadratic fit is better; 2. find the best possible regression equation; and 3. generate a meaningful estimate of experimental error and hence the confidence limits around the regression.

Sometimes the chemistry applied is not that of bond rearrangement and formation of new, colored compounds, but rather the formation of a complex which absorbs in some region of the spectrum. A well-known example of this is the formation of a brown complex between amylopectin and I_2, the basis for the Wohlgemuth assay for α-amylase. As the enzyme degrades the starch, the intensity of color due to complex formation decreases. The assay is a fixed end-point type, and the end point is reached when the color matches a certain yellow-brown standard. There are several other cases in which a polymer forms a colored complex with a reactant, but the small molecular weight products from the hydrolytic reaction do not. As an example, several different polysaccharides bind dyes such as Congo Red or Calcofluor and produce a marked intensification of absorbance as well as a shift in the wavelength of maximum absorbance. All these changes can be monitored with a spectrophotometer, but that does not mean that we have improved our understanding of the quantitative molecular basis for the absorbance. Insofar as the connection between extent of absorbance changes and number of enzymatic hydrolytic events remains tenuous, it would be advisable to use this particular form of chemistry only for fixed end-point assays.

Titrimetry.   Many chemical reactions are finished off with some sort of titration. The nitrogen in the soluble peptides from the protease reaction may be measured by the Kjeldahl method. Reducing sugars are measurable by reacting with ferricyanide, then back-titrating excess ferricyanide with thiosulfate. Hydrogen peroxide (formed by glucose oxidase reaction, or unconsumed substrate after incubation with catalase) is quantitated via an iodometric titration. Free fatty acids are extracted from a lipase incubation and titrated with alcoholic KOH. In each case the main precautionary recommendations go directly back to freshman analytical chemistry, and will not be repeated here.

Since most titrations are either acid-base or redox titrations it is possible to use some sort of automatic titrator to perform the actual titration. This usually gives more precise results than manual titrations, and frees one to do other things. With an automatic sample presenter such a device may even continue to work after everyone has gone home for the night. On the other hand, don't overlook the utility and flexibility of a simple burette and sample flask, particularly during the early stages of developing an assay, or for assays which are run infrequently. A good-quality 50 ml burette is just about one of the most inexpensive pieces of equipment in the laboratory, but the precision of results may be an order of magnitude better than with your spectrophotometer (routinely, one part per thousand versus one part per hundred).

Titration methods are not more widely used primarily because they are more time-consuming as compared to placing a cuvette in the spectrophotometer and taking a reading. If the researcher is looking at the output from 40 assays this becomes a real factor. In addition the sheer repetition fosters boredom, inattention, and mistakes. So if a titration method really seems to be the best of the available alternatives for your enzyme assay, it may be well to consider the purchase of a simple automatic titrator.

Separation of Product

Often some sort of separation procedure is applied to the enzyme reaction mixture before the product is measured. In some cases such as the Anson method for protease assay, the separation procedure defines the product: product consists of those peptides from the proteolytic reaction which are soluble in the presence of trichloroacetic acid. In many cases the initiation of the separation is also the end point for the assay, that is, the addition of methanol-chloroform to a lipase reaction or trichloroacetic acid to a protease reaction inactivates the enzyme.

Precipitation.   The peptides soluble in trichloroacetic acid constitute the product from the proteolytic reaction. After the substrate (insufficiently hydrolyzed protein) and enzyme are removed by filtration or centrifugation the peptides are quantitated by a suitable chemical method. Many reactions involving the hydrolysis of soluble polysaccharides may be stopped by the addition of alcohol or similar nonaqueous solvent. Only the smaller polysaccharide fragments remain in solution, and these are measured by a colorimetric reaction for reducing sugar end groups or the total carbohydrate is quantitated using, for example, orcinol/sulfuric acid or a similar type of reagent, depending upon the saccharide present.

Certain unreacted substrates may be precipitated by the addition of flocculants. Thus ferric ion will precipitate phytic acid remaining after the reaction with phytase is stopped. Polyamines will flocculate charged polysaccharides such as carboxymethylcellulose, and the small fragments formed by the reaction with cellulase may then be measured by some general color reaction for carbohydrate. The principal may be extended to other enzyme substrates.

Chromatography. The uses of chromatography for separating the products of enzyme reaction are numerous. The techniques may range from paper chromatography to high pressure liquid chromatography, from silica gel columns to gas phase chromatography. Specific methods will be discussed as appropriate in the next chapters in connection with particular assays.

The quantitation of product after chromatographic separation varies, depending somewhat upon the form of chromatography used. For the various kinds of columns it is usually best to have a detector in line with the output, and the detector response is continuously recorded. The detector may be a spectrophotometer if the product absorbs light. More often it is a refractive index difference detector, since this is more generally applicable to all sorts of products in liquid solution. For a gas phase chromatograph the various kinds of detectors appropriate to that method will be used.

For flat media chromatography (paper, thin layer plates) it is more common to locate the spot corresponding to product either by using a light spray of a chromogenic reagent, or by having a fluorescer incorporated in the chromatography medium and looking for the quenching due to the product. Then the product is extracted from the area containing the spot and measured in a suitable fashion. The product may be reacted directly on the plate to yield a colored spot and then quantitated by direct densitometry. Separation of radio-tagged product from substrate is often carried out in this fashion. Thin layer chromatography (TLC) is convenient; the extract from several assays can be separated simultaneously, the separation is quick (often less than an hour), and the alumina or silica gel medium is easily scraped from the plate for extraction or else placed directly in vials with scintillation medium for counting. Paper chromatography is much slower and slightly less convenient, but for certain kinds of products (e.g., the oligosaccharides from certain carbohydrase reactions) the separation is superior to that possible with TLC. Of course, high pressure liquid chromatography gives equally good separation and is much faster, but at a much higher instrument cost.

Extraction. For certain kinds of enzyme reactions it is possible to extract the product from the aqueous phase into a non-aqueous, immiscible phase. This is usually done when the substrate and/or product is a lipid (i.e., a triglyceride, fatty acid, or lecithin). In the simplest form the organic solvent, such as toluene, is added to the enzyme reaction mixture, the tube is capped and shaken vigorously, and after (when, if!) the emulsion breaks, the toluene layer is drawn off and the extracted product is measured.

Repeated extraction is required to get thorough recovery, and this often leads to dealing with a large volume of organic solvent.

A better method initially adds methanol and methylene chloride to the aqueous reaction in such a ratio that a one-phase system is obtained. Then more methylene chloride is added which moves the system into the two-phase part of the phase diagram; the reaction separates into two layers with an efficient extraction of the lipids into the lower organic phase. This may be separated and analyzed for product. This procedure is less likely to produce emulsions which are difficult to break than is the simple shaking approach. Because the extraction is more efficient it is often possible to settle for the one step, accepting a small, constant loss of product left in the methanol-water phase.

A third method is to add sufficient silicic acid to the enzyme reaction mixture to reduce it to a free-flowing powder state (silicic acid will adsorb several times its own weight of water). This is then extracted with a solvent having a polarity which will elute the desired product from the medium. Hexane, or 10% diethyl ether in hexane, will elute di- and tri-glycerides; a 50/50 mixture of the two solvents elutes monoglycerides; diethyl ether alone will bring off free fatty acids; and methanol will strip off lecithins. (It may be necessary to adjust the polarity of each solvent in order to get the kind of discrimination stated, but these are guidelines for starting your investigations.) The solvents are readily removed by evaporation and the residual lipids quantitated as desired.

Extraction is often a necessary evil when studying enzymes which catalyze certain reactions of lipids. The question is often one of the pattern of hydrolysis such as the formation of 2-monoglyceride by pancreatic lipase, or just which bond is split in hydrolysis of a lecithin by phospholipase. For high-sensitivity assay of lipase action using radio-labeled triglyceride it is often necessary to extract the lipids before they can be applied to a thin layer plate for chromatographic separation. It is usually worthwhile to expend some effort on finding an extraction method which is rapid, convenient, gives good recoveries, and above all does not produce stable emulsions. The subsequent savings in time and frustration will amply repay the effort.

Gel Diffusion

A convenient method for screening a large number of samples for enzyme activity is gel diffusion. A solution of pH buffer, a gelling agent, and substrate is placed in a container such as a Petri dish and allowed to harden. Small wells are then cut in the gel, enzyme extracts are pipetted into the wells, the dish is covered and incubated, usually overnight in a thermostatted cabinet. The next morning the dish is treated with some reagent which differentiates between unreacted substrate and product; circular zones are seen around those wells which contained active enzyme, and the diameter of the zone bears some relationship to the amount of enzyme.

The choice of substrate and reagent are interrelated, just as in any other indirect signal assay. A few examples will elucidate this.

1. For proteases, a useful gel plate incorporates casein as substrate. After incubation the plate is flooded with 5% trichloroacetic acid solution. The casein which has not been hydrolyzed precipitates in situ, giving a milky white background. A clear zone is seen around those wells which contained protease.

2. For α-amylase, a starch substrate with covalently-attached dye (amylose azure) is used. After the incubation the plate is flooded with water or salt solution which dissolves and removes the dyed oligosaccharide fragments. The result is a clear zone against a blue background.

3. For glucanase, a suitable β-D-glucan is dissolved in the buffered gel solution, along with a dye such as Congo Red. After the enzyme incubation is complete the plate is flooded with 1M NaCl; in the presence of this concentration of salt the glucan binds the dye to form a blue background. Light red zones indicate enzyme activity.

4. For esterase, the substrate used is an appropriate ester of α-naphthol. After incubation the plate is flooded with a diazotizing reagent such as a solution of tetra-azo-diorthoanisidine. This reacts with free α-naphthol to give an intense red color; the background, in the absence of esterase, is clear.

The zone diameter is related to the log of the amount of enzyme activity (i.e., diameter $= a + b \log [E]$). The enzyme diffuses into the gel throughout the incubation period, and the concentration at any given radial distance from the well is the integral of the expression for diffusion rate; the value of this integral is a logarithmic function of enzyme concentration in the well. The edge of the zone represents the limit of detection of enzyme activity in the system.

The diameter of the zone depends, first, on the sensitivity of the assay system. This may be made somewhat higher by increasing substrate concentration, but the gain is not usually great. In fact, if the substrate is a gum which increases the viscosity of the aqueous medium, diffusion of enzyme may actually be slowed and zone diameter decreased. As a rule of thumb, if two alternate substrate systems are possible, that one which has the greater sensitivity in a regular assay in solution will also give larger zone diameters in a gel diffusion assay.

The second factor governing zone diameter is the diffusion rate of the enzyme into the gel. This in turn depends upon two items: 1. the molecular weight of the enzyme, which is not under experimental control; and 2. the porosity of the gel, which is manipulatable. The most usual gel medium is agarose, the non-ionic fraction obtained from agar, used at a concentration of 0.5% (w/v) in the buffered substrate solution. Decreasing the concentration slightly will increase enzyme diffusion rate, but the gel will be softer and it is more difficult to cut wells with a sharp boundary. Another possible gel medium is crosslinked polyacrylamide, often used for electrophoresis. For the present purpose the total concentration of monomer should be as high as possible, and the crosslinking monomer should be rather low. This will give an open network with adequate rigidity. The only precaution is that the substrate must be one which will not react during the free radical polymerization of the monomers.

The factors of pH and temperature must be adjusted to fit the enzyme system under investigation. While enzyme activity is higher at higher temperature, thus extending the sensitivity of the measurement, the long incubation time makes denaturation a possibility which must be considered. For screening crude enzyme extracts, in which the enzyme tends to be more resistant to inactivation but the activity level may be quite low, using a higher temperature for incubation may be useful.

The factor b is usually between 0.5 and 0.75, that is, a ten-fold increase in enzyme concentration will increase zone diameter three- to six-fold. The measurement of diameters is usually quite consistent within one plate, but somewhat less consistent between plates. These comments, then, would lead to the estimate that a measurement of $[E_t]$ by gel diffusion has a reliability at best of $\pm$ 10%. Nevertheless, for many purposes this is quite adequate, and the ability to assay literally hundreds of samples per day with rather simple, inexpensive materials makes this technique a valuable one in the enzymologist's repertoire.

Part II PRACTICAL

8   Peptide Hydrolases

The earliest enzyme studies were concerned with digestion and with brewing. Thus it is not surprising to find early reports on the action of proteases (pepsin in stomach juices, Spallanzani in 1784) and amylases (diastases from malt, Berthelot in 1833). Given this early start it is also not surprising to note that studies of proteases and amylases have generated a greater variety of assays than for any other group of enzymes. Protease and peptidase assays are the subject of this chapter, and assays for amylase and related glycosidases are discussed in the next chapter.

The physiological reaction of interest is the hydrolysis of the peptide bond in food protein: $R-C=O-N(H)-R' + H_2O \rightarrow R-C(=O)OH + H_2N-R'$. From this comes the generic term "peptide hydrolases" (sometimes peptidohydrolases) or peptidolytic enzymes. The usual determinants of specificity are the side chains R and R' of the amino acids which comprise the peptide bond, although prior or successor amino acid members in the primary sequence of the protein chain being split may also be involved. In addition to the peptide bond, some peptide hydrolases will also catalyze hydrolysis of esters of compounds which are structurally related to the physiological amino acid substrates. Specific assays using some of these ester substrates will be discussed below.

The term "protease" is understood to apply to an enzyme having endo-peptidolytic activity (i.e., one which cleaves peptide bonds internal to a polypeptide protein chain). A "peptidase" is an enzyme with exo-peptidolytic activity, cleaving peptide bonds either at the N-terminal end of the polypeptide (aminopeptidase) or at the C-terminal end (carboxypeptidase). In general proteases will not hydrolyze a peptide bond if the R moiety has a free amino group or if the R' moiety has a free carboxyl group, whereas a peptidase requires that the appropriate group is freely ionizable.

Proteases may be broadly divided into four classes, based upon the mechanism of their enzymatic action: aspartic (or acidic), serine, sulfhydryl, and metallo (or neutral) proteases (Stauffer 1987). There are a few examples of proteases (low molecular weight proteases, Ca^{++}-activated protease) which do not fall into these classes, but more than 95% of the proteases of interest to food enzymologists are one of these four types. This classification becomes significant when choosing a synthetic small molecule substrate, as in the use of a particular substituted peptide for the assay of metallo-proteases (see the FAGLA assay, below).

With some 20 naturally-occurring amino acids, it is possible to have 400 different side chain configurations around the peptide bond being hydro-lyzed by a protease. The specificity of a protease consists both of a binding constant and a rate factor. Trypsin will only bind to a peptide which has either a lysine or an arginine residue in the R position, but in addition the rate of hydrolysis of a lysine-alanine peptide bond is quite different from the rate of hydrolysis of a lysine-phenylalanine bond. Other proteases have different specificity patterns. Substrate proteins also vary in the frequency and distribution of the different possible peptide bonds. The upshot is that in a protein-based assay (e.g., the digestion of hemoglobin), if two protease preparations give identical rates (say, 1000 Anson units per mg) it is almost a certainty that they will have different rates when assayed on a different protein such as casein.

This fact leads to the conclusion that while protein-based assays are most generally applicable to all proteases, they are of severely limited value in comparing two different protease preparations being considered for a given application. A hemoglobin unit (HU) assay is useful for standardizing several different *Aspergillus niger* preparations, when the protease will be used for modifying fish protein; the HU value may be useless for comparing the *A. niger* protease to a *Bacillus subtilis* preparation, with the view of obtaining equivalent degrees of fish meal protein hydrolysis.

SUBSTRATES

Proteins

Hemoglobin.   Many of the early protease assays were based upon the digestion of hemoglobin, in part because it is a rather homogeneous material, readily isolatable from blood obtained from slaughterhouses. Hemoglobin powder for assays is available from several biochemical supply houses; care must be taken that the material is readily and completely soluble in buffer at a neutral pH, leaving no insoluble residue. If a good grade of material is not available, and if red blood cells can be had, hemoglobin may be readily purified as follows.

Preparation.   Blood is collected from a slaughterhouse and treated with sodium citrate (10 g per liter) to prevent clotting. The red cells are col-lected by centrifugation, and washed twice (suspend and centrifuge) with cold 1.5% NaCl solution to remove plasma proteins. The washed cells are suspended in an equal volume of distilled water; after hemolysis for one hour the cell wall debris is removed by centrifugation. The concentrated hemoglobin solution is then dialyzed against two or three changes of cold distilled water and lyophilized.

Denaturation. Before use as an assay substrate hemoglobin must be denatured to render it susceptible to proteolytic action. In the simplest form, a solution of hemoglobin is adjusted to pH 1.7, allowed to sit for a period of time at room temperature, then adjusted to the pH to be used for the assay (AACC 1983, Method 22-62). This is sufficient if the assay pH is acidic, but if the pH is above 6, acid-denatured hemoglobin will slowly renature and become resistant to proteolysis. In this case urea denaturation must be used, as described by Anson (1938). Hemoglobin (2.2 g) is dissolved in 50 ml water. Urea (36 g) and 1 N NaOH (8 ml) are added and the solution is made to 100 ml volume. After 60 min at room temperature 10 ml 1 M KH_2PO_4 and 4 g urea are added. This substrate solution is 2% in hemoglobin, 6 M in urea, pH 7.5. Obviously the enzyme to be assayed must be stable in urea at the pH and temperature of the assay (the presence of substrate contributes to this stability).

Succinylation. When the extent of proteolysis is followed by using one of the color reagents for free amino groups (ninhydrin, trinitrobenzene sulfonic acid, fluorescamine) the presence of free ϵ-amino groups in lysine side chains contributes to a substantial blank reading. This may be obviated by succinylation (Schwabe, 1973). To 16 g hemoglobin in 100 ml water sufficient 4 N NaOH is added to make the pH 7.5. Powdered succinic anhydride (5 g) is added in several small aliquots with good stirring, and the pH is maintained at 7.5 by addition of NaOH. After the reaction is completed the solution is acidified to pH 2.5 and dialyzed against 100 volumes 0.1 N acetic acid. The solution is diluted to give a stock solution of the desired concentration, or it may be lyophilized and stored as a powder.

Casein. The second most often used protein is casein. This is available from biochemical supply houses, and should be specified "nach Hammarsten." As with hemoglobin it should be completely soluble at 5% in neutral buffer, giving a slightly translucent solution without any obvious turbidity.

Preparation. If a satisfactory grade of casein is not available it may be prepared as described by Dunn (1949). To 1.4 L skim milk diluted with 8.5 L water, sufficient 0.5 N HCl (about 150 ml) is slowly added (during 30 min) with good stirring to adjust the pH to 4.8. The suspension is allowed to settle for 1 hr and the supernatant liquid is siphoned off. The precipitate is suspended in water to 2.5 L total volume, with good stirring, allowed to settle and the supernatant is siphoned off. This is repeated 3 more times. The precipitate is collected by filtration or centrifugation (do not allow it to dry), then resuspended in 400 ml water, stirred, and recollected by filtration or centrifugation. Repeat twice, till the supernatant tests negative for chloride ion. Suspend the solids in 300 ml 95% ethanol, stir, and collect the solids. Repeat this step twice. Suspend the solids in 300 ml absolute ethanol, stir, and collect the solids. Repeat this step twice. Finally, suspend the casein in 300 ml diethyl ether, stir, recover the solids, and repeat one more time. The final product is allowed to air-dry overnight (loosely covered, in a good hood) to allow the ether to evaporate and to equilibrate with ambient humidity. From 1.4 L of skim milk the reported yield was 28 g of dry casein.

Denaturation. In most of the published assays the casein solution is used as made up, assuming that it is completely soluble. However Kunitz (1947), in making up the solution for trypsin assay in the presence of soybean trypsin inhibitor, added a heat denaturation step. He suspended 1 g casein in 100 ml 0.1 M phosphate buffer pH 7.6 and then heated this in a boiling water bath for 15 minutes. The implication was that this was necessary to obtain complete solution of casein. No data was shown to compare the enzymatic reaction on heated versus unheated casein substrate solution.

Dimethylation. When used in an assay depending upon colorimetric reaction with amino groups, the ϵ-amino groups of lysine give a high zero-time blank. As discussed above, this may be removed by succinylation. However, this reaction turns numerous trypsin-susceptible peptide bonds into bonds which are not hydrolyzed by trypsin (and other enzymes with similar specificity). A derivative which retains reactivity towards trypsin but does not react with amino colorimetric reagents is the N,N-dimethyl protein described by Lin et al. (1969). Casein (1.5 g) is suspended in 150 ml 0.1 M borate buffer, pH 9.0, warmed (if necessary) to dissolve, and then cooled to 0°C. Sodium borohydride ($NaBH_4$, 0.3 g) is added, and then 3 ml of 37% formaldehyde solution, reagent grade, is added to the stirred solution in 100 μl increments over a 30 min period. After the last addition the solution is allowed to react for 10 min, adjusted to pH 6 with 50% acetic acid, dialyzed and lyophilized. The color yield of this substrate with trinitrobenzene sulfonic acid is only 10% of that observed with the original casein. The same derivatization may be carried out with hemoglobin, gelatin, or any other protein substrate of interest.

Nonfat Dry Milk.

A number of assays for rennin and similar acid proteases depend upon the fact that casein precipitates in the presence of Ca^{++} when a certain degree of proteolysis has been reached. This test is readily carried out using fresh skim milk, but the time response varies slightly from batch to batch of milk. Hence many laboratories prepare a large lot of dry skim milk and characterize its reaction rate with a standard crystalline enzyme sample (Martin et al. 1981). This is then dissolved in the appropriate buffer as needed to provide a reproducible substrate over a period of time. As the supply of nonfat dry milk runs low a new batch must be obtained, characterized and compared to the previous "standard substrate" to retain uniformity in the clotting assay. The nonfat dry milk must be of the "low heat" type (i.e., not heated to destroy certain sulfhydryl compounds). The "high heat" nonfat dry milk normally sold to food processors (bakers, sausage makers) is not suitable because the casein proteins have been slightly and non-reproducibly denatured.

Gelatin.   Gelatin is produced either by acid hydrolysis (Type A) or alkaline hydrolysis (Type B) of animal collagen. It has a wide range of molecular weights (15,000 to 250,000). Viscosity is dependent upon the average molecule weight of the sample, but the gel strength (designated by the Bloom number) is less strongly correlated with this characteristic. Type A gelatin has a broad isoelectric range of pH 7 to 9, while Type B has a narrower range of pH 4.7 to 5.4. The ash content may vary from 0.5% to 2%. Since heterogeneity is produced by a variety of factors, especially the source of collagen (cow, pork, hides, bones) as well as small variations during the processing, it is not to be expected that lots purchased at various times will give identical results in a viscosity assay based upon gelatin. Thus it is advisable to lay in a large supply and characterize it using standard enzymes. The sample should be deionized with a strong mixed-ion exchange resin; this removes both ash and low molecular weight gelatin fractions. Deionized gelatins are available from pharmaceutical supply houses. If a source cannot be located, a 20% solution of gelatin at 60°C is poured through a mixed-bed ion exchanger, also kept at elevated temperatures. After this treatment the solution is cooled, allowed to gel, cut into small pieces, dried either by lyophilization or by blowing warm air over the pieces, and ground. Alternatively the moisture content of the gel pieces may be determined and they are stored under refrigeration and used directly to prepare substrate solutions.

Derivatized Proteins.   A number of protein derivatives are used to facilitate colorimetric measurement of the degree of hydrolysis. These may be either highly-colored soluble diazo derivatives, or insoluble dye-protein complexes from which dye is solubilized as proteolysis proceeds. A number of these items are available from biochemical supply houses under such names as azocoll, azoalbumin, hide powder Carmine, etc. While they are not expensive or difficult to make in the laboratory, one must weigh price against time in deciding whether to buy or make.

Azo-proteins. Several proteins have been reacted with diazotized sulfanilic acid (or sulfanilamide) to give a strongly-chromogenic red derivative. The protein may be soluble at all pH's (serum albumin, usually Cohn Fraction V), soluble at alkaline pH (casein) or insoluble (powdered collagen).

The procedure is as follows (Charney and Tomarelli 1947; Tomarelli et al. 1949). Protein (50 g) is dissolved or suspended in 1 L of 1% $NaHCO_3$. Sulfanilamide (or sulfanilic acid) (5 g) is dissolved in 200 ml water plus 6 ml 5 N NaOH. Then 2.2 g $NaNO_2$ is added and dissolved, followed by 18 ml 5 N HCl with good stirring. After 2 min add 18 ml 5 N NaOH, stir, and pour into the protein solution with vigorous stirring. The reaction is complete after about 5 min, and the azoprotein is isolated: azoalbumin by dialysis against several changes of distilled water, followed by lyophilization; azocasein by acidification to pH 4.5 and centrifugation or filtration; azocollagen by filtration and washing. A good preparation of soluble azoprotein will, at a concentration of 0.1 mg/ml, have an absorbance at 440 nm (the maximum wavelength) of around 0.35.

Protein-dye complexes. Several insoluble protein-dye complexes have been used for protease assays in which the dye becomes solubilized as the enzyme hydrolyzes the protein substrate. In general these complexes are formed by stirring a suspension of an insoluble protein with an aqueous solution of the dye for some time, followed by extensive washing of the solid phase with water and/or buffer to remove loosely bound dye. A few specific examples will be outlined here, although the method should be generally applicable.

Elastin, purified and powdered, is stirred overnight in a saturated aqueous solution of Congo Red (Shotton 1970). The next day the precipitate is washed several times with water until no more dye is removed, then dried with acetone followed by diethyl ether. The dry material is ground and sieved through a #120 screen. The wavelength of maximum absorbance is 495 nm.

Toasted soy meal (10 g) is added to an aqueous solution of Carmine dye, made by dissolving 50 mg dye in 25 ml 10% NH_4OH, followed by 225 ml 0.1 N HCl (Ilany-Feigenbaum 1966). The suspension is shaken for 1 hr, then centrifuged and the solid material washed several times with water. The final precipitate is dried with warm air and ground before use. The wavelength of maximum absorbance is 540 nm.

Blood fibrin powder (75 g) is mixed with 1.5 L 0.018 N HCl containing 0.5% indigo carmine (Nelson et al. 1961). The suspension is heated to 80°C and held for 30 min with stirring. Then it is cooled and the protein-dye complex collected by filtration on several layers of cheesecloth. It is washed by suspension in 0.018 N HCl and filtration, repeated 4 times. Finally it is dried, ground, and sieved, with the fine particles (through #100 sieve) being discarded. The wavelength of maximum absorbance is 620 nm.

The same authors (Nelson et al. 1961) dyed hide powder (75 g) by adding it to 1.5 L 0.1 M phosphate buffer pH 7.5 containing 3.75 g congo red dye at 60°C and stirring the suspension for 30 min. After filtration the solids are washed twice with distilled water, then with the phosphate buffer, then 4 times with 0.1 M borate buffer pH 8.5, then 2 to 4 times with the phosphate buffer pH 7.5. Finally the filter cake is dried, ground, and sieved, discarding the fine particles (through #100 sieve). The wavelength of maximum absorbance for this dye is 495 nm.

Synthetic Substrates

Peptides. Synthetic peptides of known sequence have found some use in protease assays, particularly for enzymes belonging to the aspartic protease class (pepsin, rennin, etc.). The hydrolysis reaction is usually monitored by a colorimetric reaction with the freed amino group, although pH-stat operation is also possible at the low pH of the reaction. A few peptides have been synthesized which show an absorbance change when the peptide bond is cleaved; these are usually based upon nitro-phenylalanine as the carboxy amino acid in the susceptible bond. Other means of measurement (formol titration, alcohol titration) are also possible, but are not usually used.

The main problems with routine use of peptides as substrates are supply and expense. While a few synthetic peptides are available from biochemical supply houses, the price is high, on the order of tens of dollars per milligram. The manufacture of a peptide using Merrifield solid-state synthetic techniques is certainly much simpler than it was a decade ago, but it is still not a routine technique. Making ten grams of a hexapeptide is not a simple two-day job for a laboratory technician.

On the positive side, the use of synthetic peptide substrates allows a more definitive study of the reaction characteristics of aspartic proteases than using protein (i.e., hemoglobin) substrates. If the colorimetric reaction for monitoring freed amino groups is one of the fluorometric variety the sensitivity is high, and each assay may require only a few nanomoles (one to two micrograms) of peptide. Peptide substrates are also useful for studying the activity and specificity patterns of peptidases.

Amides. Amino acid amides may be divided into three groups for purposes of discussion. The amine may be simply NH_3, a chromogenic aryl amine such as p-nitroaniline, or a fluorogenic amine such as Rhodamine B. The preparation and mode of use of each of these groups is slightly different.

Simple amides. A number of N-acyl amino acid amides are used in spectrophotometric assays for proteases. An example is N-benzoyl-L-arginine amide, a substrate often used for assaying sulfhydryl proteases such as papain. Upon hydrolysis of the amide bond the ultraviolet absorbance due to the N-benzoyl amino acid changes, much as it does when the ester derivative is hydrolyzed. A wide variety of these amides are available from biochemical supply houses.

Aryl amides. Arylamines often absorb light in the ultraviolet or visible range, but not when they are in the form of an amide. A common arylamine is p-nitroaniline, which absorbs at 440 nm ($\epsilon = 8800$), but is colorless when the amino group is acylated. An example is N-benzoyl-L-arginine p-nitroanilide (BAPA), widely used for the assay of trypsin and similar proteases (Erlanger et al. 1961). A second example is the amide formed by amino acids with β-naphthylamine. In this case the appearance is not monitored directly; rather, a chromophore is formed by the reaction of the freed β-naphthylamine with an added diazo compound such as Fast Garnet GBC, the absorbance at 525 nm is measured and the amount of hydrolysis is determined by reference to a standard curve (Kolehmainen and Mikola 1971).

Fluorogenic amides. A number of fluorescent polycyclic compounds bear an amino substituent. When this group is acylated the derivative is non-fluorescent. Examples are a bis(dipeptidyl) amide of Rhodamine (Leytus et al. 1983b) and an N-acyl-arginyl amide of 3-amino-9-ethylcarbazole (Monsigny et al. 1982). In both instances the hydrolysis of an amide bond converts a non-fluorescent compound into a highly fluorescent material; the rate of hydrolysis is easily followed in a suitable spectrofluorimeter.

Preparation. The synthesis of these amides follows a general route. The carboxyl group of the N-acylated amino acid is activated in a suitable solvent. Then the amine constituent is added, condensation is allowed to proceed, and finally the desired product is purified. As an illustration the preparation of (Cbz-Arg-)$_2$-Rhodamine (Leytus et al. 1983a) will be described. Cbz-L-arginine·HCl (4.0 g, 11.6 mmol) is dissolved in 80 ml cold solvent (dimethylformamide(DMF)/pyridine 1:1) at 4°C. Then 2 g (10.4 mmol) 1-(3-dimethylaminopropyl)-3-ethylcarbodiimide is added and stirred 5 min to dissolve. Next 150 mg (0.41 mmol) Rhodamine 110 dissolved in 1.5 ml of the solvent is added, the reaction is stirred for 2 hr at 4°C and then for 2 days at room temperature. Addition of 150 ml diethyl ether precipitates the reaction products which are collected by centrifugation. The desired product is purified by repeated solution and precipitation as follows: dissolve in 10 ml DMF, precipitate with 200 ml acetone; dissolve in 10 ml DMF, precipitate with 100 ml 1.2 N HCl; repeat that step; dissolve in 10 ml methanol, precipitate with 200 ml ethyl acetate; repeat that step twice. The final product is dried, yielding 340 mg of a pale pink powder that appears pure by analytical thin-layer chromatography (silica gel, butan-2-one/acetone/water 8:1:1).

Esters. Ester substrates for proteases and peptidases fall into four categories, the first three being analogous to the amide substrates: 1. esters of simple alcohols with amino acids; 2. aryl alcohol esters of amino acids; 3. fluorogenic alcohol esters of amino acids; and 4. esters of compounds bearing a structural similarity to amino acid substrates of the enzyme in question.

Simple esters. An example is the ethyl ester of N-acetyl-L-tyrosine, often abbreviated ATEE. This is a substrate for many serine proteases such as chymotrypsin and subtilisin. Upon hydrolysis of the ester bond the absorbance at 237 nm increases (Schwert and Takenaka 1955). Similar esters based upon lysine or arginine are used for trypsin-like proteases and also for many sulfhydryl proteases.

Aryl esters. A widely used aryl ester is the p-nitrophenyl ester of amino acids and related compounds. p-nitrophenol is a good leaving group, and the ionized form is a strong chromophore ($\epsilon = 18,300$ at 410 nm). These esters are subject to nucleophilic attack by OH^- (so the assay pH may not be too basic) and amines (don't use amine buffers such as Tris). p-Nitrophenol itself has a pK of 7.04, so to take advantage of the absorbance of the nitrophenoxide ion the assay has to be around pH 8. By analogy with aryl amides, it is possible to use α-naphthyl esters of amino acids as substrates, reacting the freed naphthol with a diazonium salt after stopping the reaction. This possibility has not been exploited for protease assays, although naphthyl esters of carboxylic acids have found application in esterase and lipase assays.

Fluorogenic esters. Fluorescein is a polycyclic fluorescent molecule which contains two hydroxyl groups. When these groups are esterified the compound is only slightly fluorescent. Several derivatives of fluorescein have been made which will bind to certain proteases and be hydrolyzed by the action of the enzyme (Liu et al. 1980). The p-guanidinobenzoyl diester is an excellent "burst" substrate for trypsin and plasmin and is sensitive to one picomole of enzyme.

Structural analog esters. Numerous esters have been synthesized which bear a structural analogy to the amino acids which are preferentially bound by various peptide hydrolases. Since most of these are intended for "burst" titrations they will be discussed in the next section. However an example of the concept is provided by hippuryl-DL-β-phenyllactic acid, a commonly-used ester substrate for carboxypeptidase A. The corresponding peptide substrate is N-carbobenzoxy-glycine-L-phenylalanine. Hippuric acid (N-benzoyl glycine) is esterified to the α-hydroxy group of phenyllactic acid, and the β-phenyl substituent is comparable to the side chain of phenylalanine. The two aromatic rings and the amide linkage are similarly oriented in the two molecules and provide the interactions for binding of the compounds to the active site of carboxypeptidase. While the detailed mechanism of hydrolysis is apparently slightly different for the two substrates (Riordan and Vallee 1963) they have both proven useful in studying this enzyme. Hydrolysis of the peptide bond may be followed by a change in absorbance at 224 nm, but ester hydrolysis is best monitored with a pH-stat system (Petra 1970).

Preparation. Synthesis of esters is similar to synthetic methods for amides. The carboxyl group is activated, it is reacted with the corresponding alcohol, and finally the ester is isolated and purified. As an example the synthesis of N-carbobenzoxy-L-tyrosine p-nitrophenyl ester (Walsh and Wilcox 1970) will be outlined. All solvents must be strictly anhydrous. Cbz-Tyr (1.5 g) is dissolved in 5 ml dioxane plus 1.2 ml tributylamine and cooled to 5°C. Then 0.65 ml ethyl chloroformate in 5 ml dioxane is added. After 30 min 0.7 g p-nitrophenol is added and the reaction mixture is stirred for 1 hr. The solvent is removed in vacuo, the oil is taken up in $CHCl_3$, washed (cold) with 1 N HCl, saturated Na_2CO_3, 1 N HCl, and water. The organic layer is dried and the solvent removed. The solid residue is triturated in methanol and crystallization is induced. The product is recrystallized from $CHCl_3$.

Titration Substrates. These compounds usually incorporate some molecular features analogous to good substrates, which lead to a fairly high affinity for the enzyme. In addition the bond reacting at the active site has a good leaving group so that the acylation step of the hydrolytic reaction occurs readily, but the acyl-enzyme formed is rather stable so the turnover reaction is slow (see Appendix G). The reactions of these titrants are almost invariably monitored spectrophotometrically, although the fluorescein diester mentioned earlier (Liu et al. 1980) allows fluorimetric monitoring.

The first report of a "burst" reaction was that by Hartley and Kilby (1954) on the stoichiometric rapid release of p-nitrophenol from p-nitrophenyl acetate by chymotrypsin. Another titration substrate for chymotrypsin and serine proteases with similar specificity requirements is N-*trans*-cinnamoyl imidazole (Schonbaum et al. 1961). Chase and Shaw (1967) synthesized a specific titrant for trypsin, p-nitrophenyl-p'-guanidinobenzoate, in which the guanidino moiety mimics the cationic side chain of arginine to contribute to binding to the trypsin active site. Kezdy and Kaiser (1970) synthesized an internal ester, a sultone, as a chymotrypsin titrant, and later the same authors (Kezdy and Kaiser 1976) published details of synthesis of a compound specific for sulfhydryl proteases, α-bromo-4-hydroxy-3-nitro-acetophenone, in which the leaving group is a bromide ion.

Many of the more useful titration substrates are available from biochemical or chemical supply houses. The preparation for the most part involves condensation of the alcohol leaving group with the acidic moiety, as outlined earlier. In a number of cases it is possible to start with the acyl chloride, reacting it with p-nitrophenol or imidazole in the presence of pyridine and obtaining a very high yield of pure product.

MEASUREMENT

Nearly all the means of monitoring progress of enzyme reactions discussed in Chapter 7 have been applied to peptide hydrolase reactions. Many assays use direct measurement, while others require that the reaction be stopped and further manipulations carried out before an instrument reading is obtained. For most of the assays discussed below the measurement method is given in the protocol and is directly related to the nature of the substrate (i.e., absorbance changes with chromogenic substrates). However there are two areas in which several options for measurement are available, and are worth discussing without considering the specific conditions for the enzyme reaction. These relate to the quantitation of protein fragments soluble in trichloroacetic acid (TCA) (the Anson method) and the measurement of amino groups freed during the hydrolysis of peptide bonds.

Protein Fragment Quantitation

There are basically four methods available for determining the amount of protein which has been rendered TCA-soluble by proteolysis. These are: 1. Kjeldahl analysis; 2. ultraviolet absorbance; 3. Folin-Lowry color; and 4. color development with bicinchoninic acid.

Kjeldahl.   Historically this is the oldest method, used in the original development of the Anson method (Anson 1938). An aliquot of the filtrate from the TCA-treated reaction mixture is analyzed for nitrogen by the standard Kjeldahl method. While this is an accurate method it requires a rather large amount of filtrate and is time-consuming. It is still the official method for proteolytic activity in foodstuffs (AACC 1983, Method 22-60).

UV Absorbance.   The aromatic amino acids in proteins absorb light in the ultraviolet range. With hemoglobin as a substrate the absorption band is centered at 275 nm, due mostly to tyrosine. The filtrate from the TCA-treated reaction mixture is read directly in a UV spectrophotometer against an appropriate blank (AACC 1983, Method 22-62). While the units are often referred against a known amount of tyrosine, this serves more as a check on the instrument rather than having any intrinsic significance.

Folin-Lowry Color.   The protein fragments may be quantitated with the commonly-used Lowry colorimetric reaction for proteins (AACC 1983, Method 22-63). A number of slight variations on the original Lowry protocol (Lowry et al. 1951) have been published. The one given here has served the author well for many years (Stauffer 1975). To 1 ml of the TCA-filtrate add 1 ml alkaline buffer (1 M Na_2CO_3, 0.25 M NaOH) and 0.4 ml copper reagent

(0.1% $CuSO_4 \cdot 5H_2O$, 0.2% NaK tartrate), mix and allow to stand 10 min. Then add 0.75 ml diluted phenol reagent (Folin-Ciocalteau reagent diluted with 3 vol H_2O), mix, and allow 10 min for the color to develop. Read absorbance at 700 nm, using an appropriate reagent blank. The plot of log absorbance versus log protein is linear over the range 3 to 400 micrograms protein.

Bicinchoninic Acid Color.   Recently a new colorimetric method for protein measurement has been described (Smith et al. 1985). The chemistry is a two-step reaction: Cu^{++} is reduced by protein in an alkaline buffer to Cu^+; the chelator bicinchoninic acid (4,4'-dicarboxy-2,2'-biquinoline) forms a specific purple complex with cuprous ion with an absorbance maximum at 562 nm. Reagent A contains 1% bicinchoninic acid disodium salt, 0.16% disodium tartrate, 3.25% $Na_2CO_3 \cdot H_2O$, 0.09% $NaHCO_3$, pH adjusted (if necessary) to 11.25 with NaOH. Reagent B contains 4% $CuSO_4 \cdot 5H_2O$. The working reagent is 100 parts reagent A plus 2 parts reagent B, and is stable indefinitely at room temperature. To 100 μl of sample add 2 ml of working reagent. Allow color development for 2 hr at room temperature or 30 min at 37°C. Read absorbance at 562 nm against an appropriate reagent blank. The authors show that the color development is somewhat sensitive to pH and is optimum at 11.25. If this procedure is adopted for use with TCA filtrates an amount of NaOH to neutralize the TCA and any other buffer should be added before the working reagent.

Free Amino Group Quantitation

Any peptidolytic reaction may be quantitated by measuring the increase in amount of free amino groups. This can be done directly on aliquots of the enzyme reaction mixture, and is equally applicable to protease and peptidase reactions. Three kinds of measurement will be discussed: 1. colorimetric reactions; 2. fluorometric reactions; and 3. titration.

Colorimetric Reactions.   While a large number of compounds will react with primary amino groups to give colored products, two in particular have been applied to amino acids and peptides: ninhydrin and trinitrobenzene sulfonic acid (TNBS).

   Ninhydrin. As developed for use with amino acid analyzers the ninhydrin reagent is rather touchy to make and difficult to store due to the requirement for keeping it out of contact with oxygen (Reimerdes and Klostermeyer 1976). However, for enzyme assay purposes this requirement is not necessary, and the ninhydrin reagent used by the AOAC (AOAC 1980) is quite satisfactory. The reagent contains 10 g $Na_2HPO_4 \cdot 10H_2O$, 6 g KH_2PO_4, 0.5 g ninhydrin and 0.3 g fructose in a total volume of 100 ml, pH 6.6 to 6.8.

This solution is stable for 2 weeks if kept in an amber bottle, out of the light. For the reaction 2 ml of sample (the enzyme reaction mixture may require dilution) is mixed with 1 ml ninhydrin reagent, heated 15 min in a boiling water bath, cooled to 20°C and diluted with 5 ml of water/ethanol 60:40. The absorbance at 570 nm is read within 30 min. A standard reference curve is constructed using 2 to 20 μg of glycine.

TNBS. The chromogenic reaction of TNBS with primary amines has been studied by several workers (Mokrasch 1967; Fields 1971; Snyder and Sobocinski 1975; Adams et al. 1976). A major problem is that at the alkaline pH necessary for the reaction with amino groups TNBS is also hydrolyzed by OH^- to picric acid, giving a significant blank. Adlers-Nissen (1979) investigated the relationship between the rate of TNB-amine color formation, the rate of picric acid formation, pH, and temperature, to optimize the system. He also incorporated 1% sodium dodecyl sulfate (SDS) in the flow chart to solubilize partially hydrolyzed proteins; this would not be necessary if the substrate were completely soluble. For the color reaction 0.25 ml of suitably diluted enzyme reaction mixture (0.05 to 0.5 μmoles free amine) is added to 2 ml of 0.2125 M phosphate buffer pH 8.2, 2 ml of freshly made 0.1% TNBS in water (protect from light) is added and the mixture is incubated in the dark for 1 hr at 50°C. Then 4 ml 0.1 N HCl is added, the reaction is cooled to room temperature and absorbance at 340 nm is read. Many workers recommend reading the color at 420 nm. While the TNB-amine does have a higher absorption at this wavelength, picric acid also absorbs much more strongly; the difference (TNB-amine minus blank) absorbance is about 1.5 to 1.8 times greater at 340 nm. The absorbance is converted to free amino equivalents by reference to a standard curve constructed with leucine.

Fluorometric Reactions.   Two fluorogenic reagents, initially developed for amino acid analyzer applications, have also been suggested for monitoring peptidolytic reactions. They increase the sensitivity of the assay by about 100-fold as compared to the colorimetric reagents.

Fluorescamine. This material is not readily soluble in water, so the reagent solution is 0.1 mg fluorescamine per ml in acetone. The reaction is with unprotonated amines, so the buffer should be in the pH 8 to 9.5 range, either borate or phosphate. The standard curve with leucyl-leucine is linear over the range of 0 to 10 nanomoles. Schwabe (1973) found that hemoglobin at the amounts used for these assays (about 300 μg) is itself fluorescent and exerts a quenching effect, but succinylated hemoglobin does not. Apparently the free ϵ-amino groups of lysine are responsible for most of this quenching effect as well as producing a high blank. The reaction between fluorescamine and amino groups is rapid, and the fluorescence (excite 390 nm, emit 475 nm) may be read as soon as the diluted enzyme digest and the reagent are mixed. Preston (1975) used fluorescamine to measure the amino groups in the dialyzate from a proteolysis of hemoglobin in an automated AutoAnalyzer adaptation of the method of Schwabe.

o-Phthalaldehyde. Another fluorogenic reagent originally designed for use with high-sensitivity amino acid analysis is o-phthalaldehyde. It is claimed (Benson and Hare 1975) that this reagent is up to ten times more sensitive than fluorescamine. It has the advantage of being stable in aqueous solution, and a stock solution is made by adding 800 mg (dissolved in 10 ml ethanol) to 1 L 0.4 M borate buffer pH 9.7, containing 2 ml of 2-mercaptoethanol. This reagent is mixed with effluent from the amino acid analyzer column on an equal volume basis and the fluorescence (excite 340 nm, emit 455 nm) may be read immediately. As with any of the amino group reagents, the ϵ-amino groups of lysine in a protein substrate will contribute to a rather high blank.

Titration.   In the presence of 6 to 9% formaldehyde the apparent pK of amino groups is lowered by about 2 pH units and they are readily titrated with NaOH; this is termed "Formol Titration" and was widely used during the '20s and '30s for monitoring protease reactions. The various recommendations by different workers for carrying out this technique were critically evaluated by Taylor (1957) who recommended the direct titration procedure. To a 5 ml aliquot of the enzyme reaction mixture add 1 ml of 40% formaldehyde solution (previously neutralized to pH 8.5) and titrate with NaOH to pH 9 using a pH meter. This is done at intervals during the reaction. A blank titration is run starting with just the buffer and substrate and adding the enzyme shortly before reaching the end point. The difference between the blank titration and the titration of the reaction is the moles of amino groups freed by the reaction; the titration appears to measure 93% $\pm$ 10% of the free amino groups. Formol titration is two orders of magnitude less sensitive than the colorimetric reactions and four orders of magnitude less sensitive than the fluorometric reactions. It is mentioned for its historical interest, and there may be occasions when it will be useful in spite of its low sensitivity.

SPECIFIC PROTEASE ASSAYS

In this section experimental details for several specific protease assays will be discussed. The reader should be aware that in most cases many of the parameters given—volumes, pH, time and temperature of incubation—may be adjusted to fit particular needs. The amount of enzyme preparation used, of course, is always subject to variation, depending upon the level of activity present. The following assays will provide a useful starting point for the investigation in hand; improvements for specific purposes are expected.

Protein-Based Assays

Solubilization in Trichloroacetic Acid. As discussed earlier, many assays measure the amount of protein which is soluble in the presence of trichloroacetic acid (TCA) after incubation with protease. An example is the casein assay described by Kunitz (1947). The substrate is 1 g of casein suspended in 100 ml 0.1 M phosphate buffer, pH 7.6, heated in a boiling water bath 15 min and then cooled to 30°C. For the assay 1 ml enzyme solution (at 30°C) is added to 1 ml substrate and the mixture is incubated 20 min at 30°C. Then 3 ml of 5% (w/v) TCA is added, the contents of the tube are mixed and allowed to stand 1 hr for precipitation. The tube is centrifuged and the absorbance at 280 nm of the supernatant is read. The blank comprises 1 ml substrate mixed with 3 ml TCA, followed by addition of 1 ml enzyme solution, 1 hr rest period and centrifugation. This procedure is similar to Method 22-62 of the AACC (AACC 1983), which uses acid-denatured hemoglobin at pH 4.7.

The final concentration of TCA does not seem to be critical. In the assay described, it is 3%; in the AACC method, it is 6.4%; in Miller's study of the Ayre-Anderson method (Miller, 1947), it is 6%; and in an assay recommended by Hagihara, it is 3.2%.

A superior precipitating reagent is the buffered TCA recommended by Hagihara et al. (1958). The reagent contains 0.11 M TCA (18 g/l), 0.22 M sodium acetate (18 g/l) and 0.33 M acetic acid (19.8 g/l). The assay is run with 1 ml of enzyme to 5 ml of 1.2% casein in 0.03 M phosphate buffer pH 7.5, and 5 ml of buffered TCA is added to stop the reaction. After 30 min the mixture is filtered and absorbance of the supernatant is read at 275 nm. A plot of A_{275} versus amount of enzyme extends accurately through the origin as it should. When unbuffered 0.44 M TCA is used to precipitate, the plot intersects the X-axis; low levels of enzyme activity are not registered accurately. The assay with casein as a substrate is improved by using buffered TCA reagent. I am not aware of any similar comparisons done on the hemoglobin-based assay.

The soluble diazoproteins (see above) may be used in place of casein or hemoglobin in these assays. The absorbancy at 440 nm is about four times as great as the absorbancy of the parent protein at 275 nm, so the use of the diazoprotein will give a four-fold increase in assay sensitivity. In other respects the assay procedure is the same.

Soluble Dye Complexes. Soluble proteins complex with the dye Coomassie Brilliant Blue G-250 in solution and produce a hyperchromic shift which is maximum at 595 nm (Bradford 1976). Hydrolysis products formed by protease action do not give this shift, and this has been used as the basis for a proteolytic assay (Saleemuddin et al. 1980). The dye reagent is made by dissolving 100 mg Coomassie Brilliant Blue G-250 in 50 ml ethanol, adding 100 ml 85% phosphoric acid, and diluting with water to 1 L. For the assay, 100 μg of substrate protein is incubated with enzyme in a total volume of 50 μl for 10 min. Then 5 ml of dye reagent is added to stop the reaction, and after 5 min the absorbance at 595 nm is read.

Different proteins produce a different amount of hyperchromic shift per unit weight, so the slope of the plot of A_{595} versus μg protein varies. Bovine serum albumin (unhydrolyzed) gives A_{595} of 0.275 per 25 μg (Bradford 1976). While the decrease in A_{595} is linear with incubation times up to 10 min, the reported linearity with respect to amount of enzyme taken (Saleemuddin et al. 1980, reported results with pepsin, trypsin, and papain) was very poor. This assay technique promises simplicity, but more work is needed to develop acceptable linearity before it may be widely used.

Decrease in Viscosity.   Estimation of protease activity by the rate of decrease of viscosity of a gelatin solution has been used for over 70 years. J. H. Northrop did much work with this assay method during the 1920s and early 1930s, and published a paper comparing results on several different proteins, using both viscosity decrease and formol titration as measures of the extent of enzymatic hydrolysis (Northrop 1933).

A solution of 2.5% (w/v) isoelectric gelatin in phosphate buffer is attemperated at 35.5°C. At zero time dry enzyme is added to the gelatin solution, stirred gently to dissolve and mix it thoroughly, and 10 ml is pipetted into an Ostwald viscometer mounted in a 35.5°C constant temperature bath. At intervals the outflow time for the solution in the viscometer is measured; the corresponding incubation time is taken to be the time at the midpoint of the flow measurement. A plot of outflow times versus incubation times is made, and the outflow time at zero time is found by extrapolation. The end point of the assay is that time required to decrease the initial viscosity by a certain percentage. Northrop used 3% decrease routinely, finding that viscosity drops of much more than 10% were less reliable. The time required to achieve a given viscosity decrease is directly proportional to the reciprocal of the enzyme activity.

Laufer (1938) showed that a plot of log [outflow time] versus incubation time is linear up to about 7.5% decrease in viscosity, then changes abruptly to a lesser slope. An advantage of the semilog plot is that the initial outflow rate is more accurately estimated, a problem which Northrop recognized. Laufer started with a more concentrated gelatin solution and added appropriately diluted enzyme solution to have a final gelatin concentration of 3% in the viscosimeter. He used a 10% decrease in viscosity as his endpoint, a questionable choice in view of the apparent change in reaction mode prior to that time.

An automated viscosity assay has been described by Kuiper et al. (1978). They used an AutoAnalyzer with a constant time of enzyme reaction, and then measured viscosity by the pressure drop across a capillary tube through which the reaction solution was pumped at a constant flow rate. They found that the reciprocal of the pressure drop was linearly related to the reciprocal of the amount of enzyme present, although the plot does not pass through the origin. For protease assays the substrate was casein at a concentration of 7.3% (w/v) after addition of the enzyme solution.

Colorimetric Measurements.   The hydrolysis of peptide bonds forms free amino groups which may be determined colorimetrically by reaction with TNBS or ninhydrin. An assay using dimethylated casein has been published by Lin et al. (1969), but the procedure for the TNBS color reaction given here is that developed by Adlers-Nissen (1979). The substrate is 0.1% (w/v) dimethylated casein in buffer. After attemperating 1 ml in a reaction tube, 0.1 ml enzyme solution is added at zero time and the reaction is incubated for any desired length of time. It is stopped by immersing the reaction tube briefly in a boiling water bath. For color development, add 1 ml 0.4 M phosphate buffer, pH 8.20, containing 0.25% (w/v) sodium dodecyl sulfate. Then add 2 ml freshly made 0.1% TNBS in water and incubate in the dark for 1 hr at 60°C. Add 4 ml 0.1 N HCl, cool to room temperature, and read absorbance at 340 nm. A color reaction containing only the buffers and TNBS gives the reagent blank (i.e., formation of picric acid by hydrolysis of TNBS) while an assay tube heated before addition of enzyme gives the zero time blank, correcting for free amino groups in the substrate and protein.

A standard curve may be constructed using 0 to 1.5 μmoles leucine (Adlers-Nissen 1979). For TNB-leucine the molar absorbancy $\epsilon = 9.1 \times 10^3$ $M^{-1}cm^{-1}$. Alternatively, Lin et al. (1969) determined the color yield for a number of different peptides and amino acids and found an overall average $\epsilon = 13 \times 10^3$ $M^{-1}cm^{-1}$, while Snyder and Sobcinski (1975) found an average molar absorbancy for several amino acids of 11×10^3 $M^{-1}cm^{-1}$. If the latter factor is used, then a reading of A_{340} of 1.0 (after appropriate blank and zero time corrections) corresponds to the formation of 0.74 μmoles of free amino groups in the original enzyme reaction mixture, with the assay and reaction volumes given.

Fluorimetric Measurements.   Schwabe (1973) reacted fluorescamine with the reaction product of protease with succinylated hemoglobin to detect the formation of free amino groups. Hemoglobin itself gives a high zero time blank due to ϵ-amino groups, and also quenches the fluorescence; these problems were greatly reduced using the succinylated derivative. A standard curve is constructed using 0 to 10 nmoles of the dipeptide L-leu-L-leu. Preston (1975) adapted this assay to an AutoAnalyzer, reacting the fluorescamine with the dialyzate from the enzyme-hemoglobin reaction. He found good linearity in the plot of fluorescence versus amount of enzyme.

The time linearity is much better than that found with Folin-Lowry quantitation of supernatants from TCA precipitation.

Fluorescamine assay (Schwabe 1973). The substrate is succinylated hemoglobin, 40 mg/ml of buffer; to 100 μl at 37°C is added 100 μl buffered enzyme solution at the same temperature. At desired intervals 10 μl aliquots are taken for analysis. The enzyme reaction may be stopped by brief heating (i.e., placing this aliquot into a heated tube), by adding to the fluorescamine reaction buffer if the enzyme is inactive above pH 7, or by adding the aliquot to the acetone which is the solvent for fluorescamine. Whichever method is chosen, fluorescence is developed by combining the aliquot with 2 ml buffer (either 0.1 M phosphate pH 7 or 0.1 M borate pH 8 is suitable) and 1 ml fluorescamine solution (0.1 mg/ml in acetone). After 5 min read the fluorescence (excite at 390 nm, emit at 475 nm).

o-Phthalaldehyde assay (Benson and Hare 1975). The buffer for the reagent is 0.4 M boric acid titrated to pH 9.7 with potassium hydroxide and containing 2 ml 2-mercaptoethanol per liter. To 1 L of buffer is added a solution of 800 mg o-phthalaldehyde in 10 ml ethanol. The aliquot from the enzyme reaction mixture is diluted into 1.5 ml ethanol/water (50/50, v/v) and 1.5 ml of the buffered fluorogenic reagent is added. Fluorescence is read (excite 340 nm, emit 455 nm). The reaction with amino groups is said to be immediate. The authors claim this reagent enhances sensitivity as much as ten-fold relative to fluorescamine.

Casein Clotting Reaction. Martin et al. (1981) describe a standardized method for measuring the time required for enzymes such as rennet to induce clotting in milk. The substrate is low-heat nonfat dry milk, 120 g per liter of 0.01M $CaCl_2$. To 10 ml in a stoppered tube, attempered at 30°C, 1 ml of enzyme solution is added. The tube is inclined in the temperature bath and slowly rotated at 4 rpm. The endpoint (clotting time, CT) is that time when flakes of precipitate appear on the side of the tube. A plot of 1/CT versus amount of enzyme is a straight line. With some batches of milk powder this line does not extrapolate through the origin of the plot, hence the authors recommend preparing a large batch of milk powder and determining the slope and intercept of the plot using a standard, pure enzyme.

Casein turbidity. A related assay has been described by Skelton et al. (1976) for assaying papain activity. The substrate is 1% casein in 0.1 M citrate buffer, pH 6.5. Papain is activated by dissolving in the same buffer containing 1 mM glutathione. To 2 ml of enzyme solution at 30°C is added 2 ml of substrate at the same temperature. The solution is inserted in a spectrophotometer and the "absorbance" at 425 nm is followed. The time for this measurement to increase from 0.2 to 0.5 is measured. The plot of 1/t versus amount of enzyme shows marked concavity, hence this assay has limited usefulness as presently described.

Solubilization of Dye.   The release of dye from an insoluble dye-protein complex is a quick, easy assay method (Nelson et al.   1961; Ilany-Fei-genbaum 1966; Shotton 1970). The linearity with respect to enzyme activity is usually quite limited, but the assay is extremely useful for monitoring production processes (i.e., in the manufacture of detergents containing protease, or in protease production by biofermentation). Shotton (1970) described a method of following dye solubilization which is more time-consuming than a simple fixed-time assay, but gives much better linearity with respect to enzyme activity.

The substrate, elastin dyed with Congo Red, is suspended in 0.02 M borate buffer pH 8.8, at a concentration of 1 mg/ml. 7 ml of this suspension in a stoppered tube is centrifuged and the absorbance of the supernatant is read directly at 495 nm. Then 1 ml of enzyme is added, the tube is shaken to resuspend the substrate, and the tube is incubated at constant temperature. Periodically the tube is centrifuged, A_{495} is read on the supernatant, the tube is shaken to resuspend the substrate and incubated further. The plot of A_{495} versus incubation time is sigmoid in shape. The endpoint is that time required to solubilize 50% of the dye, determined by interpolation of experimental data around that point. Complete solubilization is determined by incubation for a time long enough to produce no further change in absorbance. The plot of endpoint time versus reciprocal of enzyme amount shows good linearity.

Synthetic Substrate Assays

A large number of amino acid derivatives have been synthesized for use as substrates. These allow better definition of the enzyme-substrate system than is possible with the use of protein substrates (i.e., with respect to molar concentration of substrate, moles of reaction per unit time, etc.). Assays with most of these substrates are based upon direct spectrophotometric measurements so they support continuous monitoring of increase in product concentration. A few assays are based upon other techniques and will also be described here.

Absorbance Changes.   Substrates which contain an aromatic ring somewhere in their molecular structure absorb at ultraviolet wavelengths. Hydrolysis of a bond in the substrate changes the environment of the ring structure, leading to a shift in the wavelength of maximum absorbance. The peak wavelength in the difference spectra may be used to monitor the extent of hydrolysis. A number of such substrates, along with the maximum wavelength and the absorbance change upon hydrolysis of a 1 mM solution of substrate, are listed in Table 8-1.

Several of these substrates have limited solubility in aqueous buffers, so they are often prepared as a concentrated stock solution in a dry organic

Table 8–1.   Absorbance Changes for Some Synthetic Substrates

Substrate	Acronym	nm	$\Delta\epsilon$, mM	Reference
N-acetyl-L-Tyrosine-	ATEE	237	+0.23	Schwert and Takenaka, 1955
Ethyl Ester		275	−0.17	Schwert and Takenaka, 1955
N-Benzoyl-L-Arginine-	BAEE	253	+1.15	Schwert and Takenaka, 1955
Ethyl Ester				
N-Benzoyl-L-Arginine-	BAPA	410	+8.8	Erlanger et al. 1961
p-Nitroanilide				
N-Benzoyl-L-Tyrosine-	BTEE	254	+1.03	Hummel, 1959
Ethyl Ester		256	+0.96	Walsh and Wilcox, 1970
N-Tosyl-L-Arginine-	TAME	244	+0.80	Hummel, 1959
Methyl Ester·HCl		247	+0.41	Walsh, 1970
3-(2-furylacryloyl)-	FAGLA	345	−0.32	Feder, 1968
glycyl-L-leucinamide				
N-carbobenzoxy-glycyl-	Z-Gly-Phe	224	−1.0	Petra, 1970
L-phenylalanine				
N-carbobenzoxy-L-	Z-Tyr-pNP	400	+18.8	Walsh and Wilcox, 1970
Tyrosine p-nitrophenyl				
Ester				

solvent such as methanol, acetonitrile, or dimethylsulfoxide. An aliquot of this stock is then added to buffer at the time of the assay. If the observed absorbance change is negative, the substrate absorbance is usually quite high. A 1 mM solution of Z-Gly-Phe, for instance, has A_{224} of about 1.5, and a 2.5 mM solution of FAGLA has A_{345} of 1.9. In these cases the spectrophotometer being used must be capable of accurately measuring these high optical densities, and the relatively small absorbance change as hydrolysis occurs. A few spectrophotometric assays will be described, to illustrate the use of these substrates.

TAME (Hummel 1959). The buffer for a trypsin assay is 0.04 M Tris·HCl, pH 8.1, 0.01 M in $CaCl_2$. Substrate is 39.4 mg TAME·HCl dissolved in 100 ml buffer, a concentration of 1.04 mM. To 2.9 ml substrate add 0.1 ml enzyme solution and record the increase in A_{244}. A change of 0.0008/min corresponds to a rate of 1 μmol/min.

BTEE (Walsh and Wilcox 1970). The buffer is 0.1 M Tris·HCl, pH 7.8, 0.1 M in $CaCl_2$. The substrate stock (1 mM) contains 15.7 mg BTEE dissolved in 30 ml methanol, then diluted with water to 50 ml. For the assay, put 1.5 ml substrate solution in the cuvette, followed by 1.4 ml buffer and finally 0.1 ml enzyme solution. (The blank cuvette holds 1.5 ml substrate plus 1.5 ml buffer.) The increase in A_{256} should be linear up to 0.12; a change of 0.00096/min corresponds to a hydrolysis rate of 1 μmol/min.

FAGLA (Feder 1968). This substrate is useful for metalloproteases such as thermolysin. Many of the cationic buffers may be used (Stauffer 1971), but Feder describes results with 0.05 M phosphate buffer, pH 7.2. Substrate is dissolved in this buffer at 76.9 mg per 100 ml to give a 2.5 mM solution. This concentration, near the solubility limit of FAGLA, is still well below K_M of thermolysin for this substrate, so the run is made according to first-order kinetics. One and a half ml of substrate solution is placed in the cuvette, followed by 1.4 ml buffer and 0.1 ml enzyme solution. The initial absorbance at 345 nm should be 0.96, decreasing to an infinite time value of 0.56. The data are manipulated in the usual way to obtain the first-order rate constant for the reaction.

BAPA (Kakade et al. 1969). This substrate may be used in a continuous assay, but as applied to the measurement of trypsin inhibitor in soy meal (Kakade et al. 1969, 1974; AACC 1983, Method 71-10) it is used as a fixed-time assay for multiple samples. Buffer is 0.05 M Tris·HCl, pH 8.2, containing 0.023 M $CaCl_2$. Substrate stock is 1 g BAPA dissolved in 25 ml dimethylsulfoxide. One ml of this stock is added to 100 ml buffer. The trypsin stock solution contains 4.5 mg crystalline trypsin in 100 ml 0.001 M HCl. To 7 ml of buffered substrate at 37°C is added 1 ml H_2O and, at zero time, 1 ml trypsin stock solution. After incubating 10 min at 37°C the reaction is stopped by adding 1 ml of 30% (v/v) acetic acid. The absorbance of the reaction is read at 410 nm, against an appropriate reaction blank (substrate, then water, then acetic acid, then enzyme). The trypsin Kakade et al. used was 56% active by titration with Z-Lys-pNP, and in the assay as described they obtained a value of 0.48 for A_{410}.

Hexapeptide for acid protease (Martin et al. 1980, 1981). Martin et al. synthesized the hexapeptide L-Leu-L-Ser-L-Phe(NO_2)-L-Nleu-L-Ala-L-Leu-OMe as a substrate for acid proteases such as rennet, chymosin, pepsin, and the proteases from *Mucor sp.* (The peptide is commercially available from Bachem, Inc., Torrance, CA.) Hydrolysis occurs at the bond between nitro-phenylalanine and norleucine, with a concomitant increase in absorbance at 310 nm. The difference spectrum is pH dependent, connected with the ionization of the carboxyl group of the nitrophenylalanine which has a pK of 3.5. At pH 1.5 the absorbance change is negligible, and at pH above 5.5 it is maximal, with $\Delta\epsilon = 1080\ M^{-1}cm^{-1}$. The molecular weight of the peptide is 722, and the exact concentration of a substrate solution may be determined from its absorbance at 279.5 nm ($\epsilon = 8300\ M^{-1}cm^{-1}$). The substrate solution should be freshly made and protected from light and oxygen. The assay is straightforward; 2.9 ml of a 1 mM solution of substrate in an appropriate buffer is placed in the spectrophotometer cuvette, 0.1 ml of enzyme solution is added, and A_{310} is recorded for a suitable period of time. A hydrolysis rate of 1 μmole/min will give an absorbance change of 0.00108/min if the pH is 5.5 or higher and a change of 0.00054/min at pH 3.5. The absorbance change at other pHs may be calculated from the ϵ given above and the pK of 3.5.

Fluorometric Assays.   To illustrate the application of fluorogenic substrates, an assay using (N-Cbz-Arg-NH)$_2$-Rhodamine (Leytus et al. 1983a) will be described. The synthesis of the substrate was given earlier in this chapter. The K_M for this substrate with bovine trypsin is 0.14 mM. The buffer is 0.01 M HEPES, pH 7.5, containing 10% (v/v) dimethyl sulfoxide and 0.02 M CaCl$_2$. Substrate solution (0.175 mM) is made by dissolving 16.6 mg of substrate powder in 100 ml of buffer. To 40 μl of substrate in the bottom of a plastic cuvette 10 μl of enzyme solution is added at zero time. After incubating 5 min at 22°C, 0.95 ml of 0.01 M HEPES buffer, pH 7.5, containing 15% (v/v) ethanol, is added and the fluorescence is recorded immediately (excite at 492 nm, emit at 523 nm). To convert fluorescence readings to molar concentration of product a standard curve is constructed using the monoamide Cbz-Arg-NH-Rhodamine.

These authors also examined several dipeptidyl Rhodamine compounds, and found that (Cbz-Gly-Arg-NH)$_2$-Rhodamine has a K_M of 0.065 mM and a V_{max} five times greater. The overall sensitivity in a trypsin assay would be ten-fold greater using the dipeptidyl substrate as compared to the substrate described here (Leytus et al. 1983b).

pH-stat Assays.   If the substrate being hydrolyzed is an ester the hydrolysis may be continuously monitored by recording the addition of base in a pH-stat apparatus.

BAEE (Walsh and Wilcox 1970). The substrate is 10 mM BAEE (343 mg BAEE·HCl per 100 ml) in a dilute buffer, 0.01 M Tris·HCl pH 7.75, containing 0.1 M KCl and 0.05 M CaCl$_2$. Three ml of substrate solution is placed in the thermostatted reaction vessel of the pH-stat, adjusted to pH 7.8, 100 μl enzyme solution is added, and the base uptake is recorded. Knowing the concentration of base used, the rate of hydrolysis in μmoles/min is easily determined.

SPECIFIC PEPTIDASE ASSAYS

In many instances peptidase assays use the same chemistry as protease assays, making the adjustment for the unique substrate requirements (i.e., a free carboxyl or amino group) (Hartsuck and Lipscomb 1971; Enari and Mikola 1977). The substrates are almost all derivatives of an amino acid or a dipeptide. Some specific assays follow.

Carboxypeptidase Assays

Colorimetric.   An example is similar to the assays published by Visuri et al. (1969) and Kruger and Preston (1977). The substrate is N-carbobenzoxy-L-phenylalanine-glycine (Z-Phe-Gly), as a 1 mM solution (35.6 mg per 100 ml) in 0.025 M phosphate buffer, pH 7.2. To 200 μl substrate solution is added 50 μl enzyme solution, and the mixture is incubated for 1 hr at 37°C. The reaction is stopped by briefly heating in a boiling water bath, then color development with TNBS reagent is carried out somewhat as described earlier for proteins. 2 ml 0.4 M phosphate buffer pH 8.2 is added (the SDS is not necessary in this system because there is no protein substrate to solubilize) followed by 2 ml 0.1% TNBS in water, incubation for 1 hr at 50°C in the dark, then adding 4 ml 0.1 N HCl and reading A_{340}. An accurate standard curve can be constructed for this assay with 0 to 1.5 μmoles glycine.

Spectrophotometric   (Petra 1970). The substrate used is Z-Gly-Phe, 42.0 mg dissolved in 100 ml of buffer, 0.05 M Tris·HCl, pH 7.5, containing 0.45 M KCl. At 224 nm the absorbance of this substrate solution is about 1.8. To 3 ml substrate solution in a cuvette is added 50 μl enzyme solution and the decrease in absorbance is monitored continuously. A change in A_{224} of 0.001/min corresponds to the hydrolysis of 1 μmole/min.

pH-stat   (Petra 1970). Hippuryl-DL-β-phenyllactic acid is an ester substrate for bovine carboxypeptidase A. The substrate is 10 mM HPLA (331.4 mg of the sodium salt per 100 ml) in 0.005 M sodium veronal, 0.045 M NaCl, pH 7.5. Place 3 ml of the substrate solution in the pH-stat thermostatted reaction vessel, and after recording baseline base uptake for a minute or two, enzyme solution (10 to 150 μl) is added to start the reaction. The record of the time course of base uptake gives the rate of reaction directly in molar concentration units.

Aminopeptidase Assays

Colorimetric reactions with free amino groups do not work well with aminopeptidases, because of the specificity requirement for a free N-terminal amino group in the substrate. The two types of assays usually applied either measure the carboxyl group freed, or naphthylamine liberated during hydrolysis of an amino acid naphthylamide.

Carboxyl Titration   (Hill et al. 1958; Delange and Smith 1971). The method depends upon microtitration in the presence of ethanol of carboxyl groups generated during the hydrolysis of L-leucinamide by leucine aminopeptidase. The substrate solution is 0.125 M leucinamide in water, adjusted to pH 8.5. Buffer is 0.5 M Tris·HCl, pH 8.5. Leucine aminopeptidase is activated by incubation with 0.025 M $MnCl_2$ in buffer. The reaction consists of 0.2 ml 0.025 M $MnCl_2$, 0.5 ml buffer, 1 ml leucinamide solution, and 0.55 ml water. At zero time 0.25 ml activated enzyme solution is added to begin the reaction. Periodically 0.2 ml aliquots are taken, 3 drops of thymolphthalein indicator solution is added, and the aliquot is titrated with

0.01 N alcoholic KOH to a faint blue color. Then 1.8 ml absolute ethanol is added and the titration is continued until the faint blue color reappears. The titration equivalent at each time (corrected for a zero time blank titration) represents the moles of product formed (carboxyl group freed). Thus the series of titrations represents a progress curve (see the Henri integrated rate equation section in Chapter 2) and may be analyzed to obtain K_M and V_{max} (see the data analysis section in Chapter 2).

Diazotization   (Hopsu-Havu et al. 1968; Kolehmainen and Mikola 1971; Kruger and Preston 1978). The β-naphthylamides of amino acids serve as substrates for aminopeptidases. Upon hydrolysis of the amide bond the freed β-naphthylamine is reacted with a diazonium salt and the color developed is measured. An example is the assay described by Kruger and Preston (1978).

The substrate solution is 2 mM amino acid naphthylamide in 0.01 N HCl. The buffer used is 0.025 M phosphate, pH 7.2. The color reagent is a solution of the stabilized diazonium salt Fast Garnet GBC (1 mg/ml) in 1 M acetate buffer pH 4.2, containing 10% (v/v) Tween 20. To the assay tube is added 1.6 ml buffer and 0.2 ml substrate solution, then 0.2 ml enzyme solution, and the reaction is incubated 3 hr at 35°C. At the end of that time 1 ml of color reagent is added, the coupling reaction proceeds for at least 5 min and the absorbance is read at 525 nm. A standard curve is constructed with free β-naphthylamine (Caution! Potent Carcinogen! Use extreme care in handling!) or α-naphthylamine.

As might be expected, aminopeptidases from different sources display varying specificities with regard to which amino acid naphthylamides are hydrolyzed most readily. In at least one case (Hopsu-Havu et al. 1968) the substrate was a dipeptide naphthylamide; simple amino acid derivatives were not hydrolyzed by the hog kidney aminopeptidase which they were investigating.

TITRATION ASSAYS FOR PROTEASES

Rate assays are often run to answer the implicit question "How many moles of enzyme are in this sample?" by comparison with the rate assay results obtained with some "standard, pure" enzyme. The assumption is that the standard enzyme is 100% active, an assumption which is almost certainly not true. Titration assays usually take, for their comparison basis, spectral characteristics and purity of small organic molecules, where the assumption of 100% purity is much more likely to hold, at least for the sample on which the absorption coefficients are determined.

In certain cases titration assays are used as "all-or-none" indicators to monitor some derivatization reaction of the enzyme, to find out whether the change actually inactivates the molecule, or only alters its catalytic properties (Stauffer and Etson 1969). More often they are used to establish the true purity of a standard enzyme, and subsequent rate assays are referred back to this standard to determine the actual molar concentration of protease in an experimental preparation.

Most titration assays depend on a "burst" reaction, in which an intermediate enzyme derivative is formed, and the molar extent of the "burst" is measurable (see Appendix G) by spectrophotometric or fluorimetric means. In at least one instance a high-affinity inhibitor generates a sizable difference spectra when it binds to the protease (Komiyama et al. 1975), and can be used to directly titrate the enzyme. If a high-affinity inhibitor can be purified so that its concentration can be expressed in molar units, then the enzyme can be "titrated" by a reversal of the procedure often used for measuring the amount of inhibitor in foodstuffs (e.g., AACC 1983 Method 71-10; also see the high-affinity inhibitors section in Chapter 3). However, even though several such inhibitors for the various protease classes are known, I am not aware of any published procedure that takes advantage of this possibility.

p-Nitrophenyl Esters

The first "burst" titration was the observation by Hartley and Kilby (1954) that when larger-than-usual concentrations of chymotrypsin were reacted with p-nitrophenyl acetate there was an initial rapid release of p-nitrophenol, followed by a slower, steady rate of release. The size of the initial "burst" was proportional to the moles of chymotrypsin used in the reaction. McDonald and Balls (1957) examined this same phenomenon, using some related compounds, and found that p-nitrophenyl trimethylacetate not only gave an excellent "burst," but the resulting trimethylacetyl enzyme derivative was stable and could be crystallized. Bender et al. (1967) set out the kinetics of the "burst titration" system, and gave an assay for chymotrypsin based upon p-nitrophenyl trimethylacetate.

The buffer suggested is 0.01 M Tris·HCl, pH 8.5. The substrate stock is 0.35 M p-nitrophenyl trimethylacetate (7.8 mg per 10 ml) in dry, pure acetonitrile. The enzyme solution should be on the order of 2 M, e.g., 50 mg of chymotrypsin (2 mM) dissolved in 1 ml of pH 4.6 acetate buffer. In a 1 cm pathlength cuvette in a thermostatted compartment of a spectrophotometer set at 400 nm, place 3 ml of buffer and allow it to attemperate to 25°C. Then add 50 μl substrate stock, stir, and record absorbance for a short time to establish a baseline. Add 25 μl of the enzyme solution, stir a few seconds, then record A_{400} until the change is linear, about 20 min. Extrapolation of this slope to time zero gives the absorbance change of the "burst." p-Nitrophenoxide ion has a molar absorbance of 18,800 M^{-1} cm^{-1} and a pK of 7.04. At pH 8.5, then, $\epsilon = 18,170$ M^{-1} cm^{-1}, which can be used to convert the absorbance change into molarity of enzyme in the cuvette, and thence (via appropriate dilution factors) into actual molarity of active enzyme in the enzyme stock solution.

For trypsin, a widely-used titrant is the p-nitrophenyl ester of p-guanidinobenzoic acid (Chase and Shaw 1967). The procedure is very similar to the previous one. The authors recommend the use of 0.1 M veronal buffer, pH 8.3 (containing 0.02 M $CaCl_2$) rather than an amine buffer such as Tris which causes larger blank hydrolysis. N-Cbz-L-Lys-pNP ester has also been used to titrate trypsin (Bender et al. 1966).

Papain is titrated using N-Cbz-L-Tyr-pNP ester at somewhat acidic pH (Bender et al. 1966). With this titrant the substrate concentration is not much greater than K_S so it is necessary to run the assay at several initial substrate concentrations and plot $[S]/\sqrt{B}$ versus $[S]$ (Appendix G, equation G-6c). The enzyme molarity equals $1/(\text{slope})^2$. The buffer is 0.05 M potassium hydrogen phthalate, pH 3.18. The substrate stock solution is 1.2 mM (52.4 mg per 100 ml) in dry, pure acetonitrile. The enzyme solution is activated papain, 50 mg/ml in the buffer. The procedure is much the same as that outlined for chymotrypsin above. To 3 ml buffer in the cuvette is added 50 μl substrate stock, followed by 25 μl enzyme. The absorbance is recorded at 340 nm, and the steady-state curve is extrapolated to zero time to find the size of the burst, B. To get various $[S]$, combinations of acetonitrile and substrate stock (i.e., 10 μl acetonitrile + 40 μl substrate) are used. At 340 nm the molar absorbance coefficient for p-nitrophenol (unionized) is $6.2 \times 10^3 \, M^{-1} cm^{-1}$.

α-Bromo-4-Hydroxy-3-Nitroacetophenone (BHNA)

Because of the necessity of titrating papain at several $[S]$ with Z-Tyr-pNP and extrapolating, a titrant was sought which would give complete and irreversible reaction with sulfhydryl enzymes. One such titrant is BHNA (α-bromo-4-hydroxy-3-nitroacetophenone) (Kezdy and Kaiser 1976). The material is readily synthesized by nitrating 4-hydroxyacetophenone, then brominating that product. To 53 ml of fuming nitric acid maintained at $-30°C$ 10.5 g 4-hydroxyacetophenone is added over 15 min. The mixture is stirred at $-30°C$ for 1 hr then poured into 600 ml ice water. The precipitate is collected and recrystallized twice from ethanol. For bromination 0.52 g bromine is dissolved in 11 ml $CHCl_3$ at room temperature, then 1.0 g of the nitrated product is added, the temperature is raised to 40°C and held for 15 min. The reaction mixture is washed once with water, dried over Mg_2SO_4, and the solvent removed under vacuum. The product is recrystallized twice from CCl_4 to give yellow needles, m.p. 91.5°C to 92.5°C.

The reaction with papain releases bromide ion and forms an alkylated thiol product, with a large increase in absorbance at 323 nm ($\Delta\epsilon = 6.85 \times 10^3$ $M^{-1}cm^{-1}$). The titrant is dissolved in acetonitrile, while the enzyme is in 0.067 M phosphate buffer, pH 7.0, containing 1 mM EDTA. Two ml of buffer containing enzyme is placed in the cuvette and A_{323} is measured (Ae). Then 50 μl titrant solution is added (at least a 20% excess of titrant) and the reaction proceeds for 5 min, after which the final absorbance (Aet) is read. Finally, the absorbance of 50 μl titrant in 2 ml buffer (At) is determined. The molarity of enzyme in the buffer solution is:

$$[E] = \frac{1.025(Aet - Ae - At)}{6.85 \times 10^3}$$

The best titration is obtained when $[E]$ is from 15 to 30 μM. The titrant solution concentration should be 1.6 mM, or 41.8 mg per 100 ml acetonitrile.

N-*Trans*-Cinnamoyl Imidazole (CI)

This active amide, readily available from several biochemical supply houses, reacts stoichiometrically with several serine proteases, in particular chymotrypsin and the bacterial protease subtilisin, to give the acylated enzyme. In the reaction, the strong absorbance in the 310 to 335 nm range due to the cinnamoyl amide almost disappears as substrate is converted to a cinnamate ester. This is the basis for the spectrophotometric monitoring of the reaction. The spectrophotometric assay of chymotrypsin as described by Schonbaum et al. (1961) is as follows.

Method A. This method is preferable for enzyme preparations which are at least 25% active. 3.00 ml 0.1 M acetate buffer pH 5.1 is placed in the cuvette in the thermostatted spectrophotometer compartment, and the instrument is zeroed. Then 100 μl of CI reagent solution (1 mM, 19.8 mg per 100 ml dry acetonitrile) is added and the absorbance is recorded briefly. Next 100 μl of enzyme test solution is added. This should be an amount such that half to two-thirds the CI reacts, but an excess still remains after the "burst." The absorbance is recorded for a few minutes until it is a straight line, which is extrapolated back to the time of enzyme addition. It is also necessary to obtain the absorbance of just the enzyme; for this purpose, the spectrophotometer is zeroed with 3.00 ml of buffer, then 100 μl of enzyme solution is added and the absorbance is recorded.

Method B. This method works better with enzyme preparations which have a great deal of inactive protein present. To 3.00 ml of buffer, used to zero the spectrophotometer, 100 μl enzyme test solution is added and the absorbance is recorded. Then 100 μl CI reagent solution is added, the reaction allowed to proceed as in Method A, and the final steady state is extrapolated to the time of addition. Subsequently, to 3.00 ml of buffer is added 100 μl of reagent solution, the absorbance is recorded for a time and extrapolated back to zero time.

The various experimental absorbance values needed for the calculation are indicated in Figure 8-1. In addition, certain molar absorbancies are needed. For CI, at 310 nm, $\epsilon = 23.8 \times 10^3$ M^{-1}cm^{-1}, and at 335 nm, $\epsilon = 9.37 \times 10^3$.

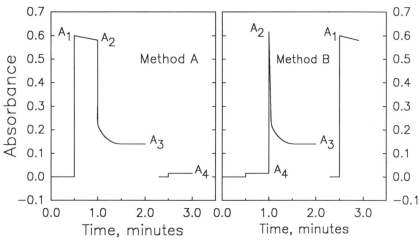

FIGURE 8-1. Titration of serine protease with N-*trans*-cinnamoylimidazole.

The change in molar absorbancy upon converting cinnamoyl imidazole to cinnamoylated enzyme $\Delta\epsilon$ is, at 310 nm, 12.85×10^3 $M^{-1}cm^{-1}$ and, at 335 nm, 8.95×10^3. In Method A, the actual size of the burst (correcting for dilutions) is given by:

$$B = 0.969A_2 + 0.969A_4 - A_3.$$

Since $B/\Delta\epsilon$ gives the operational normality of enzyme in the 3.2 ml volume of the cuvette, and this represents a 32-fold dilution of the enzyme stock, then the normality of enzyme in the stock equals $B/280$ (at 335 nm) or $B/402$ (at 310 nm).

In Method B, the actual size of the burst (correcting for dilutions) is:

$$B = 0.969A_1 + 0.969A_4 - A_3.$$

The other factors remain the same as in Method A, so B is divided by 280 (if the absorbance is recorded at 335 nm) or 402 (at 310 nm) to obtain the operational normality of the enzyme in the test solution.

Fluorescence Titration

Melhado et al. (1982) used a fluorogenic ester, fluorescein mono p-guanidinobenzoate HCl, to titrate as little as 2×10^{-11} moles of trypsin. They describe the preparation and purification of the substrate in that paper. The buffer they used was a rather complex one designed for detecting plasminogen activation in growing cells; for direct assay of trypsin or similar enzymes a simple 0.05 M phosphate buffer, pH 7.2, 0.02 M in $CaCl_2$ would undoubtedly suffice. The substrate, FMGB·HCl, is dissolved in dry dimethylformamide (0.1 mM, 5.29 mg per 100 ml). To 1 ml of the buffer in a spectrofluorimeter is added 30 μl of substrate stock, then 10 μl of enzyme solution (5 μM concentration, e.g., 0.14 mg trypsin per ml). Fluorescence is recorded (excite at 491 nm, emit at 514 nm), and extrapolated to zero time to obtain the change in fluorescence due to the "burst" reaction. A standard curve may be constructed with a weighed amount of fluorescein in order to convert a fluorescence change number into a molar concentration. This assay is undoubtedly the most sensitive titration yet devised for serine proteases.

Difference Spectra

Metalloproteases, in particular the bacterial protease thermolysin, are inhibited by the tight-binding inhibitor phosphoramidon, N-(α-Rhamno-pyranosyloxy-hydroxyphosphinyl)-L-Leucyl-L-Tryptophan. At pH 7 this microbial inhibitor has a value of K_i of 2.8×10^{-8} M (Komiyama et al. 1975). However, it has an even more interesting property, in that when it binds to the enzyme, the tryptophanyl residue enters a hydrophobic region in the protein and a large difference spectrum is generated with its peak at 295 nm. Komiyama et al. (1975) found $\Delta\epsilon = 2.04 \times 10^3$ $M^{-1}cm^{-1}$ for this change, a value which would allow this interaction to become the basis for a reasonably sensitive assay for metalloproteases ("neutral proteases") in food-related materials. A large number of commercial protease preparations based upon microbial fermentations are mixtures of serine and metalloproteases. Since the binding of phosphoramidon is the only titration assay available for metalloproteases, it would seem to merit more attention than it has received thus far.

9 Glycoside Hydrolases

The enzymology of amylases has a history at least as long as that of proteases, stemming from the early work on the "diastase" enzymes found in barley malt. Thus it is not surprising to find that some of the assays for α-amylase and β-amylase have roots in work done in the nineteenth century. Unlike the protease story there has not been such a proliferation of small synthetic substrates for glycosidases. At the present time there are no titration substrates available for any of these enzymes. Most of the assays reflect closer ties to practical applications than for proteases, perhaps because of the great economic importance of starch-derived materials (e.g., corn syrups, ethanol) and potential importance of such processes as the conversion of cellulose to glucose. In the clinical area the detection of pancreatic α-amylase is of diagnostic use, and the development of assays for this enzyme has contributed to assays for glycosides in the food field. A number of authors have compared various methods for the detection of amylase in foods (Mathewson et al. 1981; D'Appolonia et al. 1982; Hsu and Varriano-Marston 1983), and the assay of amylase in the clinical chemistry field has also been recently reviewed (Lorentz 1979; Kaufmann and Tietz 1980). In addition, a review of methods for measuring cellulase activity has also been published (Canevascini and Gattlen 1981).

SUBSTRATES

Polymeric Substrates

Amylose, Amylopectin.   Starch is a mixture of linear-chained amylose and the highly-branched amylopectin. Many amylase assays simply specify "Starch" or "Soluble Starch"; the amylose/amylopectin ratio is unknown, and may well affect the parameters of enzyme action. The colored complex formed with iodine has a different wavelength of maximum absorption for each of these starch species, which complicates the measurement of assays using this signal. For these reasons it would seem advisable to use one or the other material as an amylase substrate.

The separation of the two species is easily carried out by the method of Street and Close (1956), which is based upon an earlier publication by

Haworth et al. (1946). Eight hundred ml of water containing 1 g NaCl is brought to a boil. Ten g starch is slurried into 100 ml water at room temperature, and this slurry is slowly stirred into the boiling water. An additional 100 ml water is used to rinse the beaker. The suspension of gelatinized starch is cooled to room temperature and 1.3 g powdered thymol is added. The amylose-thymol complex precipitates completely in 48 hr. It is collected by filtration and washed with several liters of thymol-saturated water, followed by absolute ethanol to remove the thymol. The moist filter cake is dissolved in water and lyophilized, giving amylose which is readily water soluble. If the filter cake is simply air dried, 0.01N NaOH is required to redissolve the amylose. From the initial filtrate from the precipitation reaction nearly pure amylopectin may be isolated by lyophilization followed by washing with absolute ethanol to remove residual thymol. The yield of amylose depends, of course, upon the amylose content of the starting material; 10 g of a high-amylose corn starch will give 7 to 8 g amylose, while from a waxy maize starch less than 1 g would be obtained. Purified amylose and amylopectin is available from many biological supply houses.

Limit Dextrin.   Several assays for α-amylase, notably the Wohlgemuth method, differentiate the endohydrolase activity from the exohydrolase (β-amylase) activity by first treating a starch suspension with an excess of β-amylase to remove all the molecular species which react with this enzyme. The resulting substrate, limit dextrin, is essentially amylopectin in which the α (1-4) poly-glucose chains have been hydrolyzed from the non-reducing ends to the α (1-6) branch points. Upon treatment with amylase enzymes, any further changes in the properties (e.g., viscosity, iodine complex formation) of this substrate are due to the action only of α-amylase. Limit dextrin is made as follows (AACC 1983, Method 22-01; Perten 1966). Ten g (dry weight) of soluble starch ("Lintner starch," from Merck & Co.) is suspended in 25 ml water and poured slowly, with stirring, into 300 ml boiling water, using a wash bottle to complete the transfer. The suspension is boiled for 2 min, then cooled to room temperature, and 25 ml buffer (4M acetate, pH 4.7) is added, followed by a suspension of β-amylase (0.2 mg of crystalline sweet potato enzyme, or 250 mg of a special "2000°L" malt enzyme as per the AACC method) in 5 ml water and a few drops of toluene (to prevent microbial growth). The volume is made to 500 ml with water, and the solution is incubated at room temperature for at least 20 hr before use as an α-amylase substrate. The solution is usable for 3 to 5 days if stored in the refrigerator. Alternatively, the digested solution may be boiled briefly to inactivate the enzyme, dialyzed against distilled water for 2 days, and lyophilized to give a dry, stable substrate (Marchylo and Kruger 1978). Note that the molecular weight of limit dextrin depends upon the "core size" of the amylopectin used. Limit dextrins from root starches (potato, tapioca) are smaller than limit dextrins from cereal grain starches.

Reduced Starch.   If the assay employs chemical quantitation of reducing groups to measure the extent of hydrolysis, unmodified starch will give a rather large blank. To overcome this Strumeyer (1967) reduced the initial substrate with borohydride. A suspension of 2.5 g starch in 10 ml water is boiled for 1 min, then diluted to 45 ml. It is cooled in an ice bath and a cold solution of 75 mg $NaBH_4$ in 5 ml water is stirred in. The mixture is held at room temperature 1 hr, then stored in the refrigerator. For use, 0.2 ml acetone is added dropwise to 10 ml of the stock solution with good shaking to destroy excess borohydride. After 20 min at room temperature the pH is adjusted to 7 with 1 N acetic acid and the stock is diluted to 50 ml with an appropriate buffer to give a 1% solution of reduced starch. This treatment reduces the color yield of the substrate blank to less than 3% of that of unmodified starch.

α-(1-4)-D-Glucosyl Oligomers.   Some recent assays use short chain length linear oligomers, typically containing 5 to 10 glucose residues per molecule. These are obtained by size-exclusion chromatography of dextrin mixtures. The dextrins may be obtained by partial hydrolysis of amylose with α-amylase, or as corn syrup solids from a commercial supplier (Grain Processing Corporation, for instance, sells three grades M200, M250 and M255, all of which contain over 30% of their weight as oligomers with DP between 5 and 8). Chromatography is carried out on Bio-Gel P2, 200 to 400 mesh, at 65°C (jacketed column) with water as the solvent (John et al. 1969). Separation of the oligomers up to DP8 is excellent, and the separated fractions may be readily recovered by lyophilization. The use of these oligomers in particular assay schemes will be discussed later.

β-D-Glucans.   A number of gums from cereal grains (oats, barley) have been isolated for assays relating to the commercial use of these grains. The glucan from oat and barley has been purified by Wood (Wood et al. 1977; Wood and Weisz 1987). The grain is ground into a flour, which is suspended in 10 times its weight of water. The suspension is adjusted to pH 10 with 20% Na_2CO_3 solution and stirred 30 min at 45°C. The insoluble material is removed by centrifugation and extracted a second time in the same manner. The extracts are combined, adjusted to pH 4.5 with 2N HCl, and centrifuged at 4°C. To the cold supernatant an equal volume of cold isopropanol is slowly added and the precipitate is allowed to settle overnight. The impure gum is collected by centrifugation, resuspended in isopropanol, collected on a suction filter and air dried. The dried material (5 g) is dissolved in 1 L water and precipitated by addition of 250 ml saturated ammonium sulfate solution. The precipitate is collected, redissolved in water, and the procedure is repeated. After the second precipitation the material is dissolved in cold water and an equal volume of cold isopropanol is added. The precipitate is collected on a filter, washed with isopropanol, and air-dried with gentle warming.

Soluble Cellulose Derivatives. Cellulose gums have been an item of commerce for a number of years. They are available with a variety of degrees of substitution (DS), and in a range of molecular weights. Almin and Eriksson (1968) examined the influence of DS on suitability of carboxymethyl cellulose (CMC) in a viscosimetric assay for cellulase, and concluded that it was best to have DS between 0.8 and 1.0. At DS > 1 enzyme activity was reduced, and at DS < 0.5 they experienced difficulty in reproducing experimental results. The wide variety of CMCs available make it difficult to make a general characterization of this substrate. It is best to find one product designation from a reputable source, determine the molecular weight and DS of that product, and stick with that for all experimentation. In any case, each new lot of substrate must be characterized viscosimetrically (see Appendix H). One difficulty with CMC is that it is ionic in nature, thus its properties change as the pH and ionic strength of the reaction mixture is varied. To overcome this, nonionic cellulose gums such as hydroxyethyl cellulose or hydroxypropyl cellulose have been recommended (Iwasaki et al. 1964; Child et al. 1973). Again, many of these are available from suppliers such as Dow Chemical Co., Midland, Michigan. While viscosimetric characterization of these substrates have not been as extensive as that of CMC (Manning 1981) the report by Child et al. (1973) did demonstrate good linearity of response with respect to time and enzyme concentration. These substrates would seem to merit more attention from enzymologists interested in the action of cellulase.

Synthetic Small Molecules

Nitrophenyl Glycosides. Most of the glycosidases will hydrolyze the phenyl glycoside of the appropriate sugar substrate (e.g., maltase will cleave p-nitrophenyl maltoside, etc.) (Matsubara et al. 1959; Distler and Jourdian 1973; Maruhn 1976; Fretzdorff 1978). The course of the reaction is followed by measuring the amount of free phenol formed after a certain length of time. In most cases the substrate is the p-nitrophenyl (pNP) glycoside, and the assay is a fixed time assay; after a certain incubation period the reaction is stopped by making the pH alkaline and the absorbance of the freed pNP is read at 400 nm (Maruhn 1976; Mathewson and Seabourn 1983). Many of these pNP glycosides are now available from biochemical supply houses, so the synthesis will not be given here. If an unusual glycoside is required, Matsubara et al. (1959) give a general synthetic method.

While the fixed time assays are convenient for many purposes, it is sometimes desirable to measure product formation continuously. Ford et al. (1973) addressed this with respect to the pNP glycosides of glucose and N-acetyl-glucosamine. The problem is that pNP has a pK of 7.05, the ionized (high pH) form absorbs light much more than does the unionized form, but most glycosides are active in the acidic range. The isosbestic point (i.e., the wavelength at which the absorbance of the nitrophenol and the nitrophenoxide species is the same) is at 346 nm. At this wavelength the pNP-glycoside also absorbs some light, but Ford et al. found that the absorbance difference for free pNP versus pNP-glycoside is 2.6 for the glucoside and 3.6 for the N-acetyl-glucosaminide, all solutions at 1 mM concentration. Thus, although the assay is some 5 to 6 times less sensitive than when the reaction is stopped and free p-nitrophenoxide is measured, this is offset by the fact that the absorbance change can be measured continuously (initial rates can be determined) and at any desired pH for the enzyme reaction.

Fluorogenic Substrates. The use of fluorogenic substrates can increase the sensitivity of an assay by 2 to 3 orders of magnitude. For glycosides, a commonly used fluorochrome is 4-methyl umbelliferone (excite at 365 nm, emit at 460 nm) (Robinson 1956; Robins et al. 1968). The synthesis of such glycosides may be illustrated by the method for making the 4-methylumbelliferyl-glucoside (MUG). To 11.5 g 4-methylumbelliferone dissolved in 140 ml acetone is added 19 g tetra-acetyl bromoglucose, followed by 3.6 g NaOH in 90 ml water. After setting overnight at room temperature, the acetone is removed in vacuo, and the precipitate is recrystallized 4 times from ethanol (decolorize with charcoal during the first time). The tetra-acetate is dissolved in 25 ml dry methanol and a few drops of 10% barium methoxide in methanol is added to deacetylate. The precipitate is recrystallized from ethanol to give MUG hemihydrate, m.p. 211°C.

Insoluble Chromogenic Substrates

Starch-Dye Compounds. A number of procedures have been published in which a dye is attached to starch, and amylase activity is measured as the amount of dye solubilized during an incubation period (Ceska et al. 1969; Klein et al. 1969; Hall et al. 1970). In some of these methods the starch is rendered more insoluble by cross-linking (Ceska et al. 1969) and these substrates are being sold in proprietary commercial amylase assay kits. As a general indication of the way in which most of the materials are made, the synthesis of Klein et al. (1969) will be described. Amylose (500 g) is slurried into 5 L water and warmed to 50°C, then a solution of 50 g Cibachron Blue in 5 L warm water is added, followed by 1000 g Na_2SO_4 and a solution of 50 g Na_3PO_4 in 750 ml water. This is stirred at 50°C for 75 min, then the amylose-dye complex is allowed to settle. The supernatant is removed and the complex is washed several times with 5 L portions of water until the washings are colorless. The precipitate is then washed with methanol until the washings are colorless, the solids are air-dried, then ground and stored. The dye absorbs light maximally at a wavelength of 625 nm.

A fluorogenic dyed starch is described by Rinderknecht and Marbach (1970). The anthranilate released from this substrate emits at 410 nm when excited at 335 nm, and the assay is orders of magnitude more sensitive than the dye substrate, as would be expected. To make it, 50 g waxy maize starch is suspended in 100 ml 0.9% saline at 35°C. To this is added 1 g Na_2CO_3 followed by 1 g isatoic anhydride, and the reaction mixture is stirred 4 hr at 35°C. It is then diluted with 600 ml water and the starch derivative is allowed to settle in the cold. The precipitate is washed repeatedly with water until the fluorescence of the washings is constant, then it is washed twice with methanol and dried in a desiccator. For use the starch anthranilate is suspended in the desired buffer (0.5 g per 100 ml) and heated at 90°C until a colloidal solution is obtained. The authors describe a method for using this substrate in an automated (AutoAnalyzer) setup.

Dyed Cellulose, Dextran.   A few substrates relying on this same concept have been reported for cellulase and dextranase enzymes. Ng and Zeikus (1980) coupled Remazol Brilliant Blue R to cellulose using the same type of reaction as that applied by Klein et al. To 10 g Avicel microcrystalline cellulose suspended in 100 ml 50 mM Na_2SO_4 is added 100 ml of a 1% solution of the dye in water. This is stirred 30 min at 60°C, meanwhile slowly adding 20 g Na_2SO_4. The pH is adjusted to 12 by adding 5% Na_3PO_4 solution, hold for 30 min, then filter and rinse with 200 ml 1% Na_2CO_3 (at 55°C), then with warm tap water until washings are clear. Finally wash the filter cake with methanol followed by acetone, and air dry. The authors describe a method for incubating the substrate with enzyme in a stirred reaction chamber, continuously removing solution through a filter, measuring the absorption at 595 nm, and returning the liquid to the reaction chamber.

Huang and Tang (1976) modified aminoethyl-cellulose or aminoethyl-Sephadex G-200 with either trinitrobenzenesulfonic acid (TNBS) or fluorescamine to obtain chromogenic or fluorogenic substrates for cellulase or dextranase, respectively. While they describe the preparation of the AE-cellulose and AE-dextran, these items are now readily available commercially and that synthesis will not be given here. To make the trinitrophenyl derivatives, 10 g of AE-substrate is suspended in 30 ml 0.2M borate buffer pH 9, and 0.6 g TNBS is added. The mixture is adjusted to pH of 8.5 to 9.0, the reaction vessel is protected from light, and stirred for 3 hr at room temperature. Then the solid material is collected on a Buchner funnel and washed with 0.5 L borate buffer, 6 L water, and 1 L acetone. It is stored in a brown bottle at 4°C.

The fluorogenic derivative is made by suspending 10 g of AE-substrate in 30 ml of 0.2M borate buffer pH 9, adding 50 mg fluorescamine dissolved in 1.5 L acetone, stirring 40 min at room temperature, then collecting on a Buchner funnel and washing with 0.5 L borate buffer, 1 L water, and 2 L acetone. For either substrate, the authors used 2 ml of a 1% suspension in the appropriate buffer, with 5 to 10 μl enzyme. After 30 min incubation the mixture was filtered and the amount of soluble dye was measured, either by absorbance at 344 nm for the TNP derivative or in a fluorometer (excite at 390 nm, emit at 480 nm) for the fluorescamine substrate.

MEASUREMENT METHODS

Chemical Signals

Colorimetric Reactions of Reducing Sugars.   As the hydrolysis of glycosides proceeds, new reducing groups are formed, one for each enzymatic event. With the appropriate chemical reaction a colored compound can be formed which is directly related to the concentration of reducing groups in the reaction mixture. For the reaction of amylase with starch, in particular, it should be reiterated that starch itself contains many reducing groups (contributing to a high zero-time blank), but this can be substantially reduced by the method of Strumeyer (1967) described earlier.

Nelson-Somogyi reagent. The earliest reagent described which is still in use depends upon the reduction of Cu^{++} to Cu^{+} by the reducing sugar, followed by the formation of a colored complex of arsenomolybdate with cuprous ion. The stabilized copper reagent is the Somogyi reagent, while the arsenomolybdate reagent is the Nelson reagent. The recipes for the two reagents are as follows.

Somogyi Copper reagent (Somogyi 1952). 24 g anhydrous Na_2CO_3 and 12 g Rochelle salt (sodium potassium tartrate, $NaKC_4H_4O_6 \cdot 4H_2O$) are dissolved in 250 ml water. 40 ml of a 10% solution of $CuSO_4 \cdot 5H_2O$ are added with stirring, followed by 16 g $NaHCO_3$. Next, 18 g anhydrous Na_2SO_4 are dissolved in 500 ml water and boiled briefly to degas. It is cooled and added to the first solution, and the whole is made to 1 L volume with water. After a few days any slight precipitate is removed by filtration. The reagent is stable at room temperature.

Nelson Arsenomolybdate reagent (Nelson 1944). 25 g ammonium molybdate ($(NH_4)_2MoO_4$) is dissolved in 450 ml water, then 21 ml concentrated H_2SO_4 is added, followed by a solution of 3 g sodium arsenate ($Na_2HAsO_4 \cdot 7H_2O$) dissolved in 25 ml water. The reagent is mixed, held at 37°C for 24 to 48 hr, then stored in a brown bottle.

For measuring reducing sugars 1 ml of sugar solution is mixed with 1 ml Somogyi reagent and placed in a boiling water bath for 20 min. It is cooled and 1 ml Nelson reagent is added. The absorbance is read at 660 nm. The complex follows Beer's Law, so dilutions may be made as required.

3,5-Dinitrosalicylate. A second widely used reagent is the dinitrosalicylic (DNS) acid reagent described by Miller et al. (1960). This has the advantage of being a single reagent, stable and simpler to make and use than the Nelson-Somogyi color reaction. It has the disadvantage that the color yield is dependent upon the chain length of the oligosaccharide carrying the reducing end (Robyt and Whelan 1965), and the presence of Ca^{++} also influences the color yield. Thus DNS reagent is not good for following the progress of depolymerizing reactions (i.e., amylase or cellulase reactions), but it is readily applicable to reactions such as those of maltase or sucrase, when the appropriate compounds are used to construct the standard curve.

DNS Reagent. In 500 ml of water are dissolved 10 g NaOH, 10 g dinitrosalicylic acid, 2 g phenol, 0.5 g Na_2SO_3, and 200 g Rochelle salt. After everything is dissolved the solution is diluted to 1 L and stored in a brown bottle. For the color reaction, to 2 ml of reducing sugar solution is added 3 ml DNS reagent, and this is heated in a boiling water bath for 15 min. It is cooled, and absorbance is read at 640 nm.

Neocuproine Reagent. A third reagent uses neocuproine (2,9-dimethyl-1,10-phenanthroline·HCl) to develop a color with cuprous ion (Dygert et al. 1965). A disadvantage of this reagent is that while it will accurately measure a wide range of reducing sugar concentrations, the amount of reagent added must be adjusted to approximate the amount of reducing sugar present (see table on following page). Thus if the reaction has proceeded further than expected, a second measurement might be needed to obtain an accurate result. On the other hand, the color yield is independent of oligosaccharide chain length, so this reaction is recommended for depolymerase reactions.

Reagent A. In 600 ml water dissolve 40 g anhydrous Na_2CO_3 followed by 10 g glycine and 0.45 g $CuSO_4 \cdot 5H_2O$. Make to 1 L.

Reagent B. Dissolve 0.12 g Neocuproine·HCl in 100 ml water. Store in a brown bottle. If the reagent becomes yellow discard it and make a new solution.

To use, to 1 ml of reducing sugar solution, add equal volumes of Reagent A and Reagent B according to the expected amount of equivalent glucose (see table). Heat in a boiling water bath 15 min, cool, and dilute to 10 or 25 ml as needed to get a suitable absorbance reading at 450 nm.

Glucose equiv., μ g	ml Reagent A and B
5 − 25	1 + 1
26 − 50	2 + 2
51 − 75	3 + 3
76 − 100	4 + 4
101 − 125	5 + 5

The same result could be obtained, of course, by routinely using 1 + 1 ml of reagents and taking smaller aliquots of the reducing sugar solution to contain the equivalent of 5 to 25 μg glucose. Strumeyer (1967) used this color reagent to demonstrate the linearity of an amylase reaction using his reduced starch substrate.

Enzymatic Reactions of Sugars.   A number of schemes have been developed for linking the products of glycosidase reactions to oxidation-reduction reactions involving NAD^+ (or NADH). The resulting change in absorbance at 340 nm is indicative of the rate of product formation. These are auxiliary enzyme assays, and their design should keep in mind the principles discussed in earlier chapters. All these reactions work upon glucose as the initial substrate. There are two principal types of reactions: 1. direct reduction of NAD^+ to NADH catalyzed by glucose dehydrogenase (van Leeuwen 1979; Holm 1980; Vanni and Hanozet 1980); and 2. formation of glucose-6-phosphate (hexokinase, ATP) followed by reduction of $NADP^+$ to NADPH catalyzed by G6P dehydrogenase (Liu and Walden 1969; Guilbault and Rietz 1976; Larsen 1983).

Glucose dehydrogenase reaction. Vanni and Hanozet pointed out that this enzyme is only active with β-glucopyranoside, so if the enzyme of interest forms α-glucose (i.e., glucosidase or invertase), the enzyme mutarotase must be included in the reaction mixture to rapidly convert the α to the β isomer. The reaction with glucose dehydrogenase is straightforward, and if the proper amount of this enzyme is used, the rate of increase of absorbance at 340 nm is directly related to the rate of the initial (e.g., invertase) reaction, after the lag period. The actual amounts used by these authors will be given in the later discussion on specific assays.

Glucose-6-phosphate dehydrogenase reaction. The phosphorylation of glucose by hexokinase using ATP as the donor is not affected by the anomeric state of glucose, so mutarotase is not required. The reaction is of the two auxiliary enzyme type:

$$\text{Glucose} + \text{ATP} \xrightarrow{\text{ATP hexokinase}} \text{G–6–P} + NAD^+ \xrightarrow{\text{G6P dehydrogenase}} \text{NADH}$$

The source of the G6P dehydrogenase must be taken into account, since some of them use NAD^+ as a cofactor, while others require $NADP^+$ as a cofactor. Normally the measurement is of the rate of increase of absorbance at 340 nm, but Guilbault and Rietz (1976) enhanced the sensitivity of the assay by measuring the increased fluorescence due to NADH (excite at 365 nm, emit at 455 nm) directly in the reaction cuvette.

Starch-Iodine Complex Formation.   The formation of a colored starch-iodine complex is the basis for one of the oldest quantitative amylase assays (Wohlgemuth 1908). It was slowly modified as the understanding of α- and β-amylase activities developed, but was usually a fixed end point type of assay, in which the end point was the time required to obtain a certain decrease in the color level seen when a sample of reaction mixture was mixed with an iodine solution (AACC 1983, Method 22-01). If the starch species involved is amylopectin, the color is a reddish-brown, with an absorbance maximum at 570 nm. If the complexation is with amylose, the color is blue, absorbing maximally at 620 nm.

The classical form of the Wohlgemuth assay involved beginning a reaction of amylase with starch, taking aliquots at various time intervals, adding these to some standardized iodine solution, and comparing the depth of color to a set of standards (AACC 1983). One update consists of reading the absorbance of the complex in a spectrophotometer (Street and Close 1956) after a fixed time of reaction. The advantages of the fixed end point assay may be combined with the advantages of spectrophotometric measurement (Perten 1966), in which the enzyme rate is calculated from the rate of absorbance change around the time which gives the optimal amount of hydrolysis. A continuous assay is described by Marshall and Christian (1978), who include iodine directly in the reaction mixture and follow the change in absorbance at 570 nm.

All of these assays suffer from the theoretical disadvantage that the change in complex formation (i.e., absorbance) is not linearly related to the number of hydrolytic events which have occurred. Because of the reaction pattern of α-amylase, and because this pattern varies depending upon the source of the amylase, rates of color decrease do not correlate well with rates of appearance of new reducing ends (i.e., bond hydrolysis). Thus it is important to keep in mind that this type of assay is quite useful for measuring the amount of amylase in different samples of closely similar enzyme types (e.g., barley malts) but not for comparing hydrolytic activities of α-amylases from different sources such as bacterial, cereal, and fungal origins.

Instrumental Signals

Viscometric Measurements.   As large polymeric molecules, the substrates for many glycolytic enzymes generate a high solution viscosity at relatively low concentrations; this viscosity decreases more or less sharply as the macromolecules are cleaved by enzymatic action. This technique is applied most commonly to the action of cellulases (Almin and Eriksson 1968; Tschetkarov et al. 1967; Tschetkarov and Koleff 1969; Manning 1981), but has been applied to α-amylase (Tipples 1969) and (as discussed in Chapter 8) proteases (Laufer 1938; Kuiper et al. 1978).

The theory of viscometric measurement of enzyme activity has been developed by Manning (1981) and is presented more fully in Appendix H. With α-amylase and proteases the theoretical drawback is that bond cleavage is not random, and the molecular weight distribution of the product molecules changes as the reaction progresses. With these enzymes, if the assay is conducted as a fixed end point assay to a certain decrease in relative viscosity, it has utility for comparing samples from similar sources (i.e., a number of malts, or a series of bacterial fermentations). However, the reliability is less when comparing samples from different sources. With cereal malts, also, there is the difficulty that the level of β-amylase also affects viscosity assays for α-amylase (Tipples 1969), even though β-amylase by itself has very little effect on the viscosity of a starch solution.

The Falling Number apparatus has come into wide use for detecting the presence of amylase in cereals, particularly in areas where the grain is subject to sprouting due to adverse weather conditions during the harvest (Hagberg 1961; Perten 1964). This is an excellent tool for its intended use, but researchers have tended to push it beyond its limits of applicability. The response (FN, or the time for a weighted plunger to drop through a gelatinized starch suspension) is a rectangular hyperbola when plotted versus α-amylase measured by a Wohlgemuth method. At low or high levels of enzyme a small variation in one parameter leads to a large error in the predicted parameter. At low levels of enzyme a few percent difference in amylase changes FN by 10 to 30%, while at high enzyme levels a variation of 2% in FN may convert to a 40% difference in amylase concentration. More details on the use of the Falling Number apparatus for measuring various kinds of amylases may be found in a recent description by Perten (1984).

Many food laboratories have a Visco-Amylograph for the study of pasting properties of starches for various food applications. This instrument has been used to measure α-amylase activity (Ranum et al. 1978). As an assay the method is rather cumbersome and imprecise. But it is useful in giving a practical idea of how the amylase under study reacts with, and modifies the properties of the starch being used.

Nephelometric, Turbidimetric Measurements. The basic difference between these two types of measurements is that a Nephelometer measures the amount of light scattered from a hazy solution (the light energy emerging at right angles to the incident beam) while a Turbidimeter measures the amount of light which is not scattered (the light energy along the same axis as the incident beam). The former requires a special sort of instrument construction, while the latter measurement is usually made with an ordinary colorimeter. Many macromolecules will give hazy solutions which scatter less light as the molecules are depolymerized by the action of a hydrolase. This principal has been applied to measuring the rate of hydrolysis of starch and cellulose.

Nummi et al. (1981) ball-milled cellulose powder suspended in 94% ethanol at 20°C for 24 to 72 hr. The suspension was filtered through fritted glass filters (Schott Nos. 1 and 2) to yield a fine fraction with particle size of 90 to 150 μm. The iron dust from the steel balls used in the milling was removed by stirring the cellulose with a magnetic stirring bar. The cellulose powder gave a stable suspension when suspended in citrate buffer for the cellulase assay. A 0.03% suspension was used as the substrate in a Perkin-Elmer Model 91 Nephelometer; from the plots given of the change in nephelos versus time of enzyme reaction, it was very difficult to judge the linear part of the curve and the initial slope. Measured in a colorimeter, with a 0.04% cellulose suspension at 50°C, the rate of decrease in "absorbance" at 620 nm was somewhat easier to judge. While these authors ball-milled to get smaller particle sizes, it should be possible to use one of the small sized cellulose powders (BH-200, BW-200) available from commercial suppliers such as James River or Fiber Sales.

A similar idea was applied to starch by Rungruangsak and Panijpan (1978), who made a 2% starch suspension in Tris buffer, 0.05 M, pH 7.0, boiled for 1 min, then cooled. They added the sample containing amylase (pancreatic) and followed the change in absorbance at 650 nm in a simple spectrophotometer. The data they showed indicated a linear decrease for 2 to 3 min. Amylase activity has also been measured using nephelometry, in the commercial Perkin-Elmer Model 91 instrument (Campbell 1980; Osborne et al. 1981). While this has seemed to work well within a limited range of applications, some difficulty has been experienced in obtaining a starch substrate which gives reproducible results over a number of lots. Also, there has been a report (Asp and Midness 1983) that the two sensitivity scales of this instrument don't always overlap as they should.

Polarimetry. Certain glycosidase reactions lead to a large change in the light polarization of the substrate solution. The best example of this is invertase, or sucrose hydrolase. The progress of the invertase reaction is often followed by measuring the appearance of reducing sugars. This is complicated by the transferase activity of the enzyme (i.e., a sucrose molecule plus a glucose molecule yields a fructose plus a maltose molecule), and the expected stoichiometry (two reducing groups generated per enzymatic event) is not maintained. Polarimetry, particularly with the continuous-recording high sensitivity instruments available today, sidesteps this problem and gives a more accurate measurement of the course of the reaction during the early stages (Bowski et al. 1971). It is the method of choice for monitoring the invertase reaction.

Gel Diffusion.   The appearance of a clear zone as a polymeric substrate embedded in a gel is hydrolyzed has been applied to several glycosidases. Ceska (1971) used Cibacron Blue starch as the substrate; as it was hydrolyzed by amylase the blue dye diffused away into a more generalized concentration, and a clear zone resulted. Mestecky et al. (1969) used soluble starch in the agar layer. After the incubation with amylase the gel surface was flooded with an aqueous solution of iodine, to reveal the clear hydrolysis zones in contrast to the blue unreacted areas. Wood and Weisz (1987) used gel plates to measure β-glucanase in various cereals. The substrate was oat glucan in buffered agarose gel. After the desired incubation period a solution of Congo Red was layered onto the plate to form a colored complex with the unhydrolyzed glucan.

In all cases, the diameter of the clear zone is plotted versus the logarithm of enzyme concentration to get a linear plot. The rationale for this was discussed earlier (in the indirect methods section of Chapter 7). The advantage of a gel diffusion assay is that it will usually cover more than three orders of magnitude of enzyme concentration, so that it is unusual to have too little for detection or too much to measure. The disadvantage is that the incubation period may range from 18 to 72 hours, although many enzymes can be reacting during this same time. It is more applicable as a screening device than as a tool for, say, quick production control tests.

SPECIFIC ASSAYS

Most of the assays which will be described in this section are readily adaptable to other glycoside hydrolases. Thus, taking the description of the DNS method for β-amylase, by making the required obvious changes in buffer pH, substrate, and incubation time and temperature, one can also have an assay for cellulase, dextranase, invertase, etc. This is not to say that such adaptation is without pitfalls. There can always be unsuspected quirks in the nature of the reaction of the new enzyme which can lead to difficulties; any adapted assay still needs to be rigorously validated according to the principles discussed in Chapter 6. The point is that even if an assay for the particular glycosidase of interest to you is not described here, you should still be able to adapt one or more of these to suit your purposes.

Colorimetric Chemical Reactions

β-Amylase.   This assay is the one described by Bernfeld (1955), based upon the formation of maltose from soluble starch by β-amylase action, followed by color development with DNS reagent. In this case the color reaction is valid because it occurs with only one reactant (maltose). Substrate is made by suspending 1 g soluble starch in about 10 ml of water, then slowly adding this to 50 ml boiling water, completing the transfer with a wash bottle. After 2 min of additional gentle boiling the colloidal solution is cooled, 10 ml of 0.16M acetate buffer, pH 4.8 is added, and the volume is made to 100 ml. The substrate is attempered at 25°C, as is the solution of enzyme. At zero time 0.5 ml of substrate is mixed with 0.5 ml enzyme solution in a test tube. After exactly 3 min at 25°C 1 ml DNS reagent is added. The mixture is placed in a boiling water bath for 5 min, then cooled and 10 ml water is added. The absorbance is read at 540 nm, using a blank made with 0.5 ml water in place of the enzyme.

α-Amylase. A similar reaction is carried out, using reduced starch, and the neocuproine reagent for color development (Strumeyer 1967). To 1 ml of the 1% reduced starch solution in 0.02M acetate buffer, pH 4.7 containing 1mM $CaCl_2$, at the desired temperature (usually 25°C) is added 1 ml of enzyme solution at the same temperature. After incubating for a set time, 4 ml of Reagent A and 4 ml of Reagent B of the neocuproine reagent is added, the mixture is heated in a boiling water bath for 12 min, then made to 25 ml total volume and the absorbance is read at 450 nm. The blank is made using 1 ml water in place of the enzyme.

α-Amylase may also be assayed using the Somogyi-Nelson color reaction and purified amylose (Fuwa 1954). To dissolve the dried amylose (if it is not in the readily soluble lyophilized form) 200 mg is suspended in 4 ml cold 1N NaOH and held overnight. This is diluted to 80 ml with water, the pH adjusted to about 8 with 1N acetic acid and the volume made to 100 ml. The substrate stock, the reaction buffer (0.5M acetate, pH 5.0, containing 2mM $CaCl_2$), and the enzyme test solutions are attempered at 37°C. For the assay 0.5 ml of reaction buffer is added to the desired amount of enzyme solution and water is added to a total volume of 1.5 ml. At zero time 1.5 ml substrate solution is added and the reaction is incubated 30 min. Then 3 ml Somogyi reagent is added, the mixture is heated 30 min in a boiling water bath, cooled, and 3 ml Nelson reagent is added. The reaction is diluted to an appropriate volume (Fuwa used 100 ml) and absorbance is read at 500 nm. A standard curve is constructed with 0 to 0.5 mg glucose.

Cellulase, Dextranase. Miller et al. (1960) used the DNS reagent to follow the hydrolysis of carboxymethylcellulose (CMC). The substrate was 44.8 g Hercules 50T CMC dissolved in 2.8 L water with stirring. This was buffered to pH 5.0 by adding 28 g citric acid monohydrate and 78.4 g sodium citrate dihydrate, 0.4 g merthiolate, 0.4 g glucose, and making to 4 L total volume. The substrate was stored at 4°C until used. For the assay, substrate and enzyme solution were attempered at 50°C. At zero time 1 ml substrate was mixed with 1 ml enzyme and incubated 20 min. Then 3 ml DNS reagent was added, it was held in a boiling water bath 15 min, cooled, and the absorbance at 540 nm was read. The assay as reported was nonlinear; perhaps this would be rectified if the color reaction were done using the neocuproine reagent.

Koh and Khouw (1970) reported a similar assay for dextranase. The substrate was a 2% solution of Dextran 500 (Pharmacia AB) in 0.1M phosphate buffer, pH 6.0. To 1.9 ml of substrate was added 0.1 ml enzyme solution, and the mixture was incubated 30 min at 37°C. One ml of this was added to 1 ml DNS reagent, heated 15 min in a boiling water bath, cooled, 10 ml of water was added, and absorbance was read at 540 nm. A standard curve was constructed using 0 to 0.3 μM maltose. Again, linearity was not as good as it might have been, a circumstance which might be helped by using the Somogyi-Nelson or the neocuproine reagent.

Synthetic Small Molecules

p-Nitrophenyl Glycosides.   Assays using these substrates are usually of the fixed time variety, stopped by adding an alkaline buffer, and measuring the absorbance of the p-nitrophenoxide ion at 405 nm. A 1 mM solution of this material has an absorbance of 18.3. Distler and Jourdian (1973) described such an assay for β-galactosidase. The reaction mixture had a total volume of 0.1 ml. It comprised: 25 μl 0.1 M citrate buffer, pH 4.3; 25 μl of a 0.4% solution of bovine serum albumin in water; 25 μl of 20 mM pNP-β-galactoside; and 25 μl of enzyme solution. This was incubated 30 min at 37°C, then 1 ml 0.25 M glycine, pH 10, was added and the absorbance was read. Maruhn (1976) used a similar protocol for assaying β-galactosidase and N-acetyl-β-glucosaminidase activities. He made the following adjustments. Total reaction volume was 0.4 ml, with 0.1 M citrate buffer pH 4.0, no bovine serum albumin, and a 15 min incubation. The stop reagent was 0.2 ml of 0.75 M 2-amino-2-methyl-propan-1-ol pH 10.25. The substrate concentrations were 5 mM for the galactoside and 10 mM for the N-acetyl-glucosaminide. Both of these assays use semi-micro cuvettes for the absorbance measurement. If you have only the standard 3 ml cuvettes to work with (requiring about 2.5 ml for a good measurement) the volumes given above will have to be adjusted accordingly.

For a continuous measurement of the hydrolysis of pNP glycosides which is insensitive to the pH of the assay, the report of Ford et al. (1973) should be consulted (see section on substrates at the beginning of this chapter). The increase in absorbance as hydrolysis proceeds is followed at 346 nm, the isobestic point for p-nitrophenol/p-nitrophenoxide. The difference spectrum is such that complete hydrolysis of 1 mM pNP-glucoside will give an increase in absorption of 2.6, while the hydrolysis of 1 mM pNP-N-acetyl-glucosaminide will give an increase of 3.6. The difference is due to the different absorbancies of the two substrates at this wavelength.

MUG. The assay with 4-methyl-umbelliferyl-glucoside (MUG) was described by Robins et al. (1968). The substrate is dissolved in the desired buffer at any desired concentration, and 25 μl of this is placed in a fluorimeter tube. After the addition of 5 μl enzyme and incubation for 1 hr at 37°C, 1 ml of 0.02M 2-amino-2-methyl-propan-1-ol buffer pH 10.3 is added, and the fluorescence is measured (excite 365 nm, emit 460 nm). The usual care required for any fluorescence assay (the direct methods section of Chapter 7) must be observed for good results. They reported excellent linearity with β-galactosidase, β-glucuronidase, and β-glucosidase in this assay.

Dyed Insoluble Substrates

The earliest version of an α-amylase assay using these substrates (Ceska et al. 1969) is the basis of a kit sold for clinical chemistry applications ("Phadebas," Pharmacia AB). The directions for using the kit are included, and will not be repeated here. This has found some application in food laboratories.

Synergistic α-, β-Amylase Assay. The general mode of use of these dyed starch materials has been indicated earlier, where the synthesis of the substrates was given, and no further elaboration should be necessary. However, the findings of Doehlert and Duke (1983) relative to the release of color from a starch-azure substrate by α-amylase in the presence of β-amylase should be mentioned. By itself, β-amylase will not release color from this substrate. However, when some α-amylase is present, the endo-hydrolysis performed by this enzyme exposes parts of the starch molecules which are susceptible to attack by β-amylase, and the color release is greater than that due to the α-amylase action alone. They found that the synergistic effect is maximal at a β-amylase concentration of 1000 units/ml (approximately 1 mg crystalline sweet potato amylase per ml). For the substrate, 2 g starch azure is suspended in 100 ml 0.1 M potassium succinate buffer, pH 6.0, containing 3 mM $CaCl_2$ and brought to a boil, then cooled (readjust volume if necessary). For the assay, to 5 ml of substrate is added 1 ml of enzyme (α-amylase plus β-amylase) and this is incubated at 30°C. At intervals an aliquot is removed, centrifuged, and the absorbance at 595 nm of the supernatant is measured. Because of the synergistic action of the β-amylase the assay is several times more sensitive for α-amylase than in the absence of β-amylase.

Fluorimetric α-Amylase Assay.   Marchylo and Kruger (1978) used the action of α-amylase on a fluorigenic limit dextrin as the basis of a sensitive assay. The limit dextrin was made essentially as described earlier (see section on polymeric substrates), but the substrate for the β-amylase activity was a starch anthranilate made by the method of Rinderknecht and Marbach (1970) (see the section on substrates). The limit dextrin anthranilate was isolated and lyophilized. Marchylo and Kruger described the use of this substrate for an automated method, in which, after incubation with enzyme, the solubilized anthranilate was separated from the large molecules via dialysis and measured fluorimetrically. To make the substrate solution 60 ml 0.2 M acetate buffer pH 5.5, containing 1 mM $CaCl_2$ is brought to a boil and 0.5 g of the limit dextrin anthranilate is added. After this suspension is homogeneous it is added to 100 ml 0.2M acetate pH 5.5 at 90°C, and the solution is stirred for 30 min at 90°C. Then it is cooled and the volume adjusted to 200 ml using the same acetate buffer. The Technicon procedure required a flow rate of 1 ml/min substrate, 0.42 ml/min enzyme solution, incubation at 37°C for 12.5 or 25.5 min, and dialysis against 0.2M phosphate buffer, pH 7.2, flowing at 1.6 ml/min. The dialysate was measured in a Turner Model 111 Fluorimeter. For a manual procedure the substrate could be mixed with enzyme, incubated as required, the reaction stopped by addition of the phosphate buffer, insoluble starch removed by centrifugation, and the fluorescence of the supernatant measured (excite at 335 nm, emit at 410 nm).

Cellulase, Dextranase Assays.   The substrates described by Huang and Tang (1976) work well for these enzymes. The choice between the trinitrophenyl derivative or the fluorescamine derivative depends upon the instrumentation available; the latter is preferable on the grounds of sensitivity if you have a spectrofluorimeter. The assays are simple and straightforward. One g of the desired substrate (see section on substrates) is suspended in 100 ml of an appropriate buffer. After attemperating (say, at 30°C), to 2 ml of this suspension is added 5 to 10 μl of enzyme, the reaction proceeds for 30 min, and then it is filtered or centrifuged. If the TNP derivative is used the absorbance at 344 nm of the supernatant is measured. For the fluorescamine derivative the supernatant fluorescence is determined (excite 390 nm, emit 480 nm). The sensitivity is much greater than with assays using cellulose-azure or DNS methodology.

Starch-Iodine Complex Formation

Colorimetric Wohlgemuth Assay.   The development of the original Wohlgemuth assay often used in the cereal industry is the Sandstedt-Kneen-Blish (SKB) method, in which the end point of the reaction is determined by comparing the color of a test (reaction aliquot added to iodine solution) to a standard (AACC 1983, Method 22-01). An improvement used a colorimeter to measure the absorbance at 570 nm of the tests taken at various reaction times, and calculated the rate of decrease in absorbance (Perten 1966). The substrate, a limit dextrin, is made as described above (see section on limit dextrin). For a stock iodine solution dissolve 11 g KI in a minimum of water, then add 5.5 g I_2 crystals and stir until dissolved. Dilute to 250 ml and store in a brown bottle in darkness. The test iodine solution contains 40 g KI dissolved in water, then 4 ml of stock solution is added and the volume is made to 1 L. This solution should be made fresh daily. A 0.2% solution of $CaCl_2$ is also required.

The absorbance at 570 nm is set to zero using a mixture of 2 ml $CaCl_2$ solution, 10 ml test iodine solution, and 40 ml water, the mixture being attempered at 20°C. A substrate control contains 0.5 ml substrate solution, 1.5 ml $CaCl_2$ solution, 10 ml test iodine solution, and 40 ml water, also at 20°C. The absorbance should be 0.55 to 0.60. For the assay, 15 ml of enzyme solution (e.g., malt extract or fungal amylase solution, in $CaCl_2$ solution) is attempered at 30°C, then 5 ml of limit dextrin substrate is added. At intervals of 5 or 10 min after the start of the reaction, 2 ml of the reaction mixture is added to 10 ml test iodine solution plus 40 ml water, at 20°C. The absorbance at 570 nm is measured. The enzyme concentration should be such that a time between 15 and 40 min is required to reach an absorbance between .390 and .240. It was found empirically that a plot of log [absorbance] versus incubation time was linear. The amylase activity, A, equals the slope of the plot of log [absorbance] versus time multiplied by a dilution factor (this equals 1 if 5 g of sample was extracted or dissolved in 100 ml $CaCl_2$ solution, 2 if that extract is diluted with an equal volume of $CaCl_2$, etc.) times 100. It was found that the standard SKB unit equals $A \times 0.11$.

The absorbance of the iodine-limit dextrin complex is sensitive to temperature, hence the necessity of making these measurements at 20°C. While the above procedure is the standard ICC (International Cereal Chemists) colorimetric procedure, it would seem to be advantageous to prepare a series of tubes ahead of time, each containing 1 ml test iodine solution plus 4 ml water, attempered to 20°C, and add 0.2 ml of the reaction mixture at desired times. This would speed up the rate of data acquisition, as well as conserving substrate and reagents. The reaction mixture might well comprise 3 ml enzyme extract plus 1 ml substrate.

Continuous Assay. Marshall and Christian (1978) described an assay for pancreatic α-amylase which allows continuous monitoring of the decrease in absorbance due to the starch-iodine complex. The substrate is a 1% solution of soluble starch in water (boiled briefly to completely dissolve the starch). A buffered iodine solution contains 32 mg I_2, 330 mg KI, and 23.4 mg

NaCl in 100 ml of 0.01M phosphate buffer, pH 7.0. For the assay, 2.5 ml of buffered iodine is brought to 38°C, 10 μl of enzyme is added, followed by 10 μl of starch solution. The absorbance at 570 nm is monitored continuously. The decrease is linear for the first few minutes of the reaction.

This assay has not been tested for applicability to cereal and fungal amylases, to my knowledge, but it should be easy to adapt. The buffer should be acetate at pH 4.7, and $CaCl_2$ would replace the NaCl. Also, the effect of replacing soluble starch with limit dextrin should be evaluated. This would be much simpler than the colorimetric Wohlgemuth assay, and seems to be worth investigation.

Coupled Enzyme Assays

Glucose Dehydrogenase Based.

The maltose formed during the action of α-amylase on starch may be measured using a series of three auxiliary enzymes: glucosidase, to generate α-glucose; mutarotase, to rapidly isomerize this to β-glucose; glucose dehydrogenase plus NAD, yielding glucuronate plus NADH; the increase in absorbance at 340 nm due to the appearance of the latter is monitored. van Leeuwen (1979) described such an assay for serum amylase. The substrate is 1% soluble starch in water. The buffer contains 3.06 g K_2HPO_4, 1.17 g KH_2PO_4, and 3.22 g NaCl in 1 L, pH 7.2. The auxiliary enzyme mixture contains 3.4 IU/ml mutarotase, 100 IU/ml glucose dehydrogenase, and 22 mM NAD^+. A solution of glucosidase contains 100 IU/ml. For the assay, 250 μl substrate is mixed with 250 μl of auxiliary enzyme mixture, and 20 μl serum amylase sample, then incubated at 30°C for a few minutes to remove serum glucose. Then 10 μl glucosidase is added and absorbance at 340 nm is measured. The response is reported to be linear after an overall lag period of about 20 min.

Amyloglucosidase may be measured using maltose as the substrate (Holm 1980). In this case the product is β-glucose so mutarotase is not required. The only auxiliary enzyme needed is glucose dehydrogenase plus NAD^+. The report is for an AutoAnalyzer method of performing this assay.

Invertase is readily assayed using this procedure (Vanni and Hanozet 1980). The 3 ml reaction volume contains 33 mM sodium maleate buffer pH 6.8, 33 mM sucrose, 2 mM NAD^+, 0.6 IU mutarotase, and 28 IU glucose dehydrogenase. The reaction is started by adding 10 μl invertase solution, and absorbance at 340 nm is measured. This method continuously removes the glucose as it is formed, thus preventing the transferase activity of invertase which presents a problem in other, fixed-time, types of invertase assays (Bowski et al. 1971).

Glucose-6-Phosphate Dehydrogenase Based.

Glucose resulting from a hydrolytic reaction is phosphorylated by hexokinase plus ATP, and the resulting G-6-P is dehydrogenated with concomitant reduction of NAD^+

(NADP$^+$) to give NADH (NADPH) and an increase in absorbance at 340 nm. A straightforward application of this sequence is an assay for cellobiase (Liu and Walden 1969). In a 1 ml reaction volume they used 50 mM phosphate buffer pH 7.2, 3 mM ATP, 0.3 mM NADP$^+$, 2 mM EDTA, 2 mM MgCl$_2$, 3 IU yeast hexokinase, 3 IU G-6-P dehydrogenase, and the cellobiase enzyme. After equilibrating at 21°C 25 μl of cellobiose solution (concentration as desired) is added to start the reaction. The authors report a lag period of about 1 min, and linearity out to 5 min. The stoichiometry between NADPH formed and cellobiose hydrolysis is 2:1, because each hydrolytic event forms two molecules of glucose for further reaction.

Serum α-amylase is measured using maltopentaose as a substrate (Larsen 1983). Amylase cleaves this to maltose plus maltotriose; these two materials are hydrolyzed by α-glucosidase (which reacts slowly with maltopentaose) to 5 molecules of glucose, which are then phosphorylated and dehydrogenated. Thus the amplification factor is 5 in this case. The substrate solution is 24 mM maltopentaose in water. The buffer is 0.07 M phosphate, pH 6.8, containing 51 mM NaCl, 0.6 mM CaCl$_2$, 4 mM MgCl$_2$, 2.7 mM NADP$^+$, and 1.5 mM ATP (this solution is stable for 5 months if frozen, and for 1 month at 4°C). Glucosidase solution contains 250 IU/ml suspended in 3.2 M ammonium sulfate. The suspension is centrifuged, 80% of the supernatant is removed and replaced with water to bring the enzyme into solution. The kinase/dehydrogenase solution contains 125 IU/ml hexokinase and 75 IU/ml G-6-P dehydrogenase in 50% aqueous glycerol. For the assay, 1 ml buffer, 10 μl kinase/dehydrogenase solution, 80 μl glucosidase solution and 20 μl amylase solution are mixed and brought to 37°C. To start the reaction 100 μl of substrate is added and absorbance at 340 nm is monitored. The author indicated a lag period of about 6 min before a linear trace was seen. The lag may be shortened by increasing the amount of glucosidase, but this also increases the blank rate due to its slight reaction with maltopentaose.

Guilbault and Rietz (1976) used a similar scheme, monitoring the appearance of NADH fluorometrically to increase the sensitivity somewhat. The buffer was 0.05 M Tris, pH 7.1, with 3 mM MgCl$_2$, 1.2 mM NAD$^+$, and 25 mM ATP. The substrate was starch, recrystallized from methanol/water to remove maltose, dissolved at 4 g per L in 0.05 M Tris pH 7.1. The auxiliary enzymes were added individually. The stock solution concentrations were: for α-glucosidase 138 IU/ml; for hexokinase 280 IU/ml, and for G-6-P dehydrogenase 140 IU/ml. For the assay the following volumes were added to a fluorometer tube in order: buffer, 500 μl; amylase solution, 100 μl; α-glucosidase, 50 μl; hexokinase, 5 μl; G-6-P dehydrogenase, 5 μl; and finally starch solution, 500 μl. The fluorescence was measured (excite at 365 nm, emit at 455 nm). They found a lag period of 10 to 20 min, but obtained good linearity in the range of 0.5 to 5 units of amylase per ml.

Gel Diffusion

α-Amylase.   Mestecky et al. (1969) describe a gel diffusion assay based upon the reaction of iodine with unhydrolyzed starch after the incubation period. Two g agar and 1 g soluble starch were dissolved in 100 ml 0.02 M phosphate buffer pH 7.0 by autoclaving. Into siliconized cellulose acetate boxes (64 mm square) 7 ml of this solution was poured and allowed to cool and solidify. Wells having a diameter of 2 mm were cut into the gel, and 3 μl of enzyme was placed in each well. After incubating 2 to 4 hours at 37°C the plate was flooded with a solution of 5 ml tincture of iodine dissolved in 100 ml water; after 1 min the plate was rinsed with water. The diameter of each clear zone was measured.

Ceska (1971) used a dyed starch as the substrate. One g agar was dissolved in 0.02 M phosphate buffer pH 7.0 containing 0.05 M NaCl and 0.02% NaN_3 by heating to near boiling. It was cooled to 85°C and 0.5 g Cibacron Blue starch was added. The hot solution was poured into Petri dishes to give a layer 0.8 mm thick and allowed to cool and solidify. A moist sponge was attached to the cover of each dish. For the assay 5 μl drops of enzyme solution were placed directly on the surface of the gel, then the plate was covered and incubated in a chamber at 37°C or 56°C. The increase in diameter of clear zones was linear for up to 48 hr at 37°C and up to 17 hr at 56°C. A plot of zone diameter versus log amylase activity was linear over the range of 10 to 1000 IU/L.

β-Glucanase.   Wood and Weisz (1987) describe a gel assay for this enzyme found in oats and barley. To 25 ml 0.05 M MES buffer pH 5.5 was added 12.5 mg β-glucan (see section on substrates) and 125 mg agarose, and the suspension was heated to boiling to effect solution. The hot solution was poured into Petri plates to a thickness of 4 to 5 mm and allowed to cool and solidify. Wells of 4 mm diameter were cut into the gel. To each well was added 10 μl of enzyme solution. After about 20 min a 0.1% aqueous solution of Congo Red was layered onto the plate which was then covered and allowed to incubate 18 hr at room temperature. The unreacted substrate gave a purplish-red color surrounding the clear zones of enzyme activity. Again, a plot of zone diameter versus log enzyme concentration was linear over some three orders of magnitude.

Viscosity Measurements

Cellulase.   Viscosity methods measure the endohydrolase activity found in various cellulase enzyme complexes. They show relatively little response to the C_1 activity and the cellobiase activity in crude extracts of e.g., *T. viride*. Since it is the initial depolymerization step which is usually rate limiting in the conversion of cellulose wastes to glucose, the development of endocellulase is of major commercial importance. Assay methods increase in complexity as they increase in theoretical validity. It is important to choose a method corresponding to your requirements.

Simple assay. Some of the factors to consider in designing a quick, simple viscosity test were discussed in the direct methods section of Chapter 7. The test should be a fixed end point assay (i.e., the time to achieve a certain percentage decrease in viscosity, or outflow time in the viscosimeter). This decrease should be relatively small, no more than 5%, and the experimental error in the measurement of outflow time should be small relative to this 5% decrease. A suitable capillary viscosimeter is the Ubbelohde #1, which has an outflow time of about 80 sec for water, and about 200 sec for the substrate solution described next. The measurement of outflow time is usually ±0.1 sec, and the experimental relative error in measuring a change of 10 sec is about 2% (±0.1 sec at zero time and at the ending time, for a change of 10 ±0.2 sec).

A typical design for a cellulase assay is as follows. Substrate is sodium carboxymethylcellulose (CMC) with a molecular weight in the 50,000 to 100,000 range, and a DS of 0.7 to 0.8. Five g is dissolved in 1 L of 0.05 M acetate buffer pH 5.0 by stirring overnight. The solution is frozen, then thawed and boiled under reflux for 10 min, filtered through a sintered glass filter to remove any residual particles, and frozen again (in suitable portions) for storage. This process ensures complete dispersal of the CMC and gives better uniformity from batch to batch. The outflow time for this solution should be around 200 to 225 sec. For the assay, substrate solution is thawed and attemperated at 40°C. The enzyme solution is also brought to this temperature. At zero time 3 ml of enzyme is mixed with 10 ml substrate and this is poured into the viscosimeter. Outflow times are measured every 5 min, with the relevant incubation time taken as the midpoint of the measurement interval. A plot of log [outflow time] versus incubation time will allow a linear extrapolation to get an accurate outflow time at zero incubation time (Laufer 1938). From the plot one can determine the incubation time required to reduce the outflow time to 95% of the zero time value. The enzyme dilution should be such that this time is between 15 and 30 min, for the sake of reduced experimental error. The activity of the enzyme is directly related to the reciprocal of the incubation time to reduce outflow time by 5%.

More precise assay. By some minor adjustments in experimental conditions, some initial characterization work on the substrate solution, and a slightly more complicated treatment of experimental data, it is possible to determine (to a good approximation) enzyme activities in absolute terms (i.e., micromoles of glucosidic bonds hydrolyzed per minute) (Hulme 1971). The mathematics of this procedure are given in Appendix H. Briefly, the viscosity measurements are made as described above. The enzyme dilution should be adjusted so that the decrease in outflow time is about 2 sec for each 5 min of incubation time. Then $1/\eta_{sp}$ is plotted versus time. The (linear) slope of this line times a proportionality factor gives the rate of bond hydrolysis in micromoles per minute. The proportionality factor must be determined for each batch of substrate; the method is described in the Appendix. The other crucial factor is that the substrate concentration must be such that $4K\eta_{sp} < 3$. If this constraint is met this method allows measurement of cellulase activity in fundamental units. The actual manipulation required for each assay is similar to that for the simple assay discussed earlier. The data processing is slightly more complex, and worktime is required to characterize the substrate in order to arrive at the proportionality factor to be used. This extra work is repaid by having enzyme activity data which are more meaningful in absolute terms.

Child et al. (1973) used nonionic cellulose gums for a viscometric assay. The substrate stock was a solution of 310 mg powder (e.g., hydroxyethyl cellulose) dissolved in 100 ml water by shaking overnight. For the assay, 7 ml of this stock is mixed with 1 ml 0.5 M acetate buffer, pH 4.8, and 2 ml enzyme solution, all at 25°C, then the solution is poured into a viscometer and the outflow times measured every few minutes. A plot of $(1/\eta_{sp})^{1.26}$ versus time of incubation was linear (the exponent was determined empirically), and the slope of this plot was linearly related to the amount of cellulase enzyme used. The fact that $1/\eta_{sp}$ was not directly linearly related to time of incubation indicates that for their conditions of substrate character and concentration $4K\eta_{sp}$ was greater than 3.

Most precise assay. The most complete analysis of viscometric assays of cellulase is that of Manning (1981) which has already been discussed above and is presented in Appendix H. The steps necessary to accurately convert measured changes in substrate solution viscosity to changes in product concentration $[P]$ are given there. The experimental details of an assay are the same as those discussed above. These may be manipulated as required to meet your needs; this level of sophistication for a cellulase assay is probably only needed in a detailed exploration of enzyme characteristics.

Amylase.   Tipples (1969) described an assay for α-amylase extracted from cereals which used a small-volume Ostwald viscometer. The substrate is potato starch; 3.75 g is slurried in a small amount of water then added to 200 ml boiling water, completing the transfer using a wash bottle. After boiling 4 min the solution is cooled, 5 ml of 4M acetate buffer pH 4.6 is added, and the volume is made to 250 ml. The enzyme is extracted using a solution of 2% NaCl, 0.02% calcium acetate. Substrate and enzyme are attempered at 25°C. At zero time 3 ml substrate is mixed with 2 ml enzyme solution and poured into the viscometer. Outflow times are measured periodically; the end point is the incubation time required to give a

50% reduction in outflow time. Tipples examined the influence of β-amylase on viscosity measurements and showed that it could decrease the time to reach the endpoint by as much as 25%. Whether or not this interference could be decreased by using a different substrate (e.g., a limit dextrin) was not explored. Also, the large decrease in viscosity at the chosen endpoint may introduce some errors; an endpoint of 10% outflow time decrease would be more in line with theoretical considerations.

Kuiper et al. (1978) included amylase in the battery of depolymerases which they studied with their continuous flow viscometer. Inconsistencies in the viscosity of the starch solution which they used necessitated a rather cumbersome two-stage boiling step built into their automatic analyzer. They claimed good accuracy for the assay of bacterial amylase. This arrangement might be useful in a laboratory involved in quality control of production of amylase by fermentation, where many samples of a similar nature are routinely assayed.

Pullulanase (α-Dextrin-1,6-Glucanohydrolase). Hardie and Manners (1974) reported using a viscometric assay for pullulanase type enzymes which hydrolyze the α-1,6- linkages at the branch points in amylopectin and similar glucans. They used an appropriate concentration of substrate in 0.1 M acetate buffer, pH 4.5, and followed the change in specific viscosity with time. They plotted $1/\eta_{sp}$ versus time and took the slope of the line as an indicator of enzyme activity. Since this was linearly related to the amount of enzyme taken, the conditions must have been similar to those defined by Hulme (1971) (i.e., $4K\eta_{sp} < 3$), although Hardie and Manners did not explicitly note this.

Commercial Instrument Packages

Falling Number.   This apparatus is sold as part of a package for measuring α-amylase in grain and flour, primarily for the purpose of detecting sprouted grain as it enters the commercial pipelines. It is based upon work by Hagberg (1961) and Perten (1964). Several slight modifications have been made to improve the accuracy and applicability of the method, and it is now a standard specification for most flour millers and cereal chemists. The details of operation are provided by the manufacturer, and will not be given here; if you have the machine, you have all the information for its proper use.

As originally designed, the substrate is the grain or flour which also contains the enzyme. A 7 g sample is suspended in 25 ml of water, this is heated to 100°C under closely defined conditions to gelatinize the starch, and the viscosity of this slurry is measured as the time required for a plunger to fall a set distance through it. In the presence of cereal α-amylase the gelatinized starch is partially hydrolyzed, the viscosity decreases, and the plunger falls more quickly. Fungal amylase is sometimes added to flour which is deficient in this enzyme. The standard FN method will not detect this enzyme because it is denatured by the heating step. Perten (1984) has adapted the apparatus to detect fungal amylase in flour as follows. The substrate comprises 3.5 g of flour (containing the enzyme to be measured) and 3.5 g of pre-gelatinized wheat starch. The flour is suspended in 25 ml of water at 30°C, then the pre-gelatinized starch is added and the time required for the plunger to fall through this slurry is measured, also at 30°C. This time is denoted MFN, and the amount of α-amylase added (expressed in SKB units) is linearly related to 1/MFN. The test has been shown to be sensitive to the source of the pre-gelatinized starch (Perten, personal communication), and work is presently underway to standardize this factor. Also, the proportionality factor between SKB units and 1/MFN appears to vary somewhat for amylases from different microbial sources. If these complications are kept in mind, and if your laboratory has the Falling Number equipment available, this could be a useful assay for quality control tests in the right circumstances.

Nephelometry.   An assay kit is available based upon the use of the Perkin-Elmer Model 191 grain amylase analyzer (GAA) for following the decrease in light-scattering as a suspension of limit dextrin is hydrolyzed by α-amylase (Campbell 1980). Again, if you have the machine, you have the information for operating it. There are two sensitivity ranges built into the machine. Osborne et al. (1981) and Asp and Midness (1983) reported that the two ranges did not overlap as they should. Hsu and Varriano-Marston (1983) examined the performance of the GAA in some detail, and found a great deal of variability in the limit dextrin substrate supplied commercially. Also, the linearity of machine response with respect to the amount of purified barley malt α-amylase was not good. It appears that while this instrument may have utility for quickly detecting the approximate concentration of α-amylase (i.e., in grain samples in a mill), in its present state it would be a poor research tool.

10 Ester Hydrolases

Many enzymes of importance in food systems catalyze the hydrolysis of esters. The most important of these involve esters of carboxylic acids, and phosphate esters. In each of these categories there are two groups of interest. The carboxyl ester hydrolases are specific for soluble esters (esterases) or insoluble fatty acid esters (lipases). The phosphate ester hydrolases may be involved with various kinds of water-insoluble phospholipids (phospholipases) or with soluble phosphate esters (phosphatases). We will discuss substrates and assays for these four kinds of ester hydrolase enzymes.

LIPASES

The clear-cut differentiation between lipase and esterase based upon the water-solubility of the substrate was first clearly shown by Sarda and Desnuelle (1958). They demonstrated that pancreatic lipase reacted with fatty acid esters having moderate solubility only when the solubility concentration was exceeded and a lipid/water interface was formed. Stauffer and Glass (1966) showed the same phenomenon with respect to enzymes from wheat germ. Verger (1980) has reviewed the extensive subsequent work on pancreatic and other lipases. The parameter corresponding to substrate concentration [S] in the usual kinetic equation is, in the case of lipase, the area of the lipid/water interface present in the system. The significance of this fact has been explored using monolayer film techniques to clarify the role of commonly used adjuncts which stimulate lipase activity such as calcium ion and bile salts (emulsifiers).

A number of review articles and books concerning lipases have been published. Two may be mentioned here; a book, *Lipolytic Enzymes* (Brockerhoff and Jensen 1974) and a recent review by Jensen (1983). Both should be consulted if extensive work with lipase is being planned.

Substrate

Triglycerides. The most common substrate for assaying lipase activity is the natural substrate (i.e., a triglyceride of a long chain fatty acid).

The purest such substrate is a synthetic liquid triglyceride such as triolein or trilinolein. The solid triglycerides such as tripalmitin do not react as readily with the enzyme, probably because of steric factors rather than any inherent inability to bind to the enzyme. Pure triolein is rather expensive when compared to a natural vegetable oil, so olive oil or soybean oil is often specified in some publications. These contain some amount of monoglyceride, diglyceride and free fatty acid (FFA) which should be removed before use. This is easily done by passing the oil over a column of adsorbent, either neutral alumina Activity Grade I or silica gel which has been partially deactivated by the addition of 5% water followed by equilibration. The usual ratio is about 250 ml of oil percolated over 100 g of adsorbent. Air pressure (5 psi) on the top of the column will hasten the filtration. The effectiveness of the purification may be checked by measuring the interfacial tension versus distilled water; pure triglyceride will have an interfacial tension of 30 dynes whereas ordinary vegetable oil usually gives a value of 20 to 23 dynes.

Fluorogenic Substrates. Fatty acids which yield a fluorescent product upon hydrolysis are sometimes used to achieve greater assay sensitivity. A popular substrate is the ester with 4-methylumbelliferone, or 4-MUA. These are readily synthesized as per the method of Jacks and Kircher (1967). To 2 g 4-MU dissolved in 10 ml dry pyridine is added a 100% molar excess of acyl chloride. This is stirred at 70°C for 1 hr, cooled, diluted with 20 ml water, and poured into 100 ml diethyl ether. After a brief shake the layers are allowed to separate, the ether layer is dried and then evaporated. The crude product is recrystallized from methanol (for saturated fatty acid esters) or a methanol/acetone mixture for esters of unsaturated fatty acids (e.g., the oleate). For use the 4-MUA may be dissolved in a water miscible solvent such as methyl cellosolve, which is then diluted in water to give a homogeneous emulsion.

Fatty Acid Quantitation

When triglyceride is the substrate, the extent of hydrolysis is most conveniently followed by measuring the amount of FFA formed. This may be done by simple titrimetry (i.e., running the assay in a pH-stat), or, less conveniently, by stopping the reaction after a set time, extracting the lipid phase, and titrating the FFA with base. In some cases the FFA has been isolated by chromatography and measured by methylation and gas chromatography (Drapron and Sclafani 1969) or by using a C^{14}-labeled fatty acid triglyceride and quantitating the isolated FFA in a scintillation counter (Matlashewski et al. 1982).

The most convenient method, applicable in most instances, is to extract the FFA into an organic phase, form the copper soap (which remains soluble in the organic solvent), and measure the light absorption of the copper soap itself (Kwon and Rhee 1986), or of a colored complex of the copper with diethyldithiocarbamate (DDC) (Myrtle and Zell 1975).

Isooctane is used to extract the FFA, by simply shaking the reaction mixture (e.g., a lipase assay which has been stopped by the addition of N HCl), separating the phases by centrifugation, and drawing off the upper organic layer. To 5 ml of the organic layer is added 1 ml of copper reagent (5% cupric acetate in water, adjusted to pH 6.1 by adding pyridine), this is mixed vigorously for 90 sec, then the phases are allowed to separate. The absorbance at 715 nm of the upper phase is read; the concentration of copper soap is determined from an appropriate standard curve. This procedure is good for measuring up to 20 μmoles of FFA, but emulsions form at higher concentrations of soap.

For increased sensitivity, to 3 ml of the organic phase containing the copper soap, add 0.1 ml of NaDDC reagent (100 mg sodium diethyldithiocarbamate in 100 ml n-butanol). After mixing, read the absorbance at 435 nm of the copper diethyldithiocarbamate complex. This increases the measurement sensitivity by about ten-fold.

Lipase Assays

Titrimetric Assays. The simplest form of a titrimetric assay is to incubate lipase with a buffered substrate emulsion for a period of time, stop the reaction by the addition of acid, extract the FFA with isooctane, and titrate the FFA to a phenolphthalein endpoint using 0.01N KOH in methanol. With appropriate blank corrections, and with a moderately active lipase preparation, this method can give a good measure of relative enzyme activity and can serve as a guide in developing more convenient assays. Hockeborn and Rick (1982) published a thorough critique of titrimetric lipase assays, and showed that due to lag periods and similar experimental artifacts, the two-point (zero time and final time) assay as described above is not reliable.

A more reliable assay is the continuous titrimetric (pH-stat) assay, where the pH of the reaction mixture is monitored and base such as 0.01 N NaOH is added to keep the pH constant. The rate of addition of base is a direct measure of the rate of FFA formation. The reaction is usually carried out at a basic pH, around 8.5 to 9.0. To prevent the absorption of CO_2 from the air (which would require more base addition) purified N_2 is layered over the surface of the reaction mixture. Good stirring is also essential, and may be provided with either a small mechanical stirrer or a magnetic stirrer. The reaction should be carried out in a thermostatted container for temperature control. Several commercial instruments are now available which will provide all these refinements, and give the data in the form of a strip chart record from which the rate may be read directly. However, a continuous assay may be performed simply with a magnetic stirrer, pH-

is filtered through a 0.22 micron Millipore filter to give a clear suspension which contains about 11 μM 4-MU ester. For the assay 60 μl substrate, 20 μl of the desired buffer, and 0 to 20 μl enzyme solution is incubated for 10 min at 37°C. The reaction is stopped by the addition of 3 ml of cold 1M Tris buffer pH 7.5, and fluorescence is measured (excite at 365 nm, emit at 450 nm). A background blank value is obtained by using 20 μl 150 mM NaCl in place of the enzyme solution. Roberts did not report monitoring the fluorescence of the reaction mixture continuously, but with an optically clear suspension, this should work, and would be preferable to the fixed time assay described.

Low-Moisture Assays.   Lipase activity in flours and meals of various kinds is often of practical importance. A number of workers have investigated the action of the enzyme in these systems under conditions where the only moisture is that adsorbed from the surrounding atmosphere (Caillat and Drapron 1974; Drapron and Sclafani 1969) or is just sufficient to form a damp paste (Martin and Peers 1953; Matlashewski et al. 1982). As an example of this sort of assay, Matlashewski et al. mixed 90 mg triolein with 0.5 g oat flour, then added just enough buffer (0.05 M Tris·HCl, pH 7.5, with 1% Triton X100 detergent) to give 40% final moisture. After stirring to homogenize this paste, it was incubated at 37°C. At intervals a sample was removed, acidified with 0.1 ml 1 N HCl, and the FFA extracted with 10 ml chloroform/heptane/methanol 49/49/2. The extract was reacted with aqueous copper reagent (see above), the absorbance was read at 715 nm, and the concentration of fatty acid soap was determined from a standard curve.

ESTERASES

Substrates

Esterase substrates are, by definition, soluble in water. The nature of the substrate depends upon the specificity of the enzyme being studied: a carboxy esterase may be most active on ethyl propionate, pectin esterase requires methoxylated pectin, cholesterol esterase acts on O-acetyl cholesterol, etc. In addition, a wide variety of alcohol moieties are chosen to suit the assay chemistry: fluorescein diacetate for a fluorometric assay, α-naphthyl acetate for an assay in which freed α-naphthol is diazotized, salicyl acetate for a UV-spectrophotometric assay, etc. p-Nitrophenyl acetate is sometimes used as a chromogenic substrate for simple esterases, but has the drawback that it is rather readily hydrolyzed at pH's above 8, so the blank value for these determinations tends to be high. However, it does make a good "quick and easy" esterase substrate.

Esterases are often used as markers in studying plant or insect genetics. Extracts of a tissue are run on an electrophoresis slab, the developed electrophoretogram is incubated with a solution of α-naphthyl acetate for a while, then sprayed with a solution of some diazotized salt which reacts with the α-naphthol which has been freed by esterase action, depositing a colored diazo dye in the zone of active enzyme. This technique often reveals a large number of esterase isozymes, for example the 17 found in wheat (Cubadda and Quatirucci 1974).

Esterase Assays

Titrimetric Assays.
The easiest, most broadly applicable, and fundamentally soundest esterase assay is the continuous pH-stat assay. This is conducted as discussed earlier in connection with lipases, with the added advantage that since the substrate is water soluble one does not have to be concerned about the state of the substrate emulsion. This assay is also useful because the pH of the reaction can be set wherever desired. A point to remember is that if the reaction pH is in the region of, or below, the pK of ionization for the acid which is being freed by the enzyme reaction, the addition of less than 1 mole of base per mole of acid is required to keep the pH constant.

Chromogenic pH-Drop Assay.
If an esterase reaction takes place in the presence of a suitable pH indicator dye, then as acid is freed and the pH changes the absorbance of the dye solution changes. This concept has been frequently applied in stopped-flow enzyme kinetic studies to monitor rapid pH changes when two solutions are quickly mixed. It can also be used as the basis for an enzyme assay in a more usual time frame. The reaction mixture is lightly buffered so that the pH change upon the formation of a small amount of acid is relatively limited. Also, the ionization pK of the indicator dye must be in the same range of pH as the reaction. Finally, the absorbance change is "calibrated" by titrating the indicator dye/buffer system with a solution of a known concentration of the acid which is freed in the esterase reaction.

Hagerman and Austin (1986) describe such a system for assaying pectin methylesterase. The substrate is a solution of 0.5% citrus pectin in water, adjusted to pH 7.5. The indicator/buffer system is 0.01% bromthymol blue in 0.003 M phosphate buffer, pH 7.5. The enzyme sample is also adjusted to pH 7.5 before the reaction is started. For the assay, to 2 ml of pectin solution is added 0.15 ml of indicator/buffer and 0.83 ml water. The absorbance at 620 nm should be about 0.28, and is recorded briefly to establish a baseline pH change. Then 20 μl of enzyme is added to initiate the reaction, and the decrease in absorbance at 620 nm is recorded. By titrating a similar system with a solution of galacturonic acid, the authors found that a change of 0.0328 in absorbance at 620 nm corresponded to the formation of 0.1 μmole of galacturonic acid, with a linear response up to a maximum absorbance change of 0.1. The authors also carried out a pH-stat assay using five times the volume of reagents given above, and found good correspondence between the rate of base uptake in this system and the rate of acid release inferred from the spectrophotometric run.

Continuous Azo Dye Assay.   Johnston and Ashford (1980) took advantage of the strong color absorption of diazo dyes to develop a sensitive spectrophotometric esterase assay. The substrate is α-naphthyl acetate (8 mg/ml in acetone/water, 1/1 v/v). They used a number of 0.1 M citrate, phosphate, Tris, and glycine buffers to cover the pH range 3.0 to 9.5, but the Teorell-Stenhagen Universal buffer discussed in Chapter 4 would also serve. Water-soluble diazonium salts are included in the reaction mixture. As esterase hydrolyses the ester, the freed α-naphthol reacts with the diazo salt to form a highly-colored azo dye (which is kept in solution by the inclusion of bovine serum albumin (BSA) in the reaction). The authors investigated the pH-dependence of the rate of dye formation of a large number of diazonium salts. They found that in the pH range of 3.0 to 7.0 the best salt was p-nitrobenzenediazonium tetrafluoroborate, p-NBDTFB. For the pH range 7.0 to 9.5, the best diazo salt was Fast Violet B (FVB) salt (4'-amino-6'-methyl-m-benzanisidine), CI no. 37165.

The stock solutions are: Buffer, 0.1 M specific or Universal Buffer; Diazo salt, in cold distilled water, p-NBDTFB, 0.2 mg/ml, and FVB, 0.8 mg/ml; BSA, 10 mg/ml in water; Substrate, 8 mg/ml α-naphthyl acetate in acetone/water 1/1. For the assay, put 2 ml buffer, 0.2 ml BSA, 0.5 ml diazo salt, and 0.2 ml enzyme solution in a cuvette at 25°C. At zero time add 0.1 ml substrate, and record the increase in absorbance. For p-NBDTFB the absorbance of a 1 mM solution of azo dye is 26.36 at 450 nm. For FVB absorption of a 1 mM solution of the azo dye is 18.33 at 490 nm, at pH 8 or less. At higher pH the absorbance maxima shifts towards 510 nm, and the molar absorptivity also increases. The authors made a standard curve so they could convert absorbance changes at these higher pHs to molarity changes. It would be more useful to determine the isosbestic wavelength for the two ionization states of the azo dye and the molar absorptivity at that wavelength, and use this setting for assays with FVB at all pH values in the basic range.

PHOSPHOLIPASES

Phospholipases (ester hydrolases acting upon phospholipids) have several different types of specificities. Using phosphatidylcholine (PC) as the model substrate for this group, the phospholipases are classified as follows (Acker, 1985): PL-A (phospholipase A) splits off one fatty acid to give lysophosphatidylcholine (LPC); PL-B hydrolyzes the second fatty acid from LPC to give glycerophosphatidylcholine (some plant PL-Bs can release FFA from either PC or LPC); PL-C hydrolyzes the phosphate ester linkage to the glycerol moiety of PC yielding a diglyceride and phosphocholine; while PL-D acts on the other phosphate ester link to give phosphatidic acid and free choline. Two fairly recent reviews are those by Heller (1978) and by Acker (1985), while the paper by Acker (1969) discussing the activity of PL-B and PL-D in low water flour systems should also be mentioned.

PL-B and PL-D are the two types found in foodstuffs. PL-A is found primarily in various reptile and insect venoms, while PL-C is produced by a number of different bacterial species. The assay of PL-B and PL-D depends upon measuring the appearance of FFA or choline, respectively.

Substrate

Lecithin, the substrate for phospholipase, may contain four main components: phosphatidylcholine, phosphatidylethanolamine, phosphatidylserine, and phosphatidylinositol. It is usually obtained from one of two sources, either egg yolk or as a byproduct from the production of soybean oil. While crude lecithin could serve as a substrate, a more reproducible assay is based upon the use of purified PC. This is available from biochemical supply houses, but may also be prepared by the method of Singleton et al. (1965).

Liquid egg yolk (500 g) is blended with 1 L of acetone and the acetone-insoluble phosphatides allowed to settle for 1 hr at room temperature before being collected by filtration. The solids are washed with three 200 ml portions of cold acetone, then extracted with 1 L 95% ethanol. After stirring for 1 hr the solids are collected and re-extracted with 500 ml ethanol. The extracts are combined and the ethanol removed under vacuum. This residue is extracted twice with 300 ml of petroleum ether, the extracts combined, the volume reduced to 200 ml in vacuo, and then poured into 1 L of acetone at 15°C. After standing 1 hr the supernatant is decanted and the solids are washed with cold acetone. The extraction with petroleum ether and precipitation with acetone is repeated once more. Finally, the solvent is removed in vacuo and the phosphatides are stored at −20°C. The yield is about 36 g.

For chromatographic purification, a column of aluminum oxide (Merck, "For Chromatographic Absorption") is used. The bed dimensions are 4 cm diameter, 50 cm height, prepared from a chloroform slurry and washed with $CHCl_3$. Twenty-five g of crude phosphatides are dissolved in 500 ml $CHCl_3$ and applied to the column. After they have entered the column it is washed with a further 500 ml $CHCl_3$ at a flow rate of 10 ml/min. Elution of pure PC is carried out with a mixed solvent $CHCl_3$/methanol, 9/1, v/v. After about 600 ml of solvent has entered the column the eluate begins to appear hazy; this signals the elution of PC. The eluate is collected as long as significant haziness appears (about 350 to 400 ml total volume). The solvent is removed in vacuo. The purity of the resulting PC may be checked by thin layer chromatography on silica gel with a developing solvent of $CHCl_3$/ethanol/water, 2/5/2. The PC should be stored cold, preferably under nitrogen, to prevent autoxidation of the polyunsaturated fatty acids present.

Assays

Phospholipase B.

A procedure for assaying PL-B in malt extracts was given by Acker and Geyer (1968). Enzyme solution (40 ml) in pH 5.8 buffer is mixed with 10 ml of substrate solution (80 mg PC, 30 μl toluene, 10 ml water), this is treated with an ultrasonic mixer for 50 min, 60 μl toluene is added, and the suspension is incubated at 25°C. The reaction is stopped by adding 100 ml isopropanol and heating for 15 min on a steam bath. The free fatty acids are extracted with petroleum ether and quantitated by titration.

It would appear that this scheme could be simplified in some points, perhaps by extracting the FFA with isooctane directly from the reaction mixture and quantitating them with some variation of the copper reagent method as discussed above in connection with lipase. Also, a mixed micelle substrate solution using PC and sodium taurocholate (Roberts 1985) might obviate the need for the extensive ultrasonication treatment. Such developments in the assay for PL-B will probably wait upon a higher level of research interest in the properties of this enzyme.

Phospholipase D.

The measurement of choline freed by the action of PL-D on PC has, until recently, been a tedious procedure based upon precipitation from aqueous solution with Reinecke's Salt and the absorbance at 562 nm of the precipitate dissolved in acetone (Glick 1944; AACC 1983, Method 86-45). This is complicated by the necessity of first isolating the choline from the enzyme reaction mixture by ion exchange chromatography (Nolte and Acker 1975). The quantitation of choline based upon enzymatic reaction with choline oxidase (Imamura and Horiuti 1978) or choline kinase (Carman et al. 1981) is a preferable alternative.

Choline oxidase quantitation (Imamura and Horiuti 1978). Substrate solution is 10 mM PC (7.86 g/L) in water, sonicated for 10 min to give an emulsion. The buffer is 0.05 M acetate pH 5.5, 0.015 M in $CaCl_2$. To 0.35 ml of buffer is added 0.1 ml of substrate emulsion and 0.05 ml enzyme solution (PL-D in 0.01 M dimethylglutarate buffer, pH 7, containing 0.1% BSA) to start the reaction. The authors showed that the BSA was necessary to preserve the time linearity of the reaction. After 10 min at 25°C the reaction is stopped by adding 0.2 ml of 1 M Tris·HCl pH 8 containing 50 mM EDTA. The analyzing enzyme mixture contains 5 U/ml choline oxidase, 0.67 U/ml peroxidase, 5 mM 4-aminoantipyrine and 7 mM phenol in 10 mM Tris buffer, pH 8. To the stopped phospholipase reaction mixture is added 0.3 ml of this mixture, and the reaction is allowed to proceed for 20 min at 37°C. It is stopped by adding 2 ml of 1% Triton X-100 in water, and the absorbance at 500 nm is read. A standard curve covering the range of 0 to 0.2 μmoles choline is made by using 0.05 ml of dilutions of a stock choline solution in place of the PL-D enzyme solution.

The choline oxidase reaction is:

$$\text{Choline} + 2\ O_2 + H_2O \rightarrow \text{betaine} + 2\ H_2O_2.$$

The peroxidase reaction is:

$$2\ H_2O_2 + \text{phenol} + \text{4-aminoantipyrine} \rightarrow \text{quinoneimine} + 4\ H_2O.$$

Thus, one mole of choline will yield one mole of quinoneimine; the absorbance of a 1 mM solution of this dye at 500 nm is 12.0. A unit of choline oxidase oxidizes one μmole of choline per min. The assay mixture for choline oxidase consists of 0.1 M Tris·HCl buffer, pH 8, containing 0.15 M choline hydrochloride, 0.5 mM 4-aminoantipyrine, 2.1 mM phenol, and 0.4 U/ml peroxidase. To 3 ml of this mixture at 37°C is added 0.05 ml choline oxidase solution and the increase in absorbance at 500 nm is followed. The rate of change of absorbance is converted into molar units using the millimolar absorptivity value of 12.0.

Choline kinase quantitation (Carman et al. 1981). While the choline oxidase methodology is designed as a fixed-time assay, the choline kinase assay is a continuous coupled-enzyme assay which connects the liberation of choline from PC to the oxidation of NADH via the following reactions:

$$\text{Choline} + \text{ATP} \xrightarrow{\text{(choline kinase)}} \text{Phosphoryl choline} + \text{ADP}$$

$$\text{ADP} + \text{Phosphoenolpyruvate} \xrightarrow{\text{(pyruvate kinase)}} \text{ATP} + \text{pyruvate}$$

$$\text{Pyruvate} + \text{NADH} \xrightarrow{\text{(lactate dehydrogenase)}} \text{NAD}^+ + \text{lactate}$$

The formation of free choline in the reaction mixture is signaled by a decrease in the absorbance at 340 nm.

The total assay volume is 0.2 ml, so a special small volume cuvette is required. The concentrations of reactants in this final volume are: 50 mM Tris·maleate buffer pH 6.5; 0.5 mM PC; 1 mM Triton X-100; 40 mM $CaCl_2$; 10 mM $MgCl_2$; 10 mM KCl; 1 mM ATP; 1 mM phosphoenolpyruvate; 0.3 mM NADH; 10 units choline kinase; 100 units pyruvate kinase; and 100 units lactate dehydrogenase. The stock assay mixture is 11% more concentrated than this, so that 180 μl is placed in the cuvette and incubated briefly until the absorbance at 340 nm is stable. Then 20 μl of an appropriately diluted PL-D solution is added to start the reaction. After an initial lag of about 1 min, the authors report that the decrease in absorbance at 340 nm is linear for about 15 min. The rate of absorbance change is converted into μmoles/min, using the millimolar absorptivity of 6.22 for NADH.

PHOSPHATASES

These enzymes hydrolyze the C-O-P linkage of various phosphate and phosphonate ester compounds. A common division is into acid phosphatases and alkaline phosphatases, depending upon the pH optimum of the enzyme. A common pitfall is that the researcher does not recognize the influence of substrate ionization upon enzyme rate. Thus Hickey et al. (1976) obtained a variety of pH optima for the phosphatase from wheat germ, ranging from 4.6 to 9.2, when the substrate was either a p-nitrophenyl phosphate, or a phosphonate ester. As elegantly shown by Van Etten (1982) this "pH optimum shift" is actually due to the change in ionization constants for the substrate. When the appropriate correction is made (see Chapter 4) the pH optimum is constant, regardless of substrate. It is imperative that classification of these hydrolases into "acid" and "alkaline" groups be based upon sound analysis of the effect of pH on V_{max} and K_M, not on a simple determination of the pH at which a maximum rate is seen.

Phosphate Quantitation

Chemical Determination.   When a chromogenic substrate such as p-nitrophenyl phosphate is used the rate of ester hydrolysis may be determined by simple spectrophotometry. For almost all substrates of interest, however, the rate of enzyme action is measured by the formation of inorganic phosphate ion, often denoted by P_i (with the charge state determined by the pH of the reaction). The chemical quantitation of this ion is based upon reduction of the phospho-molybdate complex to give a blue material (Fiske and Subbarow 1925) which has a broad absorption peak centered at 820 nm. The original method used aminonaphthol sulfonic acid dissolved in sodium bisulfite as the reductant, but Chen et al. (1956) showed that ascorbic acid is a more convenient reductant as well as being an order of magnitude more sensitive than the Fiske-Subbarow method. The phosphomolybdate complex is soluble in organic solvents so it can be isolated from the reaction mixture when other compounds (e.g., phytate) interfere with the color development (Cooper and Gowing 1983).

Direct color development (Chen et al. 1956). Reagent A is 10% (w/v) ascorbic acid in water. Reagent B is 4.2 g ammonium molybdate tetrahydrate dissolved in 1N H_2SO_4. The working reagent, 1 part A plus 6 parts B, is made fresh daily and kept in an ice bath until used. To 0.3 ml of the sample containing P_i is added 0.7 ml of working reagent, mixed, and held for 20 min at 45°C or 1 hr at 37°C. The absorbance is read at 820 nm. The sensitivity of this determination is high; 1 μg P_i gives an absorbance of 0.26.

Indirect color development (Cooper and Gowing 1983). To 2 ml of the phosphatase reaction mixture is added 2 ml 0.3 M trichloroacetic acid, 0.5 ml 3 M H_2SO_4, and 0.5 ml of 5% aqueous ammonium molybdate solution. This is mixed, then 2.5 ml organic solvent (n-butanol/heptane, 3/2 v/v) is added, the tube is mixed for 15 sec and then centrifuged. The upper organic layer is withdrawn and added to 4 ml of 1% aqueous ascorbic acid solution. This is thoroughly mixed, allowed to separate, and the upper organic layer is discarded. After incubation for 2 hr at 37°C or 10 min in a boiling water bath the absorbance is read at 820 nm. The sensitivity is high; 1 μg P_i gives an absorbance of 0.16. This procedure was developed for application to phytase assays, in which the presence of partially-hydrolyzed phytate interferes with the reduction of the phosphomolybdate complex.

Enzymatic Determination. Purine nucleoside phosphorylase will catalyze the phosphorolysis of inosine by P_i to give hypoxanthine plus ribose-1-phosphate. Hypoxanthine is a substrate for xanthine oxidase which, coupled with hydrogen donors, generates colored formazan products. Alternatively, the uric acid formed by xanthine oxidase action may be quantitated using its absorbance at 302 nm. These reactions may be used either to measure P_i generated in a fixed-time assay, or to make a continuous coupled-enzyme assay.

Fixed time assay (Fossati 1985). The xanthine oxidase reaction is coupled, via 1-methoxy-phenazine methosulfate (Me-PMS), with a diazonium salt, 3-(4',5'-dimethyl-2-thiazolyl)-2,4-diphenyl-2H-tetrazolium bromide (MTT), to give a formazan which absorbs at 578 nm. The reagent is 0.15 M Tris·HCl buffer, pH 7.5, containing 50 IU/L purine nucleoside phosphorylase, 6 IU/L xanthine oxidase, 0.45 mM inosine, 30 μM Me-PMS, 0.6 mM MTT, and 3 g/L Triton X-100. For the assay of P_i 20 μl of sample (0 to 5 mM in P_i) is added to 2 ml of reagent, and after 15 min standing at room temperature the absorbance at 578 nm is read. The sensitivity is similar to the chemical determinations; 1 μg P_i gives an absorbance of 0.15. The author recommends that new batches of xanthine oxidase and the phosphorylase be dialyzed overnight against distilled water to remove exogenous phosphate and ammonium sulfate before use.

Continuous assay (DeGroot et al. 1985). The XO reagent contains 104 mM KCl, 50 mM Tris·HCl pH 7.4, 1 mM inosine, 500 IU/L nucleoside phosphorylase, 50 IU/L xanthine oxidase, and is saturated with O_2. To use as a coupled-enzyme assay the XO reagent is made twice as concentrated as given, then the actual assay cuvette contains 1.5 ml of the desired phosphate ester substrate plus the phosphatase enzyme being investigated, and 1.5 ml of the double strength XO reagent. The appearance of uric acid is monitored by following the increase of absorbance at 302 nm; the stoichiometry is one mole P_i yields one mole uric acid. A 1 mM solution of uric acid has an absorbance of 8.1 at this wavelength. This is only about one-fourth as sensitive as the fixed-time determination of Fossati, but has the advantage of being applicable in a continuous assay mode.

The sensitivity of this method may be more than doubled by including 1 mM 2-p-iodophenyl-3-p-nitrophenyl-5-phenyl tetrazolium chloride (INT) in the reagent mixture (2 mM INT in the double strength reagent). One mole of P_i contributes to the formation of two moles of formazan from this diazo salt, and a 1 mM solution of the formazan has an absorbance at 546 nm of 9.2.

The sensitivity of the various P_i determinations may be compared as follows. If the concentration of P_i in the color development mixture is 0.01 mM, then the chemical method of Chen et al. (1956) gives a reading at 820 nm of 0.260; the enzyme/MTT method of Fossati (1985) gives a reading at 578 nm of 0.300; the enzyme/uric acid method of DeGroot et al. (1985) gives a reading at 302 nm of 0.081; and the enzyme/INT method of DeGroot et al. (1985) gives a reading at 546 nm of 0.184. The reagents for the chemical method are the least expensive, the method of Fossati is the most sensitive, and the methods of DeGroot et al. are applicable in a continuous assay. You should choose the methodology which best fits your particular needs.

Assays

Most published phosphatase assays are fixed-time assays, because they are based on quantitation of P_i. I will describe three particular assays, just to indicate the general approach to this topic. It is easy to adapt one of these to your particular enzyme and specificity, using the principles already discussed in this chapter.

p-Nitrophenyl Phosphate Assay. Rossi et al. (1981) followed the hydrolysis of pNPP as part of their investigation of acid phosphatase from germinating maize. To 1 ml of 0.2 M acetate buffer pH 5.4 containing 12 mM pNPP they added 1 ml of enzyme preparation (in water or dilute neutral buffer). After incubating at 37°C for a period of time the reaction was stopped by adding 1 ml 1 N NaOH. The concentration of p-nitrophenoxide was measured by absorbance at 405 nm. Unlike p-nitrophenyl carboxylate esters the phosphate ester is quite stable at basic pH, so no particular blank hydrolysis is seen due to the NaOH. The specific absorption of p-nitrophenoxide is $18.3 \times 10^3 \text{ M}^{-1} \text{ cm}^{-1}$.

Organic Phosphate Assay.   The same authors also examined the activity of the enzyme on glucose-6-phosphate. To 1 ml 0.2 M maleate buffer pH 6.7 containing 60 mM G-6-P, 8 mM KF, and 8 mM EDTA was added 1 ml of enzyme preparation. After incubating at 37°C for 10 to 30 min the reaction was stopped by the addition of 1 ml cold 10% trichloroacetic acid. The authors used an adaptation of the Fiske-Subbarow method for quantitating the P_i formed, but the chemical method of Chen et al. would be preferable.

Phytase Assay.   An example of an assay for phytase is given in the work of Singh and Sedeh (1979). To 2 ml of 3 mM sodium phytate is added 2 ml 0.1 M acetate buffer pH 5.4 and 1 ml enzyme preparation. After incubation for 2.5 hr at 45°C, 3 ml 10% trichloroacetic acid is added to stop the reaction. They used the Fiske-Subbarow method for determining the amount of P_i formed, but the method of Cooper and Gowing (1983) would be better. The much higher sensitivity of this method would markedly decrease the long incubation times necessary with the rather low enzyme activity levels found by Singh and Sedeh.

11   Oxidoreductases

A large number of enzymes catalyze various reactions which are generally considered as "oxidation-reduction" reactions. In some cases the reaction may actually involve addition of an oxygen atom or molecule to the substrate; in other cases electrons are transferred between a donor substrate and an acceptor cofactor. Although there are many of these enzymes most of them are involved with intermediary metabolism in living organisms and their role in foodstuffs is slight. However there are a few of particular interest to food scientists, and they are the topic of this chapter. Polyphenoloxidase causes enzymatic browning in many foods. Ascorbic acid oxidase influences the nutritional role of Vitamin C. Lipoxygenase plays a part in oxidative rancidity of unsaturated fats. Several dehydrogenases (alcohol, lactic, glucose-6-phosphate) have analytical applications. Peroxidase is used as an indicator in food blanching. In sum, while hydrolytic enzymes are better-known to food technologists, oxidoreductases are also important in our profession.

OXIDASES, OXYGENASES

Polyphenoloxidase (PPO)

This group of enzymes includes tyrosinase from mushrooms, laccase from poison ivy, and phenol oxidases from most plants. An excellent review of the subject is the one by Mayer and Harel (1979), in which they differentiate the types of activity.

$$\text{Cresolase: Phenol} + \tfrac{1}{2} O_2 \rightarrow \text{Catechol (1,2-dihydroxy benzene)}$$

$$\text{Catecholase: Catechol} + \tfrac{1}{2} O_2 \rightarrow \text{o-benzoquinone} + H_2O$$

$$\text{Laccase: 1,4-dihydroxy benzene} + \tfrac{1}{2} O_2 \rightarrow \text{p-benzoquinone} + H_2O$$

While tyrosinase will show both cresolase (monophenolase) and catecholase (diphenolase) activity, some diphenolases have been isolated which do not have activity on any monohydroxy substrates tried. Laccase does not act on monophenols, and oxidizes only the p-dihydroxy phenols.

The monophenols most often used are tyrosine (p-hydroxy phenylalanine), p-coumaric acid (4-hydroxy cinnamic acid), or p-cresol (p-hydroxy toluene). The corresponding diphenolic substrates are DOPA (dihydroxy phenylalanine), caffeic acid (3,4-dihydroxy cinnamic acid), and 4-methyl catechol (3,4-dihydroxy toluene). Protocatechuic (3,4-dihydroxy benzoic) acid and chlorogenic acid (caffeic acid ester of 1-carboxy-1,3,4,5-tetrahydroxy cyclohexane) are also sometimes tested when a PPO is being characterized.

The light absorption maximum for the quinones derived from these different substrates varies from 390 nm to 480 nm. PPO specificity is often tested with several substrates, monitoring all reactions by the increase in absorbance at 430 nm, a wavelength at which all the quinones will give some absorption. The mistake sometimes seen in the literature is that unit activity is defined in terms of "ΔA/min," then substrates are compared on this basis. Protocatechuic acid (absorption maximum 430 nm) and L-DOPA (absorption maximum 475 nm) may react at exactly the same rate in terms of μmoles/min, but L-DOPA would show a much smaller rate of reaction in terms of absorbance change at 430 nm. In making such specificity studies the specific molar absorptivity should be determined at the absorption maximum for each substrate, and reaction rates should be reported in terms of moles/sec (katals, or, more conveniently, in nanokatals). Reported absorption maxima for the quinones from four diphenols are: 4-methyl catechol, 395 nm; protocatechuic acid, 430 nm; L-DOPA, 475 nm; and caffeic acid, 395 nm (Interesse et al. 1980).

A straightforward spectrophotometric assay for PPO activity is given by Boyer (1977). The buffer is 0.1 M phosphate pH 6.8, and substrate is L-DOPA, 2 mg/ml in buffer. To 2.4 ml of buffer at 25°C is added 0.5 ml substrate followed by 0.1 ml of enzyme. The absorption is monitored at 475 nm. The rate of absorbance change is converted using the molar absorptivity value for the quinone of 5×10^3 M^{-1} cm^{-1}.

Keyes and Semersky (1972) differentiated the cresolase and catecholase activities of mushroom tyrosinase with the substrates p-cresol and 4-methyl catechol, taking advantage of the fact that in the conversion of 4-methyl catechol to its quinone the absorptivity of these two molecular species is the same at 291 nm, but the hydroxylation of p-cresol to 4-methyl catechol produces a change at this wavelength. To measure cresolase activity, place in a cuvette: 0.75 ml of 0.05 M acetate buffer pH 4.8, 0.2 ml 1 mM p-cresol, 0.05 ml 1 mM 4-methyl catechol, and finally 1 to 5μl of the enzyme solution. The rate of absorbance change at 291 nm is converted using an absorptivity difference value of 690 M^{-1} cm^{-1} to express activity in moles/min. For catecholase activity, the reaction mixture is 0.8 ml of acetate buffer plus 0.2 ml 1 mM 4-methyl catechol plus 1 to $5\mu l$ enzyme. The decrease in absorbance at 280 nm is monitored. This plot curves rather quickly, so the tangent at zero time is required (see the initial rate measurement section in Chapter 6 for ways to determine this). The molar rate of change is found using $\Delta\epsilon = 1.12 \times 10^3 M^{-1} cm^{-1}$.

A similar idea was used by Carmona et al. (1979) employing p-coumaric acid and caffeic acid as the substrates. To 2 ml of 0.01 M phosphate buffer pH 7.0 containing 0.3 mM p-coumaric acid and 0.015 mM caffeic acid is added 30 mU enzyme and the rate of absorbance change at 334 nm is recorded. The absorptivity difference for this hydroxylation is 4.4×10^3 $M^{-1} cm^{-1}$, a sensitivity increase of 6-fold over the p-cresol reaction. For the catecholase reaction, the 2 ml of buffer contains only 0.2 mM caffeic acid and 30 mU enzyme, and the absorbance change at 310 nm is recorded. The authors did not give the molar absorptivity for this reaction.

Mazzocco and Pifferi (1976) increased the sensitivity of the catecholase reaction 20-fold by making a colored adduct of the quinone with a hydrazone, 3-methyl-2-benzothiazolone hydrazone hydrochloride (MBTH), also known as Besthorn's hydrazone. The reaction comprises 1 ml of 0.4 M citrate-phosphate buffer pH 4.2, 0.5 ml of a 0.5 M aqueous catechol solution, 0.5 ml of 0.3% MBTH in ethanol, and 0.5 ml of enzyme solution. After 15 sec 0.5 ml of 5% H_2SO_4 is added followed by 3 ml of acetone, and the absorbance of the mixture is read at 495 nm. The absorptivity of the colored adduct is 20.16×10^3 M^{-1} cm^{-1}, and 1 μmole of the quinone in the 6 ml final volume gives an absorbance reading of 3.36.

Quinones also form adducts with 2-nitro-5-thiobenzoic acid (TNB), but in this case the strong yellow color of the TNB disappears (Esterbauer et al. 1977). The color reagent is made by reducing the disulfide (DTNB, Ellman's Reagent, 20 mg in 10 ml water) with 30 mg $NaBH_4$. After 1 hr at room temperature the reaction is complete, giving a solution of 0.1 mM TNB which is stored at 4°C until used. For the assay, 1 ml of 0.3 M citrate-phosphate buffer pH 6.0 is added to 1.5 ml of TNB solution, followed by 0.3 ml of 20 mM 4-methyl catechol (for PPO activity) or 20 mM p-dihydroxybenzene (for laccase activity). After warming to 25°C 0.2 ml of enzyme solution is added and the rate of decrease of absorbance at 412 nm is recorded. For the PPO reaction the absorptivity change is 11.0×10^3 M^{-1} cm^{-1}, while for the laccase reaction the change is slightly greater, 12.4×10^3 M^{-1} cm^{-1}. The sensitivity is only half of that with MBTH, but the method has the advantage of being a continuous assay.

These reactions may also be followed by monitoring the disappearance of one of the substrates, namely O_2. This is most conveniently done using the oxygen gas electrode. This has the advantage of being applicable in situations where turbidity makes spectrophotometric measurements difficult if not impossible. As an example, Lamkin et al. (1981) used the gas electrode to assay for catecholase activity in a number of wheat flours. To 3 ml of buffer (0.1 M phosphate, 0.037 M diethylbarbiturate, pH 8.3, saturated with air) in the reaction cuvette thermostatted at 37°C was added (with stirring) 0.2 g ground wheat. The slow background consumption of O_2 was recorded for 10 min, then 100 μl of substrate (e.g., 7.5 mM catechol in water) was added, and the increased rate of O_2 consumption was recorded for another 5 min. The rate in percent of O_2 saturation/min is converted to moles/min using the known solubility of oxygen at 37°C. This methodology is generally applicable to oxidase/oxygenase enzymes and is a vast improvement over the use of the Warburg manometer referred to in some of the older literature on these enzymes.

Lipoxygenase

This is an intriguing enzyme for a number of reasons. Its primary reaction is the oxygenation of polyunsaturated fatty acids to produce a hydroperoxy conjugated diene product. The native enzyme is catalytically inactive and is activated by reaction with the hydroperoxy product, a requirement which gives a lag period in a simple reaction mixture. Even in the absence of oxygen, the enzyme will catalyze other reactions between the fatty acid and the hydroperoxy acid, giving an oxodiene product, fatty acid dimers, and inactivated enzyme. This is not the place to discuss these characteristics, but if you are beginning to work with lipoxygenase three excellent review articles should be consulted (Axelrod et al. 1981; Veldink et al. 1977; Verhagen et al. 1978).

Linoleic acid is the substrate most often used for assaying lipoxygenase, although other polyunsaturated fatty acids such as linolenic or arachidonic also react. Upon oxygenation, the product is 13-hydroperoxy-9,11-octadecadienoic acid (or 9-hydroperoxy-10,12 dienoic acid). Lipoxygenase is a metalloenzyme with an iron atom as the prosthetic group. In the native inactive enzyme this is present as Fe^{2+}; activation occurs upon oxidation to Fe^{3+} by reaction with the hydroperoxy fatty acid. If native enzyme is added to a solution of pure linoleic acid the activation depends upon traces of autoxidized fatty acid and a lag period in the reaction is seen. This lag period is overcome by the addition of a small amount of hydroperoxy acid to the substrate solution. The synthesis is as follows (Gibian and Galaway 1976). To 1 L 0.2 M borate buffer pH 9.0 is added 56 mg (200 μmoles) linoleic acid dissolved in 38 ml ethanol. While stirring at room temperature, 4 portions, each of 10 mg, of crystalline lipoxygenase are added over a 2 hr period. The reaction mixture is stirred overnight at 4°C, then acidified to pH 1.5 and extracted with 3 portions, each of 300 ml, of ether. The extracts are dried with $MgSO_4$, the ether is removed, and the hydroperoxy product is dissolved in 25 ml ethanol and stored at 5°C. The concentration of this stock solution may be determined by measuring the absorbance at 235 nm of a diluted aliquot ($\epsilon = 23 \times 10^3$ M^{-1} cm^{-1}, Gibian and Vandenberg 1987), and should be around 8 mM.

For the assay (Gibian and Galaway 1976) the substrate stock solution contains 168 mg linoleic acid plus 6 ml of hydroperoxy acid stock solution made to 100 ml in ethanol. To 2.9 ml 0.2 M borate buffer pH 9.0 in a 3 ml cuvette is added $50\,\mu l$ of substrate stock solution. After this is thermostatted in the spectrophotometer compartment, 50 μl enzyme solution is added to start the reaction. The increase in absorbance at 235 nm is recorded. Reported values for the molar absorptivity of the hydroperoxy dienoic acid have ranged from 20.5×10^3 to 28×10^3. Gibian and Vandenberg (1987) did an extensive study and concluded that the best value was 23,000 $\pm$ 580 M^{-1} cm^{-1}. This number should be used to convert rate of absorbance change to rate of change of the concentration of the hydroperoxy acid product.

If the anaerobic reaction of lipoxygenase is of interest, the paper by Verhagen et al. (1978) should be consulted. They showed that the enzyme catalyzes a double reaction:

$$Enz(Fe^{3+}) + LH \rightarrow Enz(Fe^{2+}) + H^+ + L^\bullet$$

$$Enz(Fe^{2+}) + LOOH \rightarrow Enz(Fe^{3+}) + OH^- + LO^\bullet$$

Two of the linoleyl free radicals $L^\bullet$ combine to form a fatty acid dimer. The peroxy free radical $LO^\bullet$ disproportionates; half goes to form an oxodienoic acid, while the other half dimerizes and/or forms a number of molecular breakdown products. The appearance of the oxodienoic acid is monitored by measuring absorbance at 285 nm. A typical sort of reaction mixture might contain equal amounts of linoleic acid and hydroperoxylinoleic acid (say, 50 μM for each) in deaerated 0.1 M borate buffer, pH 10.0. The reaction is started by the addition of enzyme.

The oxygen gas electrode may also be used to monitor the progress of a lipoxygenase reaction. When the buffer is well-saturated with air (O_2 concentration is 0.26 mM) and the linoleate concentration is less than 0.2 mM, the rate as judged by increasing absorbance at 235 nm corresponds with the rate determined by uptake of oxygen, for the lipoxygenase-1 isoenzyme from soybean (Axelrod et al. 1981). However, if the oxygen concentration falls below 0.025 mM, or if a different lipoxygenase enzyme is being studied, the anaerobic reaction may come into play and the two measures of enzyme activity will not correlate. If you are using both methods of assay for lipoxygenase, be aware that this may be the reason for discrepancies which you might find (assuming you have ruled out instrument errors).

Ascorbic Acid Oxidase (AAO)

This copper metalloenzyme is apparently related to laccase, and has been found in a large number of plant foodstuffs. It is of interest because of its possible influence upon natural Vitamin C, and also because of its possible involvement in baking. A good review of the enzyme is available (Kroneck et al. 1982).

A simple spectrophotometric assay for AAO is based upon the fact that ascorbate absorbs at 265 nm while the oxidation product, dehydroascorbate, does not (Racker 1952; Tono and Fujita 1982). Surprisingly, this has not been a popular means of assaying for this enzyme. The assay as described by Racker is uncomplicated: to 2.8 ml of buffer (0.025 M citrate, 0.05 M phosphate, pH 5.6) is added 0.1 ml of substrate solution (0.05% ascorbate, 1% disodium EDTA, neutralized), 0.05 ml of 1% bovine serum albumin solution, and finally 0.05 ml enzyme solution. The rate of disappearance of the absorbance at 265 nm may be converted to molar terms using the molar absorptivity for ascorbate (at pH 4.6) of $10.0 \times 10^3 \ M^{-1} \ cm^{-1}$ (Chance 1949). Tono and Fujita (1982) checked this spectrophotometric assay against various chemically-based assays involving measurement of ascorbate with indophenol or dehydroascorbate with 2,4-dinitrophenylhydrazine and found excellent agreement.

AAO may also be assayed using the oxygen gas electrode (Dawson and Magee 1955; Strothkamp and Dawson 1977). The reaction mixture contains 2 ml 0.2 M citrate-phosphate buffer pH 5.7, 0.5 ml 0.5% gelatin, 1.5 ml water, and 0.5 ml 14 mM ascorbic acid (250 mg in 50 ml water containing 50 mg phosphoric acid). After equilibrating at 25°C in the gas electrode reaction vessel, 0.5 ml of enzyme solution is added to begin the reaction. The rate of disappearance of oxygen is monitored and converted to molar terms as usual.

Fixed-time assays have been published in which the reaction of AAO with ascorbate is terminated by the addition of acid and the concentration of substrate remaining is determined chemically using 2,6-dichlorophenol indophenol (DCPIP) (Grant 1974; Pfeilsticker and Roeung 1980; Armstrong 1964). These might have application when the matrix is such that neither gas electrode nor spectrophotometric methods will work, or the level of enzyme activity is so low that very long incubation times are required to get a measurable response. Except for such extreme cases these assays are not recommended.

Sugar Oxidases

Glucose oxidase and similar sugar oxidases give, as products, hexonic acid and hydrogen peroxide. A large number of kits are commercially available for measuring e.g., glucose by coupling the glucose oxidase reaction with a peroxidase reaction and measuring the amount of oxidized dye formed. The same chemistry may be used for a glucose oxidase assay. A difficulty is that for most of the acceptor species normally used (e.g., o-dianisidine, guaiacol, etc.) the molar absorptivity of the oxidized dye is not known, so an absorbance change can only be related to a molarity by reference to a standard curve. There are a few exceptions; in the extremely sensitive peroxidase assay described by Ngo and Lenhoff (1980) the molar absorptivity of the indamine dye formed is known to be 47.6×10^3 M^{-1} cm^{-1} (see the peroxidase discussion in this chapter). If this system were used as the auxiliary enzyme part, a sensitive glucose oxidase assay could be designed.

In addition to O_2, glucose oxidase can also use o-benzoquinone as the hydrogen acceptor. Ciucu and Patroescu (1984) based a direct spectrophotometric assay on this principle. Their stock solutions were 0.1 M citrate pH 5.0, 0.1% benzoquinone in water, and a 1 M solution of glucose which was allowed to stand overnight at room temperature to ensure anomer equilibration. With all solutions at 25°C, 2 ml of glucose, 1 ml of benzoquinone, and 0.9 ml of buffer were mixed. After the addition of 0.05 ml enzyme the increase in absorbance at 290 nm is recorded for 2 min. The rate may be related to the increase in concentration of hydroquinone through the molar absorptivity of 2.31×10^3 M^{-1} cm^{-1}.

Avigad (1978) coupled the oxidation of galactose by galactose oxidase to the oxidation of NADH by NADH-peroxidase. In this case the oxidation of one nmole/ml of galactose results in the oxidation of one nmole/ml NADH, with a resultant change in absorbance at 340 nm of 0.00622. The auxiliary enzyme NADH peroxidase is assayed in a reaction mixture of 0.06 M phosphate buffer pH 7.0 containing 60 mM sodium acetate, 0.14 mM NADH, and 25 mM H_2O_2. After adding enzyme the decrease in absorbance at 340 nm is recorded. For the galactose oxidase assay the reaction mixture was 0.06 M phosphate, pH 7.0, 60 mM acetate, 0.14 mM NADH, 40 mM galactose, and 0.05 to 0.08 units NADH peroxidase per ml. After adding an appropriate amount of galactose oxidase the decrease in absorbance at 340 nm was recorded. This general system could be readily used for other oxidases which yield H_2O_2 as a reaction product.

DEHYDROGENASES

Dehydrogenases catalyze the oxidation-reduction reaction of a substrate and a cofactor, usually $NAD(P)^+$ or the reduced cofactor NAD(P)H. The reaction is reversible, and may be written, in general, as follows:

$$AH_2 + NAD(P)^+ \rightleftharpoons A + H^+ + NAD(P)H$$

At a wavelength of 340 nm $NAD(P)^+$ does not absorb light, while NAD(P)H has a molar absorptivity of 6.22×10^3 M^{-1} cm^{-1}. This property is often used to monitor the progress of the reaction. NAD(P)H can reduce a number of chromogenic molecules to form dyes which typically have absorptivities 3-fold greater than that of NAD(P)H, and with a wavelength maximum in the visible light range. This reduction is usually mediated either by phenazine methosulfate (PMS) or the enzyme diaphorase and has been used in designing several dehydrogenase assays to increase sensitivity and convenience.

Lactate Dehydrogenase (LDH)

Direct Methods. This enzyme catalyzes the reversible oxidation of L(+)lactate to pyruvate. The reaction equation is:

$$CH_3CH(OH)COO^- + NAD^+ \overset{LDH}{\rightleftharpoons} CH_3C(=O)COO^- + NADH + H^+$$

Since the reaction is reversible it will eventually come to an equilibrium state containing all four species. The equilibrium expression is (Howell et al. 1979):

$$\frac{[NAD^+][Lac]}{[NADH][Pyr][H^+]} = 2.7 \times 10^{11}$$

This may also be expressed in terms of the rate constants for the forward and backward reactions:

$$\frac{k_{pyr}}{k_{lac}} = 2.7 \times 10^{11}[H^+]$$

from which we see that while this ratio is pH dependent, the enzymatic rate constant at pH 7.4 for the enzymatic reduction of pyruvate is four orders of magnitude greater than the rate constant for the oxidation of lactate. Howell et al. (1979) pointed out that the pyruvate reduction reaction should be much more sensitive than the lactate oxidation reaction simply because the rate of change of absorbance at 340 nm will be greater for the same amount of enzyme. The concentrations of pyruvate and NADH may be each an order of magnitude or more smaller than concentrations of lactate and NAD^+ and the rate may still be greater. They demonstrated this with the following comparison experiment. Using 72 mM L(+)lactate plus 5.5 mM NAD +, or 1 mM pyruvate plus 0.15 mM NADH at pH 7.4, the same amount of LDH gave an initial absorbance increase of 0.01/min for the forward reaction, and a decrease of 0.1/min for the reverse reaction. If the reaction is monitored by the change in absorbance at 340 nm it would be preferable to follow the reduction of pyruvate. The amounts of substrate chemicals needed are much less, high quality NADH is now readily available (a situation which did not obtain some 20 years ago), and the sensitivity is superior.

This reaction is also amenable to the UV-absorptiostat methodology of Pantel and Weisz (1979) (see the direct methods section of Chapter 7). They used a 0.05 M phosphate buffer pH 7.5 containing 1.2 mM pyruvate (13.2 mg sodium pyruvate/100 ml buffer). For the titrating reagent they used 1 mg/ml NADH in the buffer, or a concentration of 1.33 mM. Addition of 0.5 ml of this to 7.0 ml of the buffer/pyruvate gives a concentration of 89 μM NADH and an absorbance at 340 nm of 0.55. The enzyme is added to start the reaction and the absorbance is kept constant by the addition of small increments of the titrating reagent. The initial amount of NADH in the reaction cell may be varied as may pyruvate concentration if you wish. The sensitivity is easily as good as the spectrophotometric assay. A rate of change in absorbance (in a 3 ml reaction volume) of 0.1/min corresponds to oxidation of 48 μmoles of NADH/min. At a concentration of 1.33 mM NADH titrant is added at a rate of 36 μl/min to maintain a constant NADH concentration. If the titrant syringe has a volume of 0.5 ml 36 μl/min corresponds to an addition rate of 7% of total syringe volume per min, a rate which is experimentally accurate and easy to work with.

Finally, it should be noted that the reaction could be run in a pH-stat in either direction. If oxidation of lactate were followed then base would be added to counteract the [H$^+$] generated; if pyruvate reduction were followed then acid would be added to keep the pH constant. I have not found any literature reports on this method of assaying LDH, but it may be useful in certain laboratory situation.

Colorimetric Methods.   NADH will react with the dye 2-(4-iodophenyl)-3-(4-nitrophenyl)-5-phenyltetrazolium chloride hydrate (INT) to give a highly colored formazan INTH which has a molar absorptivity of 19.3 $\times$ 10^3 M^{-1} cm^{-1} at 500 nm. This reaction is catalyzed by phenazine methosulfate (PMS) (Babson and Babson 1973), meldola blue (Burd and Usategui-Gomez 1973) or the enzyme diaphorase (Allain et al. 1973). Applying this chemistry increases the sensitivity of the assay 3-fold as well as making it amenable to spectrophotometry in the visible spectrum.

The PMS and diaphorase catalyzed assays are used in the continuous recording mode. For the first method (Babson and Babson 1973) the substrate is 50 mM lactate in 0.2 M Tris buffer pH 8.2. The color reagent contains 40 mg INT, 100 mg NAD$^+$ and 10 mg PMS in 20 ml water, stored in the dark at 4°C. For the assay 0.1 ml enzyme is added to 1 ml substrate and this is brought to 30°C. Then 0.2 ml of the color reagent is added and absorbance at 503 nm is followed for about 2 min. (When the enzyme source is serum, a blank reaction without lactate gives an absorbance change of around 0.02/min which must be substracted from the complete rate.) The diaphorase assay medium (Allain et al. 1973) contains 104 mM DL lactate, 3 mM NAD^+, 100 mM glycine, 12 mM Na_2CO_3, 0.6 mM INT, 152 mM KCl, 83.3 IU/l diaphorase, pH 8.55. At 37°C to 3 ml of this medium is added 20 μl enzyme and absorbance is followed at 500 nm.

PMS tends to be somewhat unstable so Burd and Usategui-Gomez (1973) suggested the used of meldola blue (9-methylaminobenzo-α-phenazonium chloride) as a more stable electron transducer. They describe a fixed time assay for LDH. The substrate is 0.2 M L(+)lactate in 0.4 M Tris pH 8.2. The color reagent is 50 mg INT, 100 mg NAD^+ and 12.5 mg meldola blue in 20 ml water, sonicated for 1 min and stored in the dark. For the assay 0.05 ml enzyme is added to 0.25 ml substrate at 37°C. After 20 sec 0.2 ml color reagent is added and incubation is continued for 12 min, at which time 5 ml 0.1 N HCl is added and the absorbance at 510 nm is read. A blank reaction uses 0.05 ml of 0.2% potassium oxalate plus 0.2% EDTA in buffer in place of the enzyme.

The water insolubility of INTH formazan sometimes causes problems, so Shiga et al. (1984) synthesized a dye 2-phenyl-3-(4,5-dimethyl-2-thiazolyl)-5-(4-pyridyl) tetrazolium bromide (PDTPT) which gives a water soluble formazan with a molar absorptivity of 18.3×10^3 M^{-1} cm^{-1} at 539 nm. For their assay, to 1 ml phosphate buffer pH 8.0 is added 100 μl 0.1 M lactate, 100 μl 5 mM PDTPT, 100 μl 0.25 mM PMS and 100 μl NAD^+. After equilibration at 37°C the LDH solution is added, the reaction is held 10 min at 37°C and stopped by the addition of 3 ml 0.1 N HCl. The absorbance at 539 nm is read.

Fluorometric Methods. Guilbault (1975) describes a fluorometric adaptation of this scheme in which resazurin is reduced by NADH to give the fluorochrome resorufin (excite at 560 nm, emit at 580 nm). The assay mixture is the same as those given above except that the 3 ml reaction volume contains 0.1 ml 0.2 mM resazurin in place of INT. Either PMS or diaphorase will catalyze the reaction between NADH and resazurin. The sensitivity is about 100-fold greater than for the colorimetric reaction, but because the excitation and emission wavelengths are so close the spectrofluorimeter must be well-designed to rigorously exclude excitation light from the detection photometer.

The cofactors involved in dehydrogenase activity can also be measured fluorometrically. NADH is naturally fluorescent (excite at 365 nm, emit at 400 nm) and using this methodology it can be continuously measured in the 10^{-5} to 10^{-7} M range of concentrations (Lowry et al. 1957). If the assay is run as a fixed time assay, then by further chemical reaction this sensitivity range can be extended to the nanomolar level (Lowry et al. 1957; Bergmeyer 1974). The reaction depends upon the formation of a highly fluorescent derivative of NAD^+ by treatment with alkali. To determine NAD^+, the NADH present at the termination of the enzyme reaction is destroyed by treatment with 0.2 N HCl for 20 min at room temperature. Then concentrated NaOH is added to make the mixture 6 M; after 30 min at 37°C the fluorescence is measured (excite 365 nm, emit 470 nm). If the concentration of NADH is to be determined, the NAD^+ present is destroyed by treatment with 0.05 N NaOH for 15 min at 50°C. Then the mixture is made 0.01% in H_2O_2 to oxidize the NADH to NAD^+, it is brought to 6 M in NaOH and held at 37°C for 30 min to develop the fluorescence. While these operations are rather more complex than most assays recommended in this chapter, they do provide a means for developing ultra-sensitive dehydrogenase assays.

Cytochrome b_2. Ordinary lactate dehydrogenase requires NAD^+ as its electron acceptor, but the enzyme from bakers yeast has a much broader acceptor specificity (Appleby and Morton 1959). A commonly used acceptor is potassium ferricyanide which has a molar absorptivity of 1.04×10^3 M^{-1} cm^{-1} at 420 nm. The assay is as follows. Mix 0.45 ml 0.2 M pyrophosphate buffer pH 8.0, 1 ml 1 M DL lactate, 1 ml 3 mM EDTA, and 0.5 ml 5 mM $K_3Fe(CN)_6$. After temperature equilibration add 50 μl enzyme and record the decrease in absorbance at 420 nm. Cytochrome b_2 will also decolorize such dyes as methylene blue and 2,6-dichlorophenol indophenol (DCPIP). The latter has a molar absorptivity of 22.0×10^3 M^{-1} cm^{-1} (Armstrong 1964) at 600 nm and would give a 20-fold sensitivity increase relative to the use of ferricyanide.

Alcohol Dehydrogenase (ADH)

This enzyme catalyzes the reversible NAD^+-dependent oxidation of ethanol to acetaldehyde:

$$\text{CH}_3\text{CH}_2\text{OH} + \text{NAD}^+ \overset{\text{ADH}}{\rightleftharpoons} \text{CH}_3\text{CH}{=}\text{O} + \text{NADH} + \text{H}^+$$

As with LDH the equilibrium point lies towards the left side of this equation and is pH dependent. A straightforward assay was described by Vallee and Hoch (1955). Buffer is 0.032 M phosphate pH 8.8. Substrate is 2 M ethanol, 12.12 ml 95% ethanol diluted to 100 ml with water. The cofactor stock is 25 mM NAD^+, 167 mg/10 ml in water. For the assay combine 1.5 ml buffer, 0.5 ml substrate, 1.0 ml cofactor and 0.2 ml enzyme solution. Measure the rate of increase in absorbance at 340 nm.

Skursky et al. (1979) increased the sensitivity of this assay 6-fold by using the ability of ADH to catalyze the reduction of p-nitrosodimethylaminiline

(NDMA) by NADH. NDMA has a molar absorptivity at 440 nm of 35.4 $\times 10^3$ M^{-1} cm^{-1}, while the reduced form is colorless at this wavelength. The NADH used to reduce NDMA is regenerated by reaction with alcohol:

$$NDMA + NADH \xrightarrow{\text{ADH}} NAD^+ + \text{reduced NDMA}$$

$$NAD^+ + \text{alcohol} \xrightarrow{\text{ADH}} NADH + \text{aldehyde}$$

The buffer is 0.1 M phosphate pH 8.5 containing 26 μM NDMA. To 1.9 ml of this is added 0.5 ml enzyme and the slow (blank) decrease in absorbance is recorded. Then 0.1 ml buffer containing 0.25 mM n-butanol and 5 mM NAD^+ is added and the rapid decrease in absorbance at 440 nm is measured. This rate minus the blank rate is the actual rate of enzyme-catalyzed NDMA reduction. The authors found that a rate of absorbance change of 0.005/min corresponded to about 1 IU/L of ADH in the enzyme sample (0.5 milliunits in the actual assay cuvette). Unfortunately, as reported, the response was not linear with respect to the amount of enzyme used.

Many of the same chemical reactions applied to LDH for amplification of the spectrophotometric absorbance could be applied to ADH if desired. The specificity of ADH is fairly broad, and a number of n-alkanols will serve as substrates. The possibilities for designing sensitive assays to fit particular needs are boundless.

Carbohydrate Dehydrogenases

Under this broad heading I will discuss three dehydrogenases of substrates which are of interest to food scientists: glucose, glycerol, and glucose-6-phosphate. The first two are NAD^+-dependent enzymes, while the last one requires $NADP^+$ as its cofactor.

Glucose dehydrogenase yields, upon oxidation of glucose, glucono-δ-lactone and NADH. An assay has been given by Banauch et al. (1975). The substrate is 0.1 M Tris pH 7.5, 168 mM in glucose (dissolve 33 g crystalline glucose monohydrate in 1 L buffer and allow it to stand overnight to attain anomeric equilibrium). To 3 ml of this is added 50 μl of 90 mM NAD^+ followed by 100 μl enzyme solution. The increase in absorbance at 340 nm is recorded.

Glycerol dehydrogenase is most active in oxidizing glycerol but also reacts well with 1,2-propandiol or 2,3-butandiol. The enzyme is activated by anions, especially sulfate, hence ammonium sulfate is included in the assay mixture. The assay as described by Lin and Magasanik (1960) begins by incubating the enzyme sample for 1 to 2 min with 0.1 ml 10 mM NAD^+

and 0.1 ml 1 M $(NH_4)_2SO_4$. Then 0.6 ml of 0.5 M carbonate buffer pH 9.0 is added, plus water to 3 ml total volume. The reaction is started by the addition of 0.3 ml 1 M glycerol (92 g/L in water) and the increase in absorbance at 340 nm is monitored.

Glucose-6-phosphate dehydrogenase from most sources requires the phosphorylated cofactor $NADP^+$, although the enzyme from a few bacterial sources can also utilize NAD^+. The product of the oxidation is the 6-phosphate ester of glucono-δ-lactone. This is the first enzyme in the phosphogluconate pathway of glucose metabolism which leads to ribose-5-phosphate, an important molecule for the synthesis of nucleotides. In food science it is most often used as an analytical tool, either in glucose analysis or in enzymatic determination of inorganic phosphate. Its assay (Noltmann et al. 1961) is similar to that of other dehydrogenases. The reaction mixture comprises 2.5 ml 0.1 M glycylglycine buffer pH 8.0, 0.1 ml 30 mM glucose-6-phosphate, 0.1 ml 10 mM $NADP^+$ and 0.2 ml 0.15 M $MgSO_4$. The reaction is initiated by the addition of 0.1 ml enzyme solution; the rate of increase in absorbance at 340 nM is measured.

Diaphorase

The food science laboratory worker would probably only want to run an assay for diaphorase in order to standardize a solution of the enzyme for other applications (e.g., as an auxiliary enzyme for a dehydrogenase assay). An excellent assay is that of Kaplan et al. (1969) in which the reduction of DCPIP by NADH is catalyzed. The reaction mixture contains 0.1 M phosphate buffer pH 7.5, 1 mM EDTA, 0.3 mM NADH and 0.09 mM DCPIP. Upon addition of enzyme the rate of decrease of absorbance at 600 nm is recorded. The molar absorptivity of DCPIP caused some confusion in the earlier literature until Armstrong (1964) clarified its extinction characteristics. The dye ionizes with a pK of 5.90. The protonated form has an absorptivity (at 600 nm) of 2.7×10^3, while the ionized form absorptivity is 22.0×10^3. The isobestic point between the two forms is at 522 nm with a molar absorptivity of 8.6×10^3 M^{-1} cm^{-1}. At pH 7.5, 97.5% of the DCPIP present is in the ionized form so the calculated molar absorptivity at 600 nm is 21.5×10^3 M^{-1} cm^{-1}. The measured rate of absorbance change in the assay should be converted to units of enzyme using this factor.

"ACTIVE OXYGEN"ASES

For lack of a better title I have chosen the above to designate those enzymes which react with chemically reactive oxygen species: peroxidase, catalase and superoxide dismutase. Hydrogen peroxide and superoxide anion, the substrates for these enzymes, are formed during the action of a number of oxidases. In vivo these species are toxic, and organisms have evolved these enzymes as detoxification mechanisms. Xanthine oxidase forms O_2^- as one of the products of its oxidation of xanthine. Superoxide dismutase catalyzes the reaction of this species with hydrogen ion:

$$2\ O_2^- + 2\ H^+ \rightarrow H_2O_2 + O_2$$

Peroxidase and catalase further lower the chemical potential of the oxygen in hydrogen peroxide as follows:

$$AH_2 + H_2O_2 \rightarrow A + 2\ H_2O$$

In the case of peroxidase AH_2 may be any of a number of organic molecules, while for catalase it is a hydrogen peroxide molecule and A is molecular oxygen. A good review of superoxide dismutase is that by Fridovich (1975). Reviews of peroxidase and catalase are too numerous to mention, but two good background papers on assay methods are those by Maehly and Chance (1954) and Chance and Maehly (1955).

Peroxidase

A major thrust in work on peroxidase assays has been to find molecular species AH_2 which give rise to A with high molar absorptivities. The general assay recipe has not changed fundamentally since that given by Chance and Maehly (1955) (which is based on still older work). The reaction mixture is 0.01 M phosphate buffer pH 7.0 containing 0.74 mM H_2O_2 (25 mg/L) and 20 mM pyrogallol (2.5 g/L). After the addition of enzyme the increase in absorbance at 430 nm is followed, due to the formation of purpurogallin ($\epsilon = 2.47 \times 10^3$ M^{-1} cm^{-1}).

Marshall and Chism (1979) examined three commonly used AH_2's. They found that o-dianisidine did not saturate the enzyme even at a level of 1 mM, thus the results of an assay would depend markedly upon the exact concentration of this donor molecule in the assay mixture (i.e., small experimental errors in weighing out o-dianisidine could lead to significant assay errors). Guaiacol and pyrogallol both seemed to approach enzyme saturation at concentrations above 0.06 M and so would be better choices for an assay. Their assays were done in 0.5 mM phosphate buffer pH 6.0 containing 0.03% (8.8 mM) H_2O_2 and the concentration of donor to be tested, monitored at 460 nm. Bovaird et al. (1982) examined the effects of various experimental parameters on the peroxidase reaction using o-phenylenediamine as the substrate. Their optimized assay mixture contained 0.1 M phosphate buffer pH 5.0, 3 mM o-phenylenediamine and 3.2 mM H_2O_2. The absorbance at 435 nm was measured. They found that 5 picomoles of horseradish peroxidase gave a rate of absorbance change of about 0.04/min.

Probably the most sensitive colorimetric assay for peroxidase is that of Ngo and Lenhoff (1980) which uses two dye precursor molecules. The reactants, 3-methyl-2-benzothiazolinone hydrazone·HCl·H_2O (MBTH) and 3-dimethylaminobenzoic acid (DMAB), upon oxidation with H_2O_2, form an indamine dye which has a molar absorptivity at 590 nm of 47.6×10^3 $M^{-1} cm^{-1}$. The assay mixture is 0.1 M phosphate buffer pH 6.5 containing 10 mM H_2O_2 (340 mg/L), 3.3 mM DMAB (545 mg/L) and 0.07 mM MBTH (15 mg/L). The reaction is initiated by addition of peroxidase and the absorbance at 590 nm is followed. A difficulty with many of the dye precursors used for peroxidase assays is that the molar absorptivities are not well defined. The good definition of the indamine dye along with its high absorbance would seem to make this the chromogenic assay of choice.

Some fluorometric assays for peroxidase have also been published. One of the most sensitive ones is that reported by Iwai et al. (1983). The precursor molecules o-dianisidine and homovanillic acid upon oxidation give rise to a highly fluorescent molecule. The assay mixture contains 2.5 ml 0.1 M Tris pH 8.5, 0.3 M o-dianisidine·2 HCl (14.5 mg/L), 0.1 ml homovanillic acid (2.5 g/L) and 0.1 ml enzyme. After attemperating this at 30°C the reaction is started by adding 30 μl 10 mM H_2O_2. The increase in fluorescence is followed (excite at 315 nm, emit at 425 nm). Unfortunately, as with most fluorometric assays, it is difficult to relate changes in fluorescence directly to changes in concentration of the fluorescing species.

Catalase

Spectrophotometric Methods.   The kinetics of the catalase reaction are such that the disappearance of H_2O_2 is always first-order. Hydrogen peroxide has a broad absorption band in the ultraviolet with the peak below 200 nm. At 200 nm, $\epsilon = 120$ $M^{-1} cm^{-1}$; and at 250 nm, $\epsilon = 30$ $M^{-1} cm^{-1}$. The assay solution is 0.05 M phosphate buffer pH 7.0 containing 60 mM H_2O_2 (6 ml 30% H_2O_2 per liter). To 2 ml of this is added 1 ml diluted catalase and the absorbance is recorded. A plot of ln [absorbance] versus time is linear; from the slope calculate the first-order rate constant k' for the reaction. Beers and Sizer (1952) found that the catalytic rate constant $k_s (= k'/[E]) = 2 \times 10^7$ $M^{-1} sec^{-1}$ for the crystalline catalase they were using. Using this method the exact wavelength at which the measurement is made is irrelevant; changes in the molar absorptivity due to wavelength variations will just displace the plot vertically without affecting the slope.

Several assays given for commercial catalase preparations convert a rate of absorbance change to a rate of concentration change at a set wavelength. An error of 1 nm in the spectrophotometer wavelength setting will make a difference of 2% in ϵ. Furthermore, these assays assume, incorrectly, that the reaction is zero order and experimental absorbance changes can be directly expressed in units of moles per second. These two factors make these assays untenable and they are not recommended. If a slightly quicker two point assay is wanted, set up your buffer pH, H_2O_2 concentration, and spectrophotometer wavelength as convenient. After adding catalase to start the reaction monitor the absorbance and measure the time T it takes to change from A_1 to A_2. Then the first order rate constant $k' = (\ln [A_1/A_2])/T$. The experimental error in this method is greater than in the method of plotting several points (on a statistical basis) but it is theoretically just as sound.

A zero-order rate can be obtained using the UV-absorptiostat method (Pantel and Weisz 1979). The buffer was 25 mM phosphate pH 7.0, the titrating reagent was 0.15 M H_2O_2 (5 g/L), and the substrate concentration in the reaction mixture was 8 mM H_2O_2. At 250 nm, using the 2 cm path length described by the authors, this gives an absorbance of about 0.5 which is convenient and spectrophotometrically accurate.

Colorimetric Method. Hydrogen peroxide reduces dichromate ion in acetic acid to give chromic acetate which absorbs light at 570 nm (Sinha 1972). The color reagent consists of 1 volume of aqueous 5% potassium dichromate mixed with 3 volumes glacial acetic. A stock 0.2 M H_2O_2 solution is made by diluting 23 ml 30% H_2O_2 to 1 L with water. The enzyme reaction mixture is 4 ml stock substrate, 5 ml 0.01 M phosphate buffer pH 7.0 and 1 ml enzyme. At intervals 1 ml aliquots of this are taken and added to 2 ml of color reagent. This is mixed, heated for 10 min in a boiling water bath, cooled, and the absorbance is read at 570 nm. A standard curve may be made with dilutions of H_2O_2 in buffer to ensure that the chromogenic reaction is linear with respect to hydrogen peroxide concentration. The enzymatic reaction is first order in H_2O_2 so the absorbances obtained at two times during the incubation are used to calculate the first-order rate constant: $k' = (\ln [A_1/A_2])/(T_1 - T_2)$. Kruger (1976) used this method to monitor catalase assays with an AutoAnalyzer and analyzed the data in the described manner.

Gas Electrode Method. Del Rio et al. (1977) describe a sensitive catalase assay using the Clark oxygen electrode to monitor the generation of O_2 during the reaction. In the reaction cell of the instrument they placed 2.95 ml of 0.05 M phosphate buffer pH 7.0 saturated with air. The recorder deflection was set to full scale. Then purified nitrogen gas was bubbled through the stirred, thermostatted (25°C) buffer until the recorder deflection was at a minimum; it was then set to zero on the scale. Next 100 μl 1 M H_2O_2 was added, and after a few minutes for temperature equilibration, 50 μl catalase solution was added. The rate of increase in O_2 saturation was recorded on the chart, and converted to molar concentrations by assuming that oxygen concentration was 0.25 mM in air-saturated buffer and 0.00 mM in the nitrogen-flushed buffer.

Superoxide Dismutase (SOD)

Superoxide anion O_2^- is the first species generated during the reduction of molecular oxygen by many intracellular oxidases, not all of which are defined. It will reduce molecules such as ferricytochrome C or nitro blue tetrazolium bromide (NBT), and will also oxidize NADH and other cofactors. These reactions are spontaneous; assays for SOD depend upon its inhibition of the rate of these other indicator reactions due to the lowering of O_2^- concentration. Three reactions are involved in a SOD assay: the reaction which generates superoxide anion; the indicator reaction; and the dismutase reaction:

$$1. \quad AH_2 + 2 O_2 \xrightarrow{\text{Oxidase}} A + 2 H^+ + 2 O_2^-$$

$$2a. \quad 2 O_2^- + NBT \xrightarrow{\text{Spontaneous}} 2 O_2 + NB \text{ Formazan}$$

$$2b. \quad O_2^- + NADH + H^+ \xrightarrow{\text{Spontaneous}} H_2O_2 + NAD^+$$

$$3. \quad 2 O_2^- + 2 H^+ \xrightarrow{\text{SOD}} H_2O_2 + O_2.$$

The generating reaction may be either enzymatic (xanthine oxidase is often used), chemical or photochemical. It is set up to be zero order (i.e., the rate of generation is constant throughout the assay). The indicating reaction is also zero order and in the absence of SOD the rate of absorbance change is termed V_0. The dismutase reaction is kinetically second order; the rate constant equals the diffusion constant for the anion and the enzyme to come together. After the generating reaction is started the two competing rates (for reaction 2 and reaction 3) establish a steady state concentration of O_2^- and a rate for the indicating reaction. The rate of absorbance change is called V_i. The actual assay for SOD consists of determining the extent to which the enzyme preparation being tested inhibits the rate of the indicator reaction. A plot of $(V_0/V_i) - 1$ versus amount of SOD is linear, and one unit of SOD is that amount of enzyme which inhibits the indicator reaction rate by 50% (Beyer and Fridovich 1987). In a typical assay protocol several reaction tubes are run containing different amounts of SOD and from the plot of $V_0/V_i - 1$ against aliquot sizes the volume containing 1 unit of SOD is determined. Three typical assays for SOD will be given here. The first assay uses an enzymatic generating reaction and a reducing indicator reaction. The second assay applies a photochemical generating reaction with a reducing indicator reaction. The third assay uses a chemical generating reaction and an oxidizing indicator reaction. For an indication of the wide variety of assay chemistries which have been used see Fridovich (1975) or Beyer and Fridovich (1987).

Enzyme generating, reducing indicator reaction, continuous assay (Beyer and Fridovich 1987). The assay reaction mixture consists of 50 mM phosphate pH 7.8, 0.1 mM EDTA, 50 μM xanthine and 10 μM ferricytochrome C. Three ml of this is incubated at 25°C and sufficient xanthine oxidase (about 6 nmoles) is added so that V_0, the rate at which the absorbance at 550 nm decreases due to reduction of ferricytochrome C, is 0.025/min. After this rate is established 10 μl of SOD solution is added and the new rate V_i is measured. This is repeated for several other concentrations of SOD and the data is processed as indicated above.

Photochemical generating, reducing indicator reaction, fixed time assay (Beyer and Fridovich 1987). A stock reaction mixture is made by mixing 27 ml 0.05 M phosphate buffer pH 7.8, containing 0.1 mM EDTA, with 1.5 ml L-methionine (30 mg/ml), 1.0 ml NBT (nitro blue tetrazolium chloride, 1.41 mg/ml) and 0.75 ml 1% Triton X-100. For the assay, to 1 ml of reaction mixture is added 10 μl riboflavin solution (4.4 mg/100 ml) and 20 μl SOD sample, the mixture is illuminated for 7 min under two 20-watt GroLux fluorescent bulbs mounted in a foil-lined box, then the absorbance at 560 nm is read. The increase in absorbance due to the formation of NBT formazan is the rate of the reaction. Again the absorbance in the absence of SOD (V_0) and in the presence of various amounts of SOD (V_i) is used to find the number of units/ml of SOD in the stock solution.

Chemical generating, oxidizing indicator reaction, continuous assay (Paoletti et al. 1986). Buffer is 0.1 M triethanolamine, 0.1 M diethanolamine adjusted to pH 7.4 with HCl. NADH solution is 7.5 mM (5 mg/ml of disodium salt in water). Ions solution contains 0.1 M EDTA and 0.05 M MnCl2 in water, neutralized to pH about 7. Mercaptoethanol solution is 10 mM in water. For each test, to 0.8 ml of buffer is added 40 μl NADH solution, 25 μl ions solution and 100 μl of the SOD solution to be tested. To start the reaction 100 μl of the mercaptoethanol solution is added, the contents of the cuvette are thoroughly mixed and absorbance at 340 nm is recorded. After a lag period of 2 to 4 min a linear decrease in absorbance occurs for 5 to 10 min. The authors report that a plot of V_i/V_0 versus log [ng SOD] is linear over the range of 4 to 40 ng SOD. It would still be preferable to make the plot discussed above which gives a wider range of linearity.

12 Miscellaneous Enzymes

There are several enzymes which don't fit into the categories of the previous four chapters but which nevertheless are of interest to food scientists. These enzymes are the topics of this chapter.

EPIMERASES

Glucose Isomerase

The discovery and exploitation of this enzyme is the basis for the High Fructose Corn Syrup Industry, one of the major food industry success stories of this century. The reaction is the isomerization of an aldohexose, D-glucose, to a ketohexose, D-fructose. The equilibrium ratio of the two sugars is roughly 45%-fructose/55%-glucose and is temperature dependent. Commercial applications today utilize immobilized glucose isomerase and a number of assays have been published for measuring the activity of the enzyme in this form. The kinetics of immobilized enzyme assays are highly dependent upon bed configurations, flow rates, solution viscosity, and other factors which are process-specific so it does not seem useful to describe such assays in this chapter. The principles of measuring the progress of isomerization in a homogeneous solution assay are also applicable to immobilized enzyme assays; proper consideration of the chemical engineering factors will have to be left to the reader.

The most straightforward solution assay uses an automatic polarimeter to follow the progress of isomerization. Lloyd et al. (1972) used a Bendix NPL automatic polarimeter with a cell having a 2 cm light pathlength (sodium line light) thermostatted at 60°C. To calibrate the instrument response they used two solutions 2.00 M D-glucose and 1.90 M D-glucose + 0.10 M D-fructose to set the recorder at 100% and 0% traverse, so a 20% recorder change corresponds to the conversion of 1% of the D-glucose substrate. The reaction mixture contains 2.0 M D-glucose, 0.02 M $MgSO_4$ and 1 mM $CoCl_2$ in 0.2 M maleate buffer pH 7.0. The enzyme sample is added to an amount of reaction mixture appropriate for the cell capacity equilibrated at 60°C and the change in polarization is recorded. The rate is linear for at least the first 2% of conversion, and the slope of the recorder trace is easily converted to μmoles glucose isomerized per minute.

Lloyd et al. (1972) also used an automated assay in which 0.22 M D-glucose (in 0.4 M maleate buffer pH 6.7, containing 0.06 M $MgSO_4$) is reacted with enzyme for a brief period of time, then the mixture is dialyzed against 0.1 M acetate buffer pH 4, the dialyzate is mixed with acid carbazole (4.4 ml of 0.5% carbazole in ethanol added to 800 ml concentrated H_2SO_4 diluted with 200 ml water), heated, and the absorbance is read. Carbazole gives a colored product only with the D-fructose, and from a standard curve made by running various combinations of glucose and fructose through the system the percent conversion can be obtained.

Mutarotase

Crystalline glucose is in the α-D-pyranose ring form. Upon solution in water the optical rotation is $+111°$. Over a period of several hours this rotation drops due to isomerization to the β-D-pyranose anomer (optical rotation $= +19°$) to a final equilibrium value of $+52.5°$. This represents a mixture of 36.4% α form and 63.6% β form. Enzymes which react with D-glucose (e.g., glucose oxidase, glucokinase) are specific for one or the other anomer, and mutarotase is often added to a solution containing glucose as a substrate in order to hasten the attainment of equilibrium (see the specific assays section of Chapter 9) in coupled enzyme assays. The assay of mutarotase itself is based upon the rate of conversion of α-D-glucose to β-D glucose and the action of an auxiliary enzyme upon the latter substrate.

Colorimetric methods use glucose oxidase to generate hydrogen peroxide from the oxidation of β-D-glucose, followed by a second auxiliary reaction of peroxidase with a chromogenic donor (Hill and Cowart 1966). A stock solution of α-D-glucose is made by dissolving 2 g crystalline anhydrous glucose in 10 ml ice cold water then immediately diluting this to 100 ml with ice cold methanol and holding this solution in an ice bath. To 2.15 ml of 0.01 M phosphate buffer pH 7.0 is added 0.1 ml of glucose oxidase (100 IU/ml), 0.1 ml horseradish peroxidase (60 IU/ml), 0.5 ml o-dianisidine (50 mg/ml), and 0.1 ml of the mutarotase sample. After this is attemperated at 25°C 0.05 ml of the cold glucose stock solution is added to start the reaction. The increase in absorbance at 460 nm is recorded. A blank reaction (water in place of the mutarotase sample) must also be run to get the rate of spontaneous mutarotation which is subtracted from the assay rate. The rate of change in absorbance per minute may be converted to molar units using the molar absorptivity of $11.3 \times 10^3 M^{-1}cm^{-1}$ for the oxidation product of o-dianisidine.

Oxygen uptake by the glucose oxidase reaction may be followed using an oxygen electrode. Miwa and Okuda (1974) describe such an assay. The α-

D-glucose stock is prepared as described above (they used 2.7 g glucose per 100 ml). In the thermostatted oxygen electrode cell they placed 1 ml 0.02 M EDTA buffer pH 7.0, 5 μl 2% glucose oxidase, 5 μl 0.2% NaN$_3$ solution (a catalase inhibitor) and the mutarotase sample. To start the reaction 10 μl of the glucose stock solution is added. The rapid consumption of oxygen is recorded for 1 to 2 min. Then 10 μl of 1.0 M phlorizin in ethylene glycol (held at 50°C to reduce viscosity) is added to completely inhibit mutarotase. The subsequent slow oxygen uptake is due to the background mutarotation and is subtracted from the total rate recorded earlier to give the rate due to mutarotase. The rate of percent of oxygen consumed per minute is converted using the value of 0.26 mM for concentration of oxygen at 100%.

Many mutarotase samples contain catalase as a contaminant which regenerates oxygen from the hydrogen peroxide product of the glucose oxidase reaction. If the amount of catalase is unknown this means that the net oxygen uptake is something less than 1 mole per mole of glucose oxidized. Miwa and Okuda used inhibition with azide to prevent catalase activity. Weibel (1976) took the opposite tack, adding excess catalase so that the stoichiometry is known to be 1/2 mole net oxygen uptake per mole of glucose oxidized. He also sidestepped the problem of preparing and holding a stock solution of α-D-glucose by generating it in situ via the hydrolysis of sucrose by invertase. To 3 ml of 0.01 M phosphate buffer pH 7.0 (containing 2 mM sucrose and air-saturated) in the oxygen electrode cell was added 30 μl 0.1 mM glucose oxidase, 2 μl catalase solution (25 mg/ml) and 20 μl invertase (20,000 IU/ml), which converts all the sucrose to fructose and α-D-glucose within a few seconds. The slow uptake of oxygen due to spontaneous mutarotation was recorded for 1 to 2 min. Then the mutarotase sample was added and the much faster oxygen consumption was recorded. The corrected rate was converted to molar rate of mutarotation using the 0.5 to 1 stoichiometry noted above.

GLUCOKINASE

This enzyme is also used as an auxiliary enzyme in the determination of glucose arising from, for example, the hydrolysis of maltose by maltase. The reaction is the phosphorylation of glucose at the 6 position by ATP giving glucose-6-phosphate and ADP. Glucose-6-phosphate is then dehydrogenated with concomitant reduction of NADP$^+$ and an increase in absorbance at 340 nm. The assay is described by Salas et al. (1963). The reaction mixture contains 0.05 M Tris buffer pH 7.4, 0.1 M glucose, 5 mM β-mercaptoethanol, 5 mM MgCl$_2$, 5 mM AT, 0.25 mM NADP$^+$, and 0.2 IU glucose-6-phosphate dehydrogenase in a total volume of 2 ml. The reaction is started by adding 25 μl of glucokinase sample and the increase in absorbance at 340 nm is recorded.

Davidson and Arion (1987) reexamined this assay and modified it slightly to increase the accuracy. Their reaction mixture comprised 0.05 M HEPES buffer pH 7.4, 0.1 M KCl, 7.5 mM $MgCl_2$, 2.5 mM dithioerythritol, 1% bovine serum albumin, 0.1 M glucose, 0.5 mM $NADP^+$ and 4 units glucose-6-phosphate dehydrogenase per ml. To 0.98 ml of this mixture was added 10 μl glucokinase sample followed by 10 μl 0.5 M ATP to start the reaction. The increase in absorbance at 340 nm was monitored.

Croxdale and Vanderveer (1986) measured the NADPH concentration fluorometrically to increase the sensitivity of the reaction. The reaction contained 0.1 M imidazole buffer pH 7.3, 0.525 mM glucose, 2.5 mM $MgCl_2$, 0.2 mg/ml bovine serum albumin, 1 mM ATP, 0.2 mM $NADP^+$, 0.2 μg/ml glucose-6-phosphate dehydrogenase, and the glucokinase sample. The reaction was carried out at 22°C and monitored in a spectrofluorometer (excite at 365 nm, emit at 400 nm).

GLUTAMATE DECARBOXYLASE

This enzyme is of interest because of its connection with incipient sprouting in seeds. The substrate, glutamic acid, is decarboxylated to give 4-amino butyric acid and CO_2. The published assays which I have found have all been based upon the Warburg manometer, measuring the rate of evolution of carbon dioxide. An example is given by Shukuya and Schwert (1960). The buffer is 0.1 M pyridine adjusted to pH 4.6 with HCl and the total Cl^- concentration adjusted to 0.1 M with NaCl (the usual buffers for this pH range such as acetate are competitive inhibitors of the enzyme) plus 10 mM glutamate. Three ml of this are placed in the manometer chamber (air gas phase), equilibrated at 36°C, and 0.2 ml of enzyme solution tipped in from the sidearm. The evolution of gas is linear for several minutes.

It should be possible to design an assay based upon the fact that the decarboxylation converts an acid with a pK of 3.2 to one with a pK of 6.2:

$$R-COO^- + H^+ + H_2O \longrightarrow R-H + H_2CO_3$$

As the reaction proceeds the pH should rise as H^+ is consumed. This should be a good candidate for either a pH-stat assay or a colorimetric assay using a lightly buffered indicator dye. Such assays may have been published but they have escaped my notice.

CARBONIC ANHYDRASE

The rapid attainment of equilibrium between CO_2 and H_2CO_3 is important in maintaining proper pH within living cells and in exhaling CO_2 through our lungs. In the food sciences it has less importance. However it has been used in the measurement of carbon dioxide in certain carbonated beverages, and is an analytical aid in AOAC method 11.062 for measuring carbon dioxide in wine. It seems worthwhile to record at least one assay for this enzyme in this chapter.

The reaction catalyzed is the hydration of CO_2 to H_2CO_3. The uncatalyzed reaction is fairly rapid (0.0041 sec^{-1} at 5°C) so a correction for the base rate is mandatory. Alsen and Ohnesorge (1973) describe a pH-stat assay for the enzyme. In the reaction cell thermostatted at 5°C is placed 2 ml of 30 mM barbital buffer pH 7.9. Gas of a known composition (3.5% to 12.5% CO_2 in N_2) is bubbled through the stirred buffer at a constant rate. The reaction is titrated with 0.1 N NaOH to keep the pH constant at 7.75. After this has proceeded long enough to establish the base hydration rate the enzyme sample is added and the reaction continued, again adding titrant at a rate sufficient to keep pH at 7.75. The difference between the two titration rates is the rate due to enzyme. By knowing the two pKs for ionization of carbonic acid ($pK_1 = 6.4$ and $pK_2 = 10.3$) we can calculate that at pH 7.75 the hydration of one mole of CO_2 yields 0.96 mole of H^+ (if the titration pH is 8.35, the stoichiometry will be exactly 1:1).

The hydration reaction has also been explored using stopped-flow techniques to deal with the rapid rates of the kinetic steps (Khalifah 1971; Rowlett and Silverman 1982) but these methods are beyond the scope of this book. The papers cited are worth mentioning because they give valuable insights into the use of indicator dyes for the spectrophotometric measurement of pH changes attendant on enzyme reactions. This topic was discussed in Chapter 7.

LYSOZYME

Lysozyme is found in many foodstuffs, particularly in milk (and derived dairy products) and eggs. It is a hydrolase, cleaving the acetal linkages in murein, a complex polymer made up of N-acetylglucosamine, N-acetylmuramic acid, and peptide side chains, found in the cell walls of certain bacteria. Hydrolysis of this structural molecule causes bacterial cells to lyse (hence the name lysozyme) and a turbid suspension of cells will slowly clear (the basis for most assays). In a formal sense lysozyme could have been included in Chapter 9, but in an operational sense it is so different from the carbohydrate hydrolases discussed there that it seems to fit better here.

A recent review by Grossowicz and Ariel (1983) discusses substrate factors, various assay methods and clinical applications of lysozyme measurements. A paper by Parry et al. (1965) describes a turbidimetric assay which is the basis for almost all assays used today. The substrate is *Micrococcus lysodeikticus* cells which have been killed by exposure to ultraviolet light and lyophilized. A suspension of 50 mg cells per 100 ml buffer, M/15 phosphate pH 6.2 is prepared, and 1.5 ml of this suspension is mixed with 0.5 ml 0.3 M NaCl and 1 ml of enzyme. The cuvette is placed in a colorimeter and % transmittance (not absorbance) is followed at 540 nm. The initial %T is about 10% and this slowly increases as cells are lysed. The plot of %T versus time is linear up to about 40% transmittance. The rate %T/min is linear with respect to enzyme for the 0 to 10 μg range; 1 μg lysozyme gives a rate of about 1%T/min.

Mörsky (1983) evaluated the effect of pH, ionic strength and substrate concentration on the reaction, and suggested an "optimized" assay. The buffer is 55 mM phosphate pH 6.2 containing 100 mg/l bovine serum albumin. The substrate suspension is 410 mg lyophilized *M. lysodeikticus* cells in 100 ml of buffer. For the assay 75 μl lysozyme sample is put into 1.0 ml buffer, warmed 5 min at 37°C, then 0.1 ml substrate suspension is added and absorbance at 700 nm is recorded for 3 min. The rate is taken from the initial linear part of the trace. Since there is no way to convert the rate of turbidity change to molar terms, Mörsky suggested rates as equivalent μg of pure crystalline egg white lysozyme. Rather the same approach was advocated by Borgen and Romslo (1977) who used the Coleman Model 91 Amylase Lipase Analyzer (a nephelometric instrument). The reaction mixture was 36 mg *M. lysodeikticus* cells in 100 ml 66 mM phosphate buffer pH 6.2. To 2.5 ml of this is added 500 ng crystalline egg white lysozyme, the solution placed in the instrument, and after the 135 sec reaction interval is passed the machine meter is adjusted to read 10. All subsequent unknowns are then read off in terms of this calibration.

Gorin et al. (1971) investigated the lysozyme reaction in some detail. They followed the absorbance of the reaction mixture over a long period of time, showed that the reaction has a kinetic order between 0 and 1, and discussed the non-correspondence between molecular enzymatic events and the rate of clearing of the cell suspension. They also noted several other peculiarities of the assay. For instance, they measured the spontaneous clearing rate of their substrate suspension (20 mg/100 ml *M. lysodeikticus* cells in 0.1 M phosphate pH 6.2, 1 mM in EDTA) and noted that while usually this was less than 0.01/hr, occasionally the rate was as high as 0.05/hr (this substrate was not used). They also mentioned the extreme variability between batches of commercial UV-killed lyophilized *M. lysodeikticus* cells (see below). The assay described by Gorin et al. is an end point assay (i.e., determining the length of time required to achieve a certain change in the substrate). The initial absorbance at 570 nm of the substrate suspension is about 0.7. To 3 ml is added 0.1 ml of lysozyme solution. A convenient absorbance value is noted within the first 15 sec of the reaction, and the length of time *t* required for the absorbance to decrease by 0.05 units from that point is measured. The lysozyme sample is adjusted to get this time to be between 30 and 150 sec. The authors defined a unit of enzyme as $1.55 \times 10^{-3}/t$; they also reported that the response was linear up to 1.8 μg lysozyme/ml.

Grossowicz and Ariel (1983) discussed the problem of variability between batches of *M. lysodeikticus* purchased commercially or prepared at different times in the laboratory. They attributed most of this to differences in the conditions used to kill the cells (usually by exposure to ultraviolet light) and lyophilize the cells for storage. Grossowicz et al. (1979) recommended that the cells not be exposed to UV light and lyophilized, but rather stored as a suspension in glycerol. Cells were grown for 18 to 20 hours in 250 ml Erlenmeyer flasks containing 100 ml Difco Brain Heart Infusion Broth, and shaken in a water bath at 37°C. The cells were harvested by centrifugation ($7500\times$ gravity) for 5 min at 4°C, washed with 60 mM Tris buffer pH 7.5, then resuspended in the same buffer. To 60 ml of this suspension was added 40 ml glycerol and the mixture was divided into aliquots and stored at -20°C. For use in assays one aliquot was diluted with 100 ml Tris buffer; 7 ml was mixed with 0.1 ml enzyme and absorbance was monitored. The authors claimed that this substrate gave reproducible results for up to 8 months of storage in the freezer. While it is certainly more convenient to order lyophilized substrate from a commercial supply house, one must weigh that convenience against the need for long term reproducibility. If the assay is calibrated in terms of crystalline egg white lysozyme, the concern becomes one of reproducibility of the standard enzyme rather than the substrate. A synthetic substrate for lysozyme is desirable. Unfortunately the few which have been tried have fallen short of the sensitivity of the *M. lysodeikticus* assay by several orders of magnitude. Until this situation is rectified we will have to use the turbidimetric assay, keeping in mind its shortcomings as well as its advantages.

Appendixes

A   Developing Rate Expressions

In Chapters 2 and 3 a number of rate expressions were used to show: 1. how enzymatic reaction rate v depends on substrate concentration $[S]$ given a certain model; 2. how the presence of an inhibitor $[I]$ affected the rate v; and 3. how these expressions could be manipulated to put them into forms convenient for the analysis of experimental data. Presenting the derivation of each expression would have consumed many pages. Instead, this Appendix sets forth the principles involved in deriving rate expressions based on one or the other of the two underlying assumptions; rapid equilibrium or steady-state intermediate concentrations.

Rapid Equilibrium Derivations

The assumption is that the various species present in the reaction are in equilibrium and that the forward and reverse reactions maintaining these equilibria are much faster than any irreversible steps such as the breakdown of ES complex to $E + P$. Each equilibrium is characterized by a *dissociation* equilibrium constant, equal to the product of the concentrations of participating species on the left hand side of the reaction, divided by the product of concentrations of species on the right hand side. This convention is generally followed by enzymologists, and while using *association* constants is mathematically correct and gives valid rate expressions, they tend to be confusing when comparing results with other reports in the literature.

The basic steps to be followed are:

1. Write down the model under consideration. Include all associating species and the rate step(s) for forming product.
2. Define each equilibrium constant in terms of the species involved. If the model involves a closed loop (i.e., different complexes at four corners of a square) there is a symmetry to the equilibrium constants which allows the four equilibria to be defined with only three constants.
3. Establish the step which defines the rate v. This will usually be a rate constant times the concentration of a particular species which is

called the "central complex." If two species contribute to product formation, as occurs in certain of the inhibition models, the "rate constant" may contain two parameters, but be related to only one central complex.

4. Write down the ratio of v to total enzyme E_t. The numerator is the rate times the central complex concentration as defined in step 3. The denominator is the sum of the concentrations of all species which contain enzyme.

5. Relate each individual species to the central complex by some combination of equilibrium constants. Substitute these into the denominator of step 4 and cancel out the concentration of central complex in the numerator and denominator.

6. Finally, set V_{max} equal to the rate constant from step 3 times $[E_t]$. Collect terms in the denominator to get separate factors for K_M and $[S]$.

To illustrate this process we will use the random-addition two-substrate model of Chapter 2. The model is:

$$E + A \rightleftharpoons EA \qquad K_A = \frac{[E][A]}{[EA]}$$

$$E + B \rightleftharpoons EB \qquad K_B = \frac{[E][B]}{[EB]}$$

$$EA + B \rightleftharpoons EAB \qquad \alpha K_B = \frac{[EA][B]}{[EAB]}$$

$$EB + A \rightleftharpoons EAB \qquad \alpha K_A = \frac{[EB][A]}{[EAB]}$$

$$EAB \longrightarrow E + P \qquad v = k_p[EAB]$$

The central complex is EAB. The equations which relate each of the other enzyme-containing species to this are:

$$[EA] = [EAB]\frac{\alpha K_B}{[B]}$$

$$[EB] = [EAB]\frac{\alpha K_A}{[A]}$$

$$[E] = K_A \frac{[EA]}{[A]} = [EAB] \frac{\alpha K_B K_A}{[A][B]}$$

The ratio equation is:

$$\frac{v}{[E_t]} = \frac{k_p[EAB]}{[E] + [EA] + [EB] + [EAB]} \qquad \text{(A-1)}$$

Substituting gives:

$$\frac{v}{[E_t]} = \frac{[EAB]k_p}{[EAB](\alpha K_A K_B/[A][B] + \alpha K_B/[B] + \alpha K_A/[A] + 1)} \qquad \text{(A-2)}$$

Canceling the *EAB* terms and collecting the terms in the denominator:

$$\frac{v}{[E_t]} = \frac{k_p}{(\alpha K_A K_B + \alpha K_B[A] + \alpha K_A[B] + [A][B])/[A][B]} \qquad \text{(A-3)}$$

Finally, setting $V_{max} = k_p[E_t]$ and multiplying through by $[A][B]$, we arrive at the final working equation:

$$v = \frac{V_{max}[A][B]}{\alpha K_A K_B + \alpha K_B[A] + \alpha K_A[B] + [A][B]} \qquad \text{(A-4)}$$

This general process is followed to derive the rate expression for any desired model. For example, the general reversible inhibition model is much more complex, but can be easily treated to arrive at a final rate equation as shown in detail in Appendix F.

Steady-State Derivations

The basic assumption of the steady-state approach is that shortly after mixing enzyme and substrate, rate equilibria are reached and maintained among all the intermediates in the system so that their concentration does not change (i.e., $d[ES]/dt = 0$). This is not strictly valid, of course, since as the reaction progresses the concentration of S decreases, hence the rate for formation of $ES, k_1[E][S]$, must also decrease leading to a slow decrease in $[ES]$. The assumption is a good approximation for most kinetic studies, because usually the measurement being made is of the initial rate v_0. Moreover the rate constants themselves are invariant, so the various enzyme parameters such as K_M, which are actually functions of several rate constants, also remain constant.

The difference between the two approaches is sometimes significant. The use of "titrating substrates" for enzymes would not be possible if we were restricted to the rapid equilibrium approach. Steady-state kinetics also makes it possible to arrange experimental conditions so as to dissect the various steps in a kinetic reaction and thus reach a deeper understanding of the energetics of enzyme catalysis. The development of an analytical understanding of the amounts of auxiliary enzymes needed for an accurate assay depends upon taking a steady-state approach, as does the method for finding inhibition rates for high-affinity inhibitors. These topics are discussed in more detail in Appendix B and Appendix G, respectively.

The original method of deriving steady-state expressions used by Briggs and Haldane quickly becomes unmanageable when one is considering a complex reaction. More recently a procedure called "Theory of Graphs," introduced by E.L. King and C. Altman, has produced several variants which make the derivation of complex rate expressions relatively simple. The technique which I recommend is the modified method described by Fromm (1970, 1975). The ordered-addition two-substrate model is used to illustrate this method.

1. The kinetic model is set up in geometric form, with each enzyme form being at a numbered *node*. Each pathway from one node to another is characterized by a rate constant which includes the concentration of non-enzyme species (substrate or inhibitor).

$$E \xrightarrow{k_1[A]} EA \qquad \text{node } E = (1)$$

$$EA \xrightarrow{k_2} E \qquad \text{node } EA = (2)$$

$$EA \xrightarrow{k_3[B]} EAB \qquad \text{node } EAB = (3)$$

$$EAB \xrightarrow{k_4} EA$$

$$EAB \xrightarrow{k_5} E + P$$

2. A determinant is written for each node. Start by writing down the *branches*. These are the single steps which lead to the node from any other node, with the associated rate constant. Thus for node 1 there are

two branches: $2 \longrightarrow 1(k_2)$ and $3 \longrightarrow 1(k_5)$. Next to each branch write the product of the nodes not involved in that particular step, i.e., $2 \longrightarrow 1(3)$ and $3 \longrightarrow 1(2)$. The number within the parentheses is the summation of rate constants leading *away* from that particular node. Thus, $(1) = k_1[A]$, $(2) = k_2 + k_3[B]$, and $(3) = k_4 + k_5$.

The determinant of the node is the sum of the products of the rate constant of the branch times the associated node product:

$$\{1\} = 2 \longrightarrow 1(3) + 3 \longrightarrow 1(2) = k_2(k_4 + k_5) + k_5(k_2 + k_3[B])$$

$$\{2\} = 1 \longrightarrow 2(3) + 3 \longrightarrow 2(1) = k_1[A](k_4 + k_5) + k_4(k_1[A])$$

$$\{3\} = 2 \longrightarrow 3(1) = k_3[B](k_1[A])$$

3. Each determinant is expanded, and forbidden and redundant terms are discarded. A forbidden term is one involving the product of the forward and backward rate constants for the same equilibrium, i.e. k_1k_2 or k_3k_4. In this particular case no such terms appear. Where the same product appears more than once in the same determinant it is retained only once. In $\{1\}$ the product k_2k_5 is redundant, and in $\{2\}$ the term $k_1k_4[A]$ appears twice. After expanding the determinants and removing redundancies, the three determinants for this model are:

$$\{1\} = k_2(k_4 + k_5) + k_3k_5[B]$$

$$\{2\} = k_1[A](k_4 + k_5)$$

$$\{3\} = k_1k_3[A][B]$$

4. Now the expression for the ratio of the rate to total enzyme is set up. The rate v_0 equals the rate constant(s) times the determinant(s) for those nodes which give product and free enzyme, and E_t is the total of all determinants.

$$\frac{v_0}{E_t} = \frac{k_5\{3\}}{\{1\} + \{2\} + \{3\}} \tag{A-5}$$

$$= \frac{k_5k_1k_3[A][B]}{k_2(k_4 + k_5) + k_3k_5[B] + k_1[A](k_4 + k_5) + k_1k_3[A][B]} \tag{A-6}$$

5. Bring k_5 to the denominator of the left side ($V_{\max} = k_5E_t$). Divide the numerator and denominator of the right side by the rate constants remaining

in the numerator (k_1k_3 in this case) to leave only substrate concentration terms in the numerator. This gives:

$$\frac{v_0}{V_{max}} = \frac{[A][B]}{k_2(k_4 + k_5)/k_1k_3 + [A](k_4 + k_5)/k_3 + [B]k_5/k_1 + [A][B]}$$

(A-7)

The rapid equilibrium expression for the ordered-addition two-substrate model is:

$$\frac{v}{V_{max}} = \frac{[A][B]}{K_A K_B + [A]K_B + [A][B]}$$

(A-8)

By comparing like coefficients of the two models we see that: $K_A = k_2/k_1$ and $K_B = (k_4 + k_5)/k_3$; the result for K_A is the same in either case, but on the rapid equilibrium assumption K_B is just k_4/k_3.

In addition, the steady-state equation has an extra term in the denominator, $[B]k_5/k_1$, which does not appear in the rapid equilibrium equation. In experimental runs when data pairs for $v, [B]$ are obtained, the Hanes plot of $[B]/v$ versus $[B]$ has a slope of $(1 + k_5/k_1[A])/V_{max}$. At different levels of $[A]$ a slight variation in the slopes might be seen. However, one would expect $k_1 \gg k_5$, and this deviation would probably not be experimentally detectable.

Our example here is a fairly simple one to work out. Fromm (1970) uses the random-addition two-substrate model for his illustration. The determinants may contain as many as 16 individual terms; keeping everything straight requires painstaking, careful work. If the nodes are increased, for instance, by including inhibition in the model, the determinants expand almost exponentially. Nevertheless, the "Theory of Graphs" method is the best one extant for developing steady-state equations. Since in many cases the steady-state expression is required to solve the experimental problem at hand, it is well to be familiar with this procedure.

B   Coupled Enzyme Reactions

The use of an auxiliary enzyme to measure the rate of formation of product in the reaction of interest is a useful technique in many situations. Several specific assumptions must be met for a coupled enzyme assay to be valid. The system under consideration may be modeled as follows:

$$A \xrightarrow{k_1} B \xrightarrow{k_2} C \xrightarrow{k_3} D$$

We will first consider the situation in which only one auxiliary enzyme is used (the appearance of C is followed) and then we consider the extension to the use of two sequential auxiliary enzymes. This analysis was first published by W. R. McClure (1969).

One Auxiliary Enzyme

The assumptions which must be met for this system to yield valid results are:

1. k_1 is constant (zero order) during the course of the assay. This means either that $[A]$ for the first reaction is a saturating concentration(s) for the enzyme so $k_1 \approx V_{max}$, or else an insignificant fraction of A is converted to B during the observation period.
2. The first reaction is essentially irreversible. This follows from the fact that B is being continuously removed.
3. k_2 is first order with respect to B. This means that $[B] << K_B$, and that any other substrates of the auxiliary enzyme are at saturating concentrations. Then the rate constant for the second enzyme reaction is given by $-d[B]/dt = (V_b/K_B)[B]$, and $k_2 = V_b/K_B$.
4. The auxiliary reaction $B \longrightarrow C$ is irreversible. This condition is met for a reversible reaction if the equilibrium is far on the side of C, or if only a small fraction of the total reaction is followed.

Having a linear rate of production of C depends upon achieving a steady-state concentration for B. This is expressed as:

$$\frac{d[B]}{dt} = k_1 - k_2[B] \ (= 0 \text{ in the steady state.}) \tag{B-1}$$

Integration gives:

$$[B] = \left(\frac{k_1}{k_2}\right)(1 - e^{-k_2t}) \tag{B-2}$$

At $t = 0$, $[B] = 0$, and as t approaches ∞, $[B]$ approaches k_1/k_2. This is the steady-state concentration of B.

In practice we want to measure the rate of formation of C sometime before ∞, so we begin the measurement of $d[C]/dt$ at a time when the rate is essentially constant at some fraction of the rate achievable at $t = \infty$. Rearranging and taking the log of equation B-2, we get:

$$1 - [B]\left(\frac{k_2}{k_1}\right) = e^{-k_2t}; \qquad \ln\left(1 - [B]\frac{k_2}{k_1}\right) = -k_2t. \tag{B-3}$$

At some time t^* we achieve a fraction F of the ultimate steady state value of $[B]$, i.e., $[B] = F(k_1/k_2)$. Substituting this into B-3 and rearranging gives:

$$t^* = \frac{-\ln(1 - F)}{k_2} \tag{B-4}$$

This is the basic equation for defining the amount of auxiliary enzyme to use in the assay. By defining the fraction of the steady-state concentration of B which is desired (i.e., the acceptable percentage error in assuming that $d[C]/dt = -d[A]/dt$), and the length of time we are willing to wait to achieve this, we define the amount of auxiliary enzyme required. If $F = 0.99$ (i.e., a 1% discrepancy between the measured rate $d[C]/dt$ and the rate $-d[A]/dt$) and we take an acceptable lag period t^* of 5 sec (about the time it takes to mix the reactants and place them in the spectrophotometer or O_2 electrode cell) then $k_2 = 0.92 \ sec^{-1}$. From runs made with the auxiliary enzyme on the substrate B under first-order conditions, and *at the same pH and temperature as the reaction of interest* ($A \longrightarrow B$), the amount of enzyme necessary to give this value of k_2 is known.

Note that the activity level of the first enzyme is irrelevant. There are many incorrect statements in the literature, such as this one: "The auxiliary enzyme E_2 should be present at levels which catalyze the conversion of its substrate at about 100 times the rate at which E_1 is acting. . . The actual excess needed depends upon the relative V_{max} and K_m value of the two enzymes." In fact, as McClure's analysis showed, the relative V_{max} and K_m value of the two enzymes is meaningless.

Two Sequential Auxiliary Enzymes

When two auxiliary enzymes are required in order to arrive at a detectable product the situation is less clear-cut. The end result of this analysis is that to achieve a given fraction of the ultimate steady-state concentration of $[C]$ at a given time t^* (with its associated error in equating $d[D]/dt$ with $-d[A]/dt$), a non-unique relationship between the two auxiliary enzyme first order constants k_2 and k_3 is obtained. From this relationship the amounts of the two enzymes needed to achieve the desired results can be specified. The assumptions are those used before, namely that k_1 is a zero-order rate constant, k_2 and k_3 are first-order rate constants, and all three reactions are essentially irreversible.

The concentration of C depends upon two rates:

$$\frac{d[C]}{dt} = k_2[B] - k_3[C] \tag{B-5}$$

Substituting for $[B]$ from equation B-2 we get:

$$\frac{d[C]}{dt} = k_1 - k_1 e^{-k_2 t} - k_3[C] \tag{B-6}$$

As t goes to ∞ and $d[C]/dt = 0$, this simplifies to:

$$[C] = \frac{k_1}{k_3} \tag{B-7}$$

When $t = 0$, of course $[C] = 0$.

Equation B-6 can be solved and the constant of integration evaluated using the two limits $[C]_0 = 0$ and $[C]_\infty = k_1/k_3$ to give:

$$[C] = \frac{k_1}{k_3} - \frac{k_1(k_3 e^{-k_2 t} - k_2 e^{-k_3 t})}{(k_3 - k_2)k_3} \tag{B-8}$$

The first term on the right hand side is $[C]_\infty$ and the rest of the expression represents the time it takes to achieve the steady state value. If, as before, we want to know the time t^* when $[C]$ equals some fraction F of the steady-state value, substitute $F(k_1/k_3)$ for $[C]$ and rearrange:

$$F\left(\frac{k_1}{k_3}\right) = \frac{k_1}{k_3} - \frac{1}{(k_3 - k_2)}\left(\frac{k_1}{k_3}\right)(k_3 e^{-k_2 t} - k_2 e^{-k_3 t})$$

$$(k_3 - k_2)(1 - F) = k_3 e^{-k_2 t} - k_2 e^{-k_3 t} \tag{B-9}$$

As mentioned earlier, there is no unique solution for this equation for any given values of F and t^*. When $k_3 >> k_2$ then the equation is the same as B-4 above, i.e., $t^* = -\ln(1 - F)/k_2$, and when $k_2 >> k_3$, $t^* = -\ln(1 - F)/k_3$. McClure evaluated this equation numerically by choosing a value for k_2, and then over a range of values of k_3, calculating the time t for $F = 0.99$. The plot was of k_3 on the ordinate, and time to reach 0.99 of the steady-state value of $[C]$ on the abscissa, and consisted of a series of lines corresponding to various values of k_2. From this plot McClure constructed a nomogram which related $1/k_2$ to $1/k_3$ at various values of t^* for $F = 0.99$. Because of the symmetrical nature of B-9, it turns out that on a graph with axes $1/k_2$ and $1/k_3$ the relationship for any given t^* describes a quarter circle with the center at the origin. In other words, $(1/k_2 + 1/k_3) = f(t^*)$. When k_2 approaches ∞, and $1/k_2$ approaches 0, the value of t^* is related to $1/k_3$ by: $t^* = -\ln(1 - F)/k_3$; for $F = 0.99$, $t^* = -\ln(0.01)/k_3 = 4.60517(1/k_3)$, and:

$$\frac{1}{k_2} + \frac{1}{k_3} = 0.21715t^* \tag{B-10}$$

If the desired lag time is 5 sec, then $1/k_2 + 1/k_3$ equals 1.0858. If an amount of the first auxiliary enzyme is taken to give $k_2 = 1 \ sec^{-1}$, then $1/k_3$ equals 0.0858, or an amount of second enzyme to give a k_3 of 11.67 sec^{-1} would be required.

The amounts of the two auxiliary enzymes used in a given assay can be adjusted depending upon availability, activity and price. If these are about the same, the most economical choice is with $k_2 = k_3(= 1.84 \ sec^{-1}$ for $t^* = 5$ sec). If one or the other enzyme is more readily available or cheaper, the quantities are adjusted accordingly. This relationship gives a rational basis for calculating the quantities of each auxiliary enzyme to use. More importantly, it once more emphasizes that the activity level of the initial enzyme in the series is irrelevant.

C Swinbourne Analysis of First-Order Reactions

When dealing with a first-order reaction it is often difficult to directly determine the concentration of the reactant of interest after the reaction has gone to completion. Sometimes this is because it is impracticable to wait until completion is attained; at other times extraneous factors (instrument drift, instability of some other species present) cause the difficulty. Regardless of the cause, experimentalists often wish to gather data during the course of a first-order kinetic run and then determine forthwith the rate constant (and possibly the final reactant concentration).

Several methods of doing this have been published, the earliest being by Guggenheim in 1926. However, a much simpler method is that published by E. S. Swinbourne (1960). This method is generally applicable to first-order reactions; Swinbourne demonstrated it with some of his data on a gas-phase reaction. We are interested in how it is used in analyzing first-order enzymatic runs, so the discussion will be couched in those terms.

Assuming that the experimental data consists of measurements of product concentration P as a function of time t, the procedure is as follows. A series of readings $P_1, P_2 \ldots P_n$ are made at times $t_1, t_2 \ldots t_n$, and a second series $P'_1, P'_2, \ldots P'_n$ are made at times $t_1 + T, t_2 + T \ldots t_n + T$. Then from the integrated first-order law,

$$P_\infty - P_n = P_\infty \exp(-k t_n) \qquad \text{and} \qquad \text{(C-1)}$$

$$P_\infty - P'_n = P_\infty \exp[-k(t_n + T)] \qquad \text{(C-2)}$$

Dividing C-1 by C-2 gives:

$$\frac{P_\infty - P_n}{P_\infty - P'_n} = \exp(kT); \qquad P_\infty - P_n = P_\infty \exp(kT) - P'_n \exp(kT)$$

and, upon further rearrangement,

$$P_n = P_\infty[1 - \exp(kT)] + P'_n\exp(kT) \qquad \text{(C-3)}$$

A straight line is obtained when readings in the first series are plotted versus readings in the second series, i.e., P_1 vs. P'_1, P_2 vs. P'_2, etc. The slope of the line is $\exp(kT)$, so:

$$k = \frac{\ln(\text{slope})}{T} \qquad \text{(C-4)}$$

At $t = \infty$, $P_n = P'_n$, and is the point at which the experimental line intersects a 45° line through all points $P_n = P'_n$. If the plot is fitted by linear least squares to get the intercept a and slope b, then the intersection point is:

$$P_\infty = \frac{a}{(1 - b)} \qquad \text{(C-5)}$$

Experimentally, P is read at a number of equally spaced times. T should be between one-half and one reaction half-lives (the half-life being that length of time to reduce $P_\infty - P$ by 50%), and may be chosen from a rough estimate of P_∞. In making the plot, remember that P' for one data pair may well be P for a subsequent point on the plot. To illustrate, the pressure-time data given by Swinbourne are reproduced here; the plot for $T = 4$ min is shown in Chapter 2, Figure 2-3.

t (min)	p (cm)	t (min)	p (cm)	t (min)	p (cm)
2	17.72	10	24.25	18	27.25
4	19.83	12	25.24	20	27.69
6	21.59	14	26.05	22	28.05
8	23.05	16	26.71	24	28.36

$P_1 = 17.72$ plots against $P'_1 = 21.59$, $P_2 = 19.83$ plots against $P'_2 = 23.05$, $P_3 = 21.59$ plots against $P'_3 = 24.25$, etc. A LLS fit of the 10 data pairs gives a slope of 1.4776, from which $k = 0.0976$ min^{-1}, and an intercept of -14.2195; the calculated P_∞ is 29.77 cm pressure.

Certain features of this graph should be kept in mind.

1. The original data are used directly. There is no need to take logarithms, differences, etc. This makes the method very convenient.
2. Readings towards the end of the reaction are "telescoped" towards the end of the graph. They have less relative weight in determining the slope than those points during the first few half-lives.

3. For best accuracy, the slope and intercept should be calculated from a least squares fit. Reading slope and the intersection point (P_∞) directly from the plot is usually less satisfactory.
4. The method is relatively insensitive to deviations from the strict first-order law. If there is any question about the order of the reaction, it should be checked independently.

If, at the beginning of an enzyme run for measuring the first-order rate constant [S] is not sufficiently smaller than K_M to make the conditions accurately first order, the Swinbourne plot will be convex. In this case the experiment should be rerun with a smaller initial value for [S].

Overall, the simplicity and versatility of the Swinbourne plot for analyzing first-order data in which P_∞ is not readily available make it the graphical method of choice.

D Statistics and Enzyme Assays

The experimental data obtained in developing an enzyme assay are used to decide whether or not certain conclusions can be drawn from the results. Examples might be: Is this response linear with respect to amount of enzyme used? or, Is the slope of the Hanes plot for the inhibited reaction identical with the slope of the plot in the absence of inhibitor (showing competitive inhibition)? One can (and should) plot the data and see how it looks. But then statistical techniques should be applied to give an objective measure of the probability of correctness of the conclusion drawn. This discussion will be limited to a few fundamental statistical ideas; for a fuller exposition consult a good text such as that by Draper and Smith (1981) and talk with a knowledgeable statistician. However, it is amazing how often even the simple ideas presented here are not used in published enzyme assay studies, to the detriment of the quality of the presentation.

Linear Least Squares Fitting

Regression equations express a fixed relationship between a known variable (X) and a response variable (Y). The underlying rationale is that if the parameters of the equation are known, then for a given value of X in an experiment the value of Y can be calculated, or predicted. The steps in setting up a regression equation are:

1. The form of the equation is established. This may be known a priori (e.g., the M-M equation) or various models may be tried to see which best fits the experimental data.
2. A series of values of X are used in experiments, and for each experiment the value of Y is measured. Examples would be $[S]$ (for X) and v (Y).
3. The data sets X, Y are used to determine the values of the parameters in the regression equation which best fit the experimental data. The meaning of the phrase "best fit" will be explored below. In the M-M equation the parameters are V_{max} and K_M.
4. The parametric values are then used in the regression equation to predict the value of Y expected at each given value of X. This is

termed $\hat{Y}$. In an enzyme experiment this would be the rate expected for a given $[S]$.

The Least Squares method of curve fitting minimizes the sum of the squares of the differences between Y and $\hat{Y}$ at each data point. The sum of squares is:

$$\text{SS} = \Sigma d_i^2 = \Sigma(Y_i - \hat{Y}_i)^2 \qquad \text{(D-1)}$$

where the Σ sign means summing the indicated quantities for each individual data point i of the N data sets. Those parametric values which minimize SS give the "best fit" mentioned in point 3 above. To find them the partial derivative of $\Sigma\, d_i^2$ with respect to each parameter is equated to zero. This gives a set of j linear equations, where j equals the number of parameters, which can then be solved by algebraic manipulation. These linear equations are known as the Normal Equations. To demonstrate the principle, we will use the simple linear equation $Y = a + bX$, but the means for expanding it to other situations will also be indicated.

A basic assumption in this derivation is that all the error in the experiment is in Y_i, and is random and normally distributed. In other words, the value taken for X_i is assumed to be inerrant, and experimental error all inheres in Y_i. This comes into play when weighting factors are used in making the calculations (Cleland 1979), but we will not go into such depth in this exposition.

For a linear fit, the predicted value $\hat{Y} = a + bX$, so equation D-1 becomes:

$$\Sigma d_i^2 = \Sigma(Y_i - a - bX_i)^2 \qquad \text{(D-2)}$$

$$\frac{\partial \Sigma d_i^2}{\partial a} = \Sigma\{2(Y_i - a - bX_i)(-1)\} = -2[\Sigma Y_i - aN - b\Sigma X_i] = 0 \quad \text{(D-3a)}$$

$$\frac{\partial \Sigma d_i^2}{\partial b} = \Sigma\{2(Y_i - a - bX_i)(-X_i)\} = -2[\Sigma X_i Y_i - a\Sigma X_i - b\Sigma X_i^2] = 0$$
$$\text{(D-3b)}$$

The factor 2 comes from the differentiation of the exponent 2 in equation D-2, the factor (-1) in D-3a comes from the differentiation of $-a$ with respect to a, and $(-X_i)$ in D-3b comes from the differentiation of $-bX_i$ with respect to b. The coefficient N for the a term in D-3a arises from the fact that the sums are taken over N terms, so summing 1 N times gives N.

Simplifying equations D-3a,b and rearranging slightly puts them into the form most often seen in statistics books:

$$aN + b\Sigma X_i = \Sigma Y_i \qquad \text{(D-4a)}$$

$$a\Sigma X_i + b\Sigma X_i^2 = \Sigma X_i Y_i \tag{D-4b}$$

This system of two linear equations in two unknowns is readily solved by simple algebraic manipulation (multiply D-4b by N; multiply D-4a by ΣX_i; subtract one from the other). We get:

$$b = \frac{N\Sigma X_i Y_i - \Sigma X_i \Sigma Y_i}{N\Sigma X_i^2 - (\Sigma X_i)^2} \tag{D-5a}$$

$$a = \frac{\Sigma Y_i - b\Sigma X_i}{N} \tag{D-5b}$$

The Normal Equations can be expanded to models with more than two parameters, i.e.,

$$Y = a + bX(1) + cX(2) + \cdots$$

The general paradigm is:

$$aN + b\Sigma X(1)_i + c\Sigma X(2)_i + \cdots = \Sigma Y_i \tag{D-6}$$

$$a\Sigma X(1)_i + b\Sigma X(1)_i^2 + c\Sigma X(1)_i X(2)_i + \cdots = \Sigma X(1)_i Y_i$$

$$a\Sigma X(2)_i + b\Sigma X(1)_i X(2)_i + c\Sigma X(2)_i^2 + \cdots = \Sigma X(2)_i Y_i \cdots$$

This set of equations can be solved either by successive elimination of constants by cross-multiplication and subtraction, or, more conveniently, by applying matrix methods. With more than three constants, matrix math is definitely the method of choice; the cross-multiplied coefficients become quite unwieldy to handle.

If $X(1)$, $X(2)$, etc. are powers of the same variable, this is just a simple polynomial expression. The rate expression for substrate inhibition, for instance, is readily solved by this expression where $Y = v$, $X(1) = [S]$ and $X(2) = [S]^2$ (see the low-affinity inhibitors section of Chapter 3). In the determination of the rate constant for denaturation of an enzyme using the integrated rate equation (see the data analysis section of Chapter 2) the independent variable is time, with $X(n) = t^n$ for n equal 2 or 3, depending on what level of truncation of the expansion is justified by the fit.

The polynomial fit to equation 2-19 has an unusual feature in that the curve must pass through the origin, i.e., $a = 0$. In setting up the algebraic equations to find b, c, etc. in this case, one less Normal Equation is needed than for the full curve fit. However, reference should be made to the

Normal Equation set, such as equation D-6, in developing the expressions for calculating d (for a third order polynomial), then c, then b. These expressions are *not* the same as for fitting a full third-order polynomial and then just setting $a = 0$ at the end of the calculation.

Goodness of Fit

The next point in the analysis is the significance of the parametric values calculated; how good is the fit? While the sum of squares of deviations may be minimized for the experimental data in hand and the model used, it is good to have a measure of the goodness of fit. One such measure is the coefficient of correlation, often referred to as r^2. For the two-parameter linear model, this is calculated as follows.

$$\Sigma x_i^2 = \Sigma X_i^2 - \frac{(\Sigma X_i)^2}{N} \tag{D-7a}$$

$$\Sigma y_i^2 = \Sigma Y_i^2 - \frac{(\Sigma Y_i)^2}{N} \tag{D-7b}$$

$$\Sigma x_i y_i = \Sigma X_i Y_i - \frac{\Sigma X_i \Sigma Y_i}{N} \tag{D-7c}$$

$$r^2 = \frac{\Sigma x_i y_i}{\sqrt{\Sigma x_i^2 \Sigma y_i^2}} \tag{D-7d}$$

An example where r^2 is useful is in the computer fitting of first-order data, where successive estimates of P_∞ are made and a fit is calculated to the regression $\ln(P_\infty - P)$ versus t (see the data analysis section of Chapter 2). The best estimate of P_∞ is that one which gives the highest correlation coefficient.

Two other pieces of information can be usefully obtained from an LLS analysis, namely the standard errors of the slope and intercept, S.E.(b) and S.E.(a) respectively. These are obtained as follows:

$$\Sigma d_i^2 = \Sigma y_i^2 - \frac{(\Sigma x_i y_i)^2}{\Sigma x_i^2} \tag{D-8a}$$

$$\sigma^2 = \frac{\Sigma d_i^2}{N - 2} \tag{D-8b}$$

$$S.E.(b) = \frac{\sigma}{\sqrt{\Sigma x_i^2}} \qquad \text{(D-8c)}$$

$$S.E.(a) = \sigma \sqrt{\frac{\Sigma X_i^2}{N \Sigma x_i^2}} \qquad \text{(D-8d)}$$

These are useful in comparing slopes or intercepts from plots of inhibitor analyses. If inhibition is competitive, for instance, then the slopes of the Hanes plots of data obtained at different $[I]$ should be the same within their respective standard errors. Also, in making replots of slopes or intercepts in the case of hyperbolic inhibition, knowing the S.E. of each point helps establish the reliability (probable error) of the secondary slope or intercept, and hence the reliability of the derived value of K_i.

The correlation coefficient r^2 gives the fraction of the total variation in the experimental values of Y which is explained by the regression on X. In other words, if you calculate a linear fit with an r^2 of 0.96, 96% of the spread of the individual Y_i values is linearly due to their association with the variable values of the X_i. The other 4% is due to other causes, such as random error in Y_i. Note that does not mean that the linear fit is necessarily the correct one! There is no specific level of r^2 which justifies such a conclusion. Other evidence is required, as we will show below.

The S.E. is analogous to the standard deviation about the mean of a list of numbers. If you calculate a linear slope b equal to 10 with S.E. equal to 1, this means that there is a 95% chance that the true slope for this data lies between 8 and 12 ($b \pm 2$ S.E.). A similar statement holds for the intercept a.

Residuals

We still have the question: Is the linear fit the correct one for this data? While there are rigorous methods for answering this question (Draper and Smith 1981) a simple method which will serve well in most situations is to plot the residuals from the regression. The residual at each point is the difference between the experimental value Y_i and the predicted value $\hat{Y}_i$. The sum of these residuals equals 0, and when plotted against X_i they should form a horizontal pattern centered about the zero axis. If the overall pattern of residuals is curved, or forms a line having a positive or negative slope, then the simple linear fit is not the correct one. To illustrate this I will show some linear fits and residual plots (Figure D-1) for assay data published by Hsu and Varriano-Marston (1983).

These workers were examining two commercial assay methods for measuring α-amylase activity. They used α-amylase isolated from malted barley

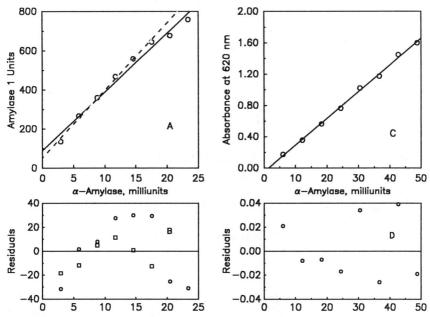

FIGURE D-1. Plots of amylase assay results. A. Nephelometric assay. B. Residuals from linear fit of nephelometric assay. C. Dyed-starch assay. D. Residuals from linear fit of dyed-starch assay.

which they standardized using the colorimetric method of Robyt and Whelan (1965). The first method is a nephelometric method which uses limit dextrin as the substrate and measures the time required to decrease turbidity by a given amount. The data are shown in Figure D-1A. The solid line is a linear fit to all eight data points and the dashed line is a fit to just the first six points. The two regression equations (with associated S.E.) are:

$$Y = 90.20 \ (24.51) + 29.97 \ (1.66) \times \alpha\text{-amylase mU}, \ r^2 = 0.982; \text{ and}$$

$$Y = 53.95 \ (13.37) + 34.44 \ (1.17) \times \alpha\text{-amylase mU}, \ r^2 = 0.995.$$

Now normally if one has a correlation coefficient of 0.982 that would be considered very good, with no further questions asked. The increase in r^2 to 0.995 does not in itself justify dropping the two top points in this run.

A plot of the residuals, however (Figure D-1B) tells a different story. The circles are residuals from the first regression (eight points) and the squares are from the second regression. Both plots have a definite hump, with the beginning and end residuals being negative and the middle residuals being positive. When you see this pattern, suspect that a quadratic term is needed for a more accurate fit. In the present case, the quadratic fit gives an r^2 of 0.998, and a plot of the residuals (not shown) is a narrower, horizontal band.

A second indication that a linear fit is inadequate is the nonzero intercept. As you know, in a properly run assay the response should be zero in the absence of enzyme. In this case the intercepts are four S.E.s away from zero, indicating a definite problem. In the quadratic fit the intercept is 6.59 (S.E. 16.83), or, statistically speaking, indistinguishable from 0. The quadratic fit is much more reliable, and the fact that the response has a curvilinear rather than a straight line dependency upon enzyme concentration is probably due to the nature of the nephelometric assay.

Hsu and Varriano-Marston also studied the Phadebas assay for α-amylase, a commercial kit which uses a dyed starch substrate, incubation for 15 min and then spectrophotometric measurement of the amount of solubilized dye. The results are shown in Figures D-1C and D-1D. The regression equation for the linear fit is:

$$Y = -0.0556 \ (0.0230) \ +0.0343(0.0007) \times \alpha\text{-amylase mU}; \ r^2 = 0.997.$$

The pattern of the residuals is basically horizontal, although the fact that it widens towards the right side indicates that the size of the random error increases as more starch is hydrolyzed. From this pattern we can conclude that a linear fit is the most appropriate one to use. The intercept of the regression line misses the origin by more than 2 S.E. indicating that there is some systematic error in the assay which tends to give low results, and which should be sought and eliminated.

By applying these few simple statistical concepts and techniques to your enzyme assay data you will greatly increase the soundness of the conclusions you draw from that data, and enhance the validity of your results for others who may need to rely upon them.

E   Utility Computer Programs

Much of the analysis of data generated by enzyme kinetic runs consists of "number-crunching"; finding the best straight line fit to a Hanes plot, obtaining the parameters for the expression for substrate-inhibited enzyme catalysis, and similar tasks. Thirty years ago this meant sitting down with the raw data, a pad of paper, and a mechanical calculator to grind out the intermediate calculations which eventually gave V_{max}, or K_i, or whatever parameter was being sought. Today the capabilities and ubiquity of the personal computer make such an approach obsolete. Generally speaking it is easier to take the raw data to a PC, along with the appropriate software programs, enter the data interactively, and record the results of the computer calculations.

The calculations are done faster, with no slips during the process. However, the basic slogan of all users of computers must be repeated:

G I G O !

The computer program HYPER, for instance, cannot give V_{max} with a relative standard error of 5% when the measurements of velocity contain a 20% variation. A LLS fit to a Hanes plot does not give good values for V_{max} and K_M from a fixed-time assay when significant product inhibition is present. If a random-addition two-substrate model is assumed when in fact the system is an ordered-addition one, the results are likely to be "hash." Remember that *a PC is a speedy, accurate idiot*. You must still apply good judgement in choosing the proper kinetic model and a reliable assay for the system, and good experimental technique in acquiring the raw data. In dealing with any unfamiliar enzyme system it is strongly recommended that appropriate graphs of the data be made and the selection of the correct model confirmed by visual inspection. While Artificial Intelligence computer programs to do this could, perhaps, be written, they are probably not available to working enzymologists; human intelligence is still more widely applicable.

The programs listed in this appendix have been written in Microsoft QuickBasic. Adaptations to other varieties of BASIC can be easily made.

The "print" statements are for writing to the monitor screen. If you prefer to have the output go to a printer, make the changes appropriate to your system. In all programs except one (M-MFIT) the fit is made via a set of Normal Equations and the calculations are done with matrix methods. These are listed separately as MATRIX subroutines, so adaptations of the programs for particular needs should be fairly straightforward.

The programs HYPER, SUBIN, BELL, HABELL, and HBBELL are rewrites in QuickBasic of FORTRAN programs published by Cleland (1979). The other three programs have been written by the present author. If the former programs are used to analyze data, Cleland's authorship should be acknowledged in any publication.

Input and Output of the Programs

These programs in general require three inputs: the number of data sets (*NP*), and, for each data set, the dependent variable (*Y*) and the independent variable (*X*). The output may be considered generally the parameters of interest with their associated standard error S.E. In addition, a table of experimental *Y* and calculated *Y* is also given, to allow detection of "outliers" by inspection.

A. M-MFIT. Fits the equation $Y = V_{max}X/(K_M + X)$. This program calculates values of V_{max} and K_M using four different approaches: the Direct Linear plot, and LLS fits to the Hanes plot, the Eadie plot, and the Lineweaver-Burk plot of the $v,[S]$ data. Since these different methods are optimally applicable to data with different error characteristics (see the discussion in Chapter 2), it is desirable to use the output which best fits your experimental situation. It is also sometimes helpful to compare these results with the results from HYPER, to get further insight into possible unsuspected error sources in the experimental system.

B. HYPER. Fits the equation $Y = V_{max}X/(K_M + X)$. Estimates of V_{max} and K_M are first made from a double reciprocal transformation of the data, then these are corrected using the Gauss-Newton method of iterative approach to the best fit (a fuller discussion is given by Cleland, 1967, 1979). The number of iterations is 3 (statement "For NT = 1 to 3") but may be changed if desired. This is probably the most useful program for routine analysis of well-behaved data.

C. POLYFIT. Fits the polynomial equation $Y = a + bX + cX^2 + \cdots$ In addition to NP and the individual data sets *Y,X* the program also prompts for the order of polynomial which is to be fit. This may be as large as 10, until you run into memory limitations on the PC or the time for solving the associated matrix becomes unacceptably long. The complete equation with $n = 2$ is used to fit substrate inhibition data (see the low-affinity inhibitors section of Chapter 3). A prompt also allows you to try different orders of the equation to fit a given set of data.

D. SUBIN. Fits the equation $Y = V_{max}X/(K_M + X + X^2/K_s)$. As for HYPER, an initial estimate of the three parameters is refined by Gauss-Newton iteration. The number of iterations is 5 ("For NT $= 1$ to 5") and may be changed. For well-behaved data this gives a better fit than does the use of the second-order polynomial method. However, substrate inhibition data tends to be more erratic, and on occasion SUBIN will "blow up," i.e., at each iteration the parameter values become more ridiculous (even though the statistical fit improves). If this occurs, consult Cleland (1979) for ways to find the problem in the data. The second-order polynomial will always give some estimates for V_{max}, K_M and K_s, even though the reliability is less.

E. EST1ST. Estimates best values for P_0, P_∞ and k' for $\ln(P_\infty - P) = -k't$. Inputs are NP and data sets P,t. An initial estimate is made of k' using the first two data points. This may prove to be too far from the true value to work well with the Gauss-Newton method. If you find that the program as written does not converge after the five iterations, you may have to write an Input statement and give an estimate of k' from a quick inspection of the data.

F. BELL. Fits the equation $\log Y = \log\{C/(1 + H/K_a + K_b/H)\}$. The data sets are Y (e.g., a rate constant),pH. The output is the maximal value of the rate constant C and the two pK's accounting for the bell-shaped dependency of C on pH. If the two pK's are less than 0.6 pH units apart the equation is not precisely correct. In this case a correction must be made in the paradigm. The program does this; the outputs then are apparent pK's which may not be precisely the same as true pK's. For a fuller discussion of this, see Cleland (1979) or Dixon and Webb (1979).

G. Sigmoid pH dependency. HABELL fits the equation $\log Y = \log\{C/(1 + H/K_a)\}$. The plot of Y versus pH describes a sigmoid curve with the kinetic constant approaching zero at low pH and a maximum value at high pH. The data sets are Y,pH. HBBELL fits the equation $\log Y = \log\{C/(1 + K_b/H)\}$. The kinetic constant approaches its maximum at low pH, and decreases to nearly zero at high pH. The data should lie in the range of 5% to 95% of the estimated maximum rate, i.e., ± 1.3 pH units of the pK. Extreme values, especially at low rates, have a marked biasing effect on the fitting algorithm; the errors may be as much as 15% in C and 0.2 units in pK due to one or two poor data points at the extremes.

Numerous other direct-fitting computer programs have been given by Cleland (1979). If the need extends beyond the programs given here, and you would like to transform his FORTRAN programs into your BASIC version, the cited reference should be consulted. Comparison of his listings (e.g., for HYPER) with the QuickBasic program listing given here should make the principles involved easy to apply.

```
Rem: QBasic Program: M-MFIT

Print "This program fits rate data to the Michaelis-Menten equation, using"
Print "the Direct Linear Plot and the Hanes, Eadie, and Lineweaver-Burk"
Print "linear transformations.": Print
DIM V(30), S(30),VM(500), KM(500),X(30), Y(30)
Input "How many data points";N
For I=1 TO N
  Input "RATE";V(I)
  Input "SUBSTRATE CONCENTRATION";S(I)
Next
Print "ITERATION NUMBER";
NN = N*(N-1)/2
  For I=1 TO NN:VM(I)=1000:KM(I)=1000:Next: IN = 0
  For J=1 TO N-1: For K=J+1 TO N
   VEST = (S(J)-S(K))/(S(J)/V(J)-S(K)/V(K))
   KEST = (V(K)-V(J))/(V(J)/S(J)-V(K)/S(K))
   IN = IN + 1
     For L = 1 TO I
     IF VEST>VM(L) Then Goto 100 Else Goto 110
100 Next L
110 For M = IN TO L+1 STEP -1
VM(M) = VM(M-1):NEXT M: VM(L)=VEST
  For L=1 TO I: IF KEST >KM(L) Then Goto 120 Else Goto 130
120 Next L
130 For M = IN TO L+1 STEP -1
KM(M) = KM(M-1):NEXT M: KM(L)=KEST
  Next K
Print J;:NEXT J: Print
T1=NN/2+.05:T2=INT(T1)
IF (T1-T2)>.2 GOTO 300
GOTO 320
300 VDM = VM(T2+1):KDM = KM(T2+1): GOTO Hanes
320 VDM = (VM(T2)+VM(T2+1))/2:KDM = (KM(T2)+KM(T2+1))/2

REM:FITS THE DATA USING THE HANES PLOT
Hanes:
For I=1 TO N: X(I) = S(I):Y(I) = S(I)/V(I):Next
  Gosub LinearLeastSquares
VHM = 1/B: KHM = A/B

REM:FITS THE DATA USING THE EADIE PLOT
For I=1 TO N: X(I) = V(I):Y(I) = V(I)/S(I):Next
  Gosub LinearLeastSquares
KEM = -1/B: VEM = KEM*A

REM:FITS THE DATA USING THE LINEWEAVER-BURK PLOT
For I=1 TO N: X(I)=1/S(I):Y(I)=1/V(I):Next
  Gosub LinearLeastSquares
VLM=1/A:KLM=B*VLM
```

```
Print "SUBSTRATE ";" EXPTL V  ";"DIR LIN V ";" HANES V  ";_
" EADIE V  ";" L-B V  "
DEVD=0:DEVH=0:DEVE=0:DEVL=O
  For I=1 TO N
   VD = VDM*S(I)/(KDM+S(I)):DEVD=DEVD+(VD-V(I))^2
   VH = VHM*S(I)/(KHM+S(I)):DEVH=DEVH+(VH-V(I))^2
   VE = VEM*S(I)/(KEM+S(I)):DEVE=DEVE+(VE-V(I))^2
   VL = VLM*S(I)/(KLM+S(I)):DEVL=DEVL+(VL-V(I))^2
    Print Using "###.####  ";S(I),V(I),VD,VH,VE,VL
  Next

DD=(INT(SQR(DEVD/(N-1))*1000000!+.5))/1000000!
DH=(INT(SQR(DEVH/(N-1))*1000000!+.5))/1000000!
DE=(INT(SQR(DEVE/(N-1))*1000000!+.5))/1000000!
DL=(INT(SQR(DEVL/(N-1))*1000000!+.5))/1000000!: Print
Print "","DIRECT LIN.  ";"HANES PLOT   ";"EADIE PLOT    ";"L-B PLOT"
Print "V MAX ",VDM,VHM,VEM,VLM
Print "KM ",KDM,KHM,KEM,KLM
Print "STD. DEV. ",DD,DH,DE,DL
End

REM: Linear Least Squares Subroutine for direct fitting of Y = A + B*X
LinearLeastSquares: G=0:H=0:L=0:M=0:O=0
  For I=1 TO N: X=X(I):Y=Y(I)
   G=G+X:H=H+X^2:L=L+Y:M=M+X*Y:O=O+Y^2: Next I
B=(N*M-G*L)/(N*H-G*G): A=(L-B*G)/N
XD=H-G*G/N: YD=O-L*L/N: XYD=M-G*L/N: CC=XYD/SQR(XD*YD)
Return
```

```
Rem: QBasic Program: HYPER, Rewritten from W. W. Cleland's Fortran program.

Print "This program fits rate data to the equation v = V*S/(Km + S)."
Print: DIM V(50), A(50), S(3,4), Q(3), SM(3), SS(3)
Input "How many data points";NP
  N=2:N1=3:N2=4
Gosub MatrixClear

For I=1 TO NP
  Input "RATE";V(I)
  Input "SUBSTRATE CONCENTRATION";A(I)
    Q(1)=V(I)^2/A(I)
    Q(2)=V(I)^2
    Q(3)=V(I)
  Gosub MatrixSetup
Next I

Gosub MatrixSolve: CK=S(1,1)/S(2,1)
Print "ESTD. K ";CK;" ESTD. V ";1/S(2,1)

For NT = 1 to 3
Gosub MatrixClear
For I=1 TO NP
  D=CK+A(I)
  Q(1)=A(I)/D
  Q(2)=Q(1)/D
  Q(3)=V(I)
  Gosub MatrixSetup
Next I

Gosub MatrixSolve: CK=CK-S(2,1)/S(1,1)
Next NT

S2=0: For I = 1 TO NP
  S2=S2+(V(I)-S(1,1)*A(I)/(CK+A(I)))^2:Next I
S1=SQR(S2/(NP - N))

Gosub MatrixVar
SL=CK/S(1,1): VINT=1/S(1,1): VK=1/SL
SEV=S1*SQR(S(1,2)): SECK=S1*SQR(S(2,3)/S(1,1))
SEVI=SEV/S(1,1)^2
S(1,3)=S1*SQR(CK^2*S(1,2)+S(2,3)+2*CK*S(1,3))
SESL=S(1,3)/S(1,1)^2: SEVK=S(1,3)/CK^2
Print "K = " CK,"S.E.(K) = " SECK
Print "V = " S(1,1),"S.E.(V) = " SEV
Print "Sigma = " S1: Print
Print "SUBSTRATE","EXPTL V","CALC V"

For I = 1 TO NP
  CV = S(1,1)*A(I)/(CK+A(I)):  Print A(I),V(I),CV
Next I
END
For the rest of this program, insert the MATRIX Subroutines.
```

```
Rem: QBasic Program: POLYFIT

Print "This Program fits a polynomial Y = A0 + A1*X + A2*X^2 + ..."
Print "up to degree 10.": Print
DIM X(50), Y(50), Q(12), A(11), SM(12), SS(12), S(12,13)
Input "How many data points ";NP
   YT=0: For I = 1 to NP
   Input "X";X(I): Input "Y";Y(I): YT = YT + Y(I): Next I
Start: Input "What degree polynomial do you want to try (1-10)";D
   N = D + 1: N1 = N + 1: N2 = N + 2
Gosub MatrixClear
For I = 1 to NP
   For J = 0 to D
   Q(J+1) = X(I)^J
   Next J
   Q(N1) = Y(I)
   Gosub MatrixSetup
Next I
Gosub MatrixSolve
Gosub MatrixVar
Print "   X(I)  ","   Y(I)  ","Pred. Y ","Residual"
For K = 1 to N: A(K-1) = S(K,1): Next K
S2 = 0: R2U = 0: R2L = 0: YA = YT/NP
For I = 1 to NP: YP = 0
 For J = 0 to D: YP = YP + A(J)*X(I)^J:  Next J
 S2 = S2 + (Y(I) - YP)^2
 R2U = R2U + (YP - YA)^2
 R2L = R2L + (Y(I) - YA)^2
 Print X(I),Y(I),YP,Y(I)-YP
Next I: S1 = SQR(S2/(NP - N))
For I = 0 to D
   Print "A";I;" = ";A(I);"      ";"Std. Error ";S1*SQR(S(I+1,I+2))
Next I
Print: Print "Sigma = ";S1; "    Correlation, R^2 = ";R2U/R2L
Print "F Statistic = ";R2U/3/S1^2; " Residual Sum of Squares = ";S2
Print: Input "Do you want to try another fit (Y/N)?";A$
If A$ = "Y" or A$ = "y" Then GoTo Start Else GoTo Final
Final: END
```

For the rest of this program, insert the MATRIX Subroutines.

```
Rem: QBasic Program: SUBIN, Rewritten from W. W. Cleland's Fortran program.

Print "This program fits rate data to the equation v = V*S/(Km + S + S^2/Ki)."
Print: DIM V(50), A(50), S(4,5), Q(4), SM(4), SS(4)
Input "How many data points";NP
N=3:N1=4:N2=5
Gosub MatrixClear
For I=1 TO NP
  Input "RATE";V(I)
  Input "SUBSTRATE CONCENTRATION";A(I)
    Q(1)=V(I)^2/A(I)
    Q(2)=V(I)^2
    Q(3)=V(I)^2*A(I)
    Q(4)=V(I)
  Gosub MatrixSetup
Next I
Gosub MatrixSolve
CK=S(1,1)/S(2,1): B=S(3,1)/S(2,1)
Print "ESTD. K " CK;" ESTD. Ki " B;" ESTD. V ";S(2,1)

For NT = 1 to 5
Gosub MatrixClear
For I=1 TO NP
    D=1+CK/A(I)+B*A(I)
    Q(1)=1/D
    Q(2)=1/D^2/A(I)
    Q(3)=A(I)/D^2
    Q(4)=V(I)
  Gosub MatrixSetup
Next I
Gosub MatrixSolve
CK=CK-S(2,1)/S(1,1): B=B-S(3,1)/S(1,1)
Next NT

S2=0
For I = 1 TO NP: S2=S2+(V(I)-S(1,1)/(1+CK/A(I)+B*A(I)))^2:Next I
S1=SQR(S2/(NP - N))
Gosub MatrixVar
SL=CK/S(1,1): VINT=1/S(1,1)
CKI=1/B: VK=1/SL
SEV=S1*SQR(S(1,2)): SECK=S1*SQR(S(2,3)/S(1,1))
SEB=S1*SQR(S(3,4)/S(1,1)): SEVI=SEV/S(1,1)^2
S(1,3)=S1*SQR(CK^2*S(1,2)+S(2,3)+2*CK*S(1,3))
SESL=S(1,3)/S(1,1)^2: SEVK=S(1,3)/CK^2: SEKI=SEB/B^2
Print "K = " CK,"S.E.(K) = " SECK
Print "V = " S(1,1),"S.E.(V) = " SEV
Print "Ki = " CKI,"S.E.(Ki) = " SEKI
Print "SIGMA = " S1: Print
Print "SUBSTRATE","EXPTL V","CALC V"
 For I = 1 TO NP: CV = S(1,1)/(1+CK/A(I)+B*A(I)): Print A(I),V(I),CV: Next
END
```

For the rest of this program, insert the MATRIX Subroutines.

```
Rem: QBasic Program: EST1ST

Print "This program fits rate data to the first order rate equation,"
Print "Y = Yinf + (Yo - Yinf)*Exp(-k't).": Print
DIM T(50), Y(50), S(4,5), Q(4), SM(4), SS(4)
Input "How many data points"; NP: N = 3: N1 = 4: N2 = 5
For I = 1 to NP
  Input "Time";T(I)
  Input "Y Variable";Y(I)
Next
Ko = Abs((Log((Y(1)-Y(NP))/(Y(2)-Y(NP))))/(T(2) - T(1)))
Print "Estimated Ko ";Ko

For NT = 1 to 5
Gosub MatrixClear
For I = 1 to NP
  Q(1) = 1
  Q(2) = Exp(-Ko*T(I))
  Q(3) = T(I)*Q(2)
  Q(4) = Y(I)
  Gosub MatrixSetup
Next I
Gosub MatrixSolve
Ko = Ko - S(3,1)/S(2,1)
Next NT

Gosub MatrixVar
YInf = S(1,1)
YZero = S(2,1) + YInf
K = Ko - S(3,1)/S(2,1)
Print "Time","Y, Data","Y, Calc."
S2 = 0
For I = 1 to NP
Ycalc = (S(1,1) + S(2,1)*Exp(-K*T(I)))
S2 = S2 + (Y(I)-YCalc)^2
Print T(I), Y(I), Ycalc
Next I:Print

S1 = SQR(S2/(NP-3))
SEInf = SQR(S2*S(1,2))
SEZero = SQR(S2*Abs(S(1,2)-S(2,3)))
SEK = SQR(S2*Abs(S(3,4)/S(2,1)))
Print "Y(zero time) = ";Yzero;" Std. Error = ";SEZero
Print "Y(infinity) = ";Yinf;" Std. Error = ";SEInf
Print "Rate Constant = ";K;" Std. Error = ";SEK
END
```

For the rest of the program, insert MATRIX Subroutines.

```
Rem: QBasic Program: BELL, Rewritten from W. W. Cleland's Fortran program.

Print "This program fits data to the curve log Y = log [C/(1 + H/Ka + Kb/H)]."
DIM V(50),A(50),Q(4),SM(4),SS(4),S(4,5)
Input "How many data points";NP:N=3:N1=4:N2=5
Gosub MatrixClear
For I = 1 TO NP
  Input "RATE";V(I)
  Input "PH";A:A(I)=1/EXP(2.3026*A)
    Q(1) = V(I)
    Q(2) = A(I)*Q(1)
    Q(3) = Q(1)/A(I)
    Q(4) = 1
  Gosub MatrixSetup
Next I
Gosub MatrixSolve
CKB = S(3,1)/S(1,1)
CKA = S(1,1)/S(2,1)
Print "ESTD. K1 ";CKA;"ESTD. K2";CKB
SCKA = CKA: SCKB = CKB

For NT = 1 to 5
Gosub MatrixClear
For I = 1 TO NP
    D = 1 + A(I)/CKA + CKB/A(I)
    Q(1) = 1
    Q(2) = A(I)/D
    Q(3) = D/A(I)
    Q(4) = LOG(V(I)*D)
  Gosub MatrixSetup
Next I
Gosub MatrixSolve
    CKA = CKA + S(2,1)*CKA^2
    CKB = CKB - S(3,1)
    IF CKA <=0 THEN CKA = SCKA
    IF CKB <=0 THEN CKB = SCKB
    CV = EXP(S(1,1))
Next NT

Gosub MatrixVar
DF = 1/LOG(10):Z = CKA^2 - 4*CKA*CKB
If Z <= 0 Then Goto ClosepK Else Goto WidepK

WidepK:Gosub SigmapK
  PKA = -DF*LOG(CKA): SEPKA = S1*DF*CKA*SQR(S(2,3))
    Print "PKA = ";PKA,"S.E.(PKA) = ";SEPKA
  PKB = -DF*LOG(CKB): SEPKB = S1*DF*SQR(S(3,4))/CKB
Print "PKB = ";PKB,"S.E.(PKB) = ";SEPKB
Print "SIGMA = ";S1
  Z = SQR(Z): TCKB = (CKA - Z)/2: TCKA = (CKA + Z)/2
  TPKA = -DF*LOG(TCKA): TPKB = -DF*LOG(TCKB)
  Print "WITH TWO GROUPS INVOLVED SEPARATED BY MORE THAN 0.6 UNITS,"
  Print "THE TRUE PKS ARE:"
  Print "TRUE PKA = ";TPKA,"TRUE PKB = ";TPKB: END
```

```
ClosepK:
Print "PKS CLOSER THAN MINIMUM SEPARATION OF 0.6 UNITS."
Print "TRUE PKS ARE IDENTICAL."
N=2:N1=3:N2=4:NT=0

For NT = 1 to 5
Gosub MatrixClear
For I = 1 TO NP
  D = 1 + A(I)/CKA + CKA/4*A(I)
  Q(1) = 1
  Q(2) = (1/4*A(I) - A(I)/CKA^2)/D
  Q(3) = LOG(V(I)*D)
  Gosub MatrixSetup
Next I

Gosub MatrixSolve
CKA = CKA - S(2,1)
Next NT
CV = EXP(S(1,1))
CKB = CKA/4
Gosub MatrixVar

Gosub SigmapK
PKA = -DF*LOG(CKA)
PKB = PKA + .60206
SEPKS = S1*DF*SQR(S(2,3))/CKA
Print "APPARENT PKS ARE CALCULATED TO BE:"
Print "APPARENT PKA = ";PKA, "APPARENT PKB = ";PKB; "S.E. (PKS) = ";SEPKS
Print "SIGMA = ";S1
END

SigmapK: S2 = 0
PRINT "   PH    ";"  EXPTL V ";" CALC V ";"   DIFF   ";"EXPTL LOGV";_
"CALC LOG V";"   DIFF   "
For I = 1 TO NP
  X = CV/(1 + A(I)/CKA + CKB/A(I))
  PH = -DF*LOG(A(I))
  X2 = DF*LOG(V(I))
  X3 = DF*LOG(X)
  DX1 = X2 - X3
  DX = V(I) - X
  PRINT USING "#####.####";PH;V(I);X;DX;X2;X3;DX1
  S2 = S2 + DX1^2
Next I
  S2 = S2/(NP-N)
  S1 = SQR(S2)
  SEV = S1*SQR(S(1,2))*CV
Print "MAXIMUM RATE = ";CV," S.E. = ";SEV
Return
```

For the rest of the program, insert MATRIX Subroutines.

```
Rem: QBasic Program: HABELL, Rewritten from W. W. Cleland's Fortran program.

Print "This program fits data to the curve log Y = log {C/(1 + H/Ka)}"
Print
DIM V(50),A(50),Q(3),SM(3),SS(3),S(3,4)
Input "How many data points";NP: N = 2:N1 = 3:N2 = 4
Gosub MatrixClear
For I = 1 TO NP
  Input "RATE"; V(I)
  Input "PH";A:A(I) = 1/EXP(2.3026*A)
    Q(1) = V(I)
    Q(2) = A(I)*Q(1)
    Q(3) = 1
  Gosub MatrixSetup
Next I

Gosub MatrixSolve
CKA = S(1,1)/S(2,1): SCKA = CKA
Print "ESTD. KA ";CKA

For NT = 1 to 5
Gosub MatrixClear
For I = 1 TO NP
  D = 1 + A(I)/CKA
  Q(1) = 1
  Q(2) = A(I)/D
  Q(3) = LOG(V(I)*D)
  Gosub MatrixSetup
Next I

Gosub MatrixSolve
CKA = CKA + S(2,1)*CKA^2: IF CKA <= 0 THEN CKA = SCKA
CV = EXP(S(1,1)): Next NT
Gosub MatrixVar
DF = 1/LOG(10): S2 = 0
Print "   PH   ";" EXPTL V ";" CALC V ";"  DIFF  ";"EXPTL LOGV";_
"CALC LOG V";"  DIFF  "
For I = 1 TO NP
  X = CV/(1 + A(I)/CKA): PH = -DF*LOG(A(I))
  X2 = DF*LOG(V(I)): X3 = DF*LOG(X)
  DX1 = X2 - X3: DX = V(I) - X
    Print USING "#####.####";PH,V(I),X,DX,X2,X3,DX1
  S2 = S2 + DX1^2
Next I

S1 = SQR(S2/(NP-2)): SEV = S1*SQR(S(1,2))*CV
Print "MAXIMUM RATE = ";CV;"  S.E. = ";SEV
PKA = -DF*LOG(CKA): SEPKA = S1*DF*CKA*SQR(S(2,3))
Print "PKA = ";PKA; "  S.E. = ";SEPKA
Print "SIGMA = ";S1
END
```
For the rest of the program, insert MATRIX Subroutines.

```
Rem: QBasic Program: HBBELL, Rewritten from W. W. Cleland's Fortran program.

Print "This program fits data to the curve log Y = log {C/(1 + Kb/H)}"
Print
DIM V(50),A(50),Q(3),SM(3),SS(3),S(3,4): N = 2:N1 = 3:N2 = 4
Input "How many data points";NP
Gosub MatrixClear

For I = 1 TO NP
  Input "RATE"; V(I)
  Input "PH";A:A(I) = 1/EXP(2.3026*A)
    Q(1) = V(I)
    Q(2) = Q(1)/A(I)
    Q(3) = 1
  Gosub MatrixSetup
Next I

Gosub MatrixSolve
CKB = S(2,1)/S(1,1): SCKB = CKB
PRINT "ESTD. KB ";CKB

For NT = 1 to 5
Gosub MatrixClear
For I = 1 TO NP
  D = 1 + CKB/A(I)
  Q(1) = 1
  Q(2) = D/A(I)
  Q(3) = LOG(V(I)*D)
  Gosub MatrixSetup
Next I

Gosub MatrixSolve: CKB = CKB - S(2,1)
IF CKB <= 0 THEN CKB = SCKB: CV = EXP(S(1,1))
Next NT

Gosub MatrixVar
DF = 1/LOG(10): S2 = 0
Print "   PH   ";" EXPTL V ";" CALC V ";" DIFF   ";"EXPTL LOGV";_
"CALC LOG V";" DIFF  "
For I = 1 TO NP
  X = CV/(1 + CKB/A(I)): PH = -DF*LOG(A(I))
  X2 = DF*LOG(V(I)): X3 = DF*LOG(X)
  DX1 = X2 - X3: DX = V(I) - X
    Print USING "#####.####";PH,V(I),X,DX,X2,X3,DX1
  S2 = S2 + DX1^2
Next I
S1 = SQR(S2/(NP-2)): SEV = S1*SQR(S(1,2))*CV
Print "MAXIMUM RATE = ";CV;"   S.E. = ";SEV
PKB = -DF*LOG(CKB): SEPKB = S1*DF*SQR(S(2,3))/CKB
Print "PKB = ";PKB; "   S.E. = ";SEPKB
Print "SIGMA = ";S1
END
```
For the rest of this program, insert the MATRIX Subroutines.

Rem: QBasic Matrix Manipulation Subroutines MATRIX

```
MatrixClear:
  FOR J=1 TO N2:FOR K=1 TO N1:S(K,J)=0:NEXT K:NEXT J:RETURN

MatrixSetup:
  FOR J=1 TO N1:FOR K=1 TO N:S(K,J)=S(K,J)+Q(K)*Q(J):NEXT K:NEXT J:RETURN

MatrixSolve:
  FOR K = 1 TO N: SM(K)=1/SQR(S(K,K)):NEXT
SM(N1)=1
  FOR J = 1 TO N1: FOR K = 1 TO N
   S(K,J)=S(K,J)*SM(K)*SM(J):NEXT K:NEXT J
SS(N1)=-1: S(1,N2)=1
 FOR L=1 TO N: FOR K = 1 TO N
  SS(K)=S(K,1):NEXT K
    FOR J = 1 TO N1: FOR K = 1 TO N
       S(K,J)=S(K+1,J+1)-SS(K+1)*S(1,J+1)/SS(1):NEXT K:NEXT J:NEXT L
 FOR K = 1 TO N: S(K,1)=S(K,1)*SM(K):NEXT
Return

MatrixVar:
  For J = 2 to N1: For K = 1 to N
   S(K,J) = S(K,J)*SM(K)*SM(J-1):
   Next K: Next J
Return
```

F General Model for Reversible Inhibition

Reversible inhibition by low-affinity inhibitors can have several different effects. The velocity at saturating substrate concentration can be lowered (noncompetitive inhibition), the concentration of substrate to produce half the maximum velocity can be increased (competitive inhibition), or both effects may appear (mixed inhibition). Further, the tertiary complex of enzyme, substrate and inhibitor (ESI) may break down to form product but at a rate which is less than that of the breakdown of the ES complex (partial inhibition).

The Rate Equation for General Hyperbolic Inhibition

The general model for reversible inhibition is depicted schematically in Figure 3-1 (Chapter 3). The rapid equilibrium model and constants are:

$$E + S \rightleftharpoons ES \qquad K_M = \frac{[E][S]}{[ES]} \qquad \text{(F-1a)}$$

$$E + I \rightleftharpoons EI \qquad K_i = \frac{[E][I]}{[EI]} \qquad \text{(F-1b)}$$

$$ES + I \rightleftharpoons ESI \qquad \alpha K_i = \frac{[ES][I]}{[ESI]} \qquad \text{(F-1c)}$$

$$EI + S \rightleftharpoons ESI \qquad \alpha K_M = \frac{[EI][S]}{[ESI]} \qquad \text{(F-1d)}$$

$$ES \longrightarrow P + E \qquad \frac{dP}{dt} = k_p[ES] \qquad \text{(F-1e)}$$

$$ESI \longrightarrow P + E + I \qquad \frac{dP}{dt} = \beta k_p[ESI] \qquad \text{(F-1f)}$$

263

The treatment is that for the Rapid Equilibrium assumption (Appendix A), with ES as the central complex.

$$[E] = \frac{[ES]K_M}{[S]} \tag{F-2a}$$

$$[EI] = \frac{[E][I]}{K_i} = \frac{[ES]K_M[I]}{K_i[S]} \tag{F-2b}$$

$$[ESI] = \frac{[ES][I]}{\alpha K_i} \tag{F-2c}$$

The ratio expression is:

$$\frac{v}{[E_t]} = \frac{k_p[ES] + \beta k_p[ESI]}{[E] + [ES] + [EI] + [ESI]} \tag{F-3}$$

Substituting for the various enzyme species, we have:

$$\frac{v}{[E_t]} = \frac{[ES](k_p + \beta k_p[I]/\alpha K_i)}{[ES](K_M/[S] + 1 + K_M[I]/K_i[S] + [I]/\alpha K_i)} \tag{F-4}$$

Cancelling the terms in $[ES]$, substituting for $V_{max} = k_p[E_t]$, and collecting the terms in K_M and $[S]$ in the denominator, we get:

$$\frac{v}{V_{max}} = \frac{(1 + \beta[I]/\alpha K_i)[S]}{K_M(1 + [I]/K_i) + [S](1 + [I]/\alpha K_i)} \tag{F-5}$$

Rearrangement gives the final form analogous to the M-M equation:

$$v = \frac{V_{max}\{(1 + \beta[I]/\alpha K_i)/(1 + [I]/\alpha K_i)\}[S]}{K_m\{(1 + [I]/K_i)/(1 + [I]/\alpha K_i)\} + [S]} \tag{F-6}$$

This is the same form as equation 3-1 (Chapter 3):

$$v = \frac{V_{app}[S]}{K_{app} + [S]} = \frac{V_{max}\{f_v\}[S]}{K_M\{f_k\} + [S]}$$

with $f_v = (\alpha K_i + \beta[I])/(\alpha K_i + [I])$ and $f_k = (\alpha K_i + \alpha[I])/(\alpha K_i + [I])$.

The significance of f_v and f_k were discussed in detail in Chapter 3, and the expressions for these terms in the various inhibitory patterns are given in Table 3-1 in that chapter.

Derivation of Delta Replot Equations

Three kinds of delta replots were recommended in Chapter 3:

1. delta intercept, $\Delta KV = K_{app}/V_{app} - K_M/V_{max}$;
2. delta slope, $\Delta V = 1/V_{app} - 1/V_{max}$; and
3. delta first order, $\Delta' = V_{max}/K_M - V_{app}/K_{app}$.

Delta Intercept. $\Delta KV = (K_M/V_{max})(f_k/f_v - 1)$

$$\frac{f_k}{f_v} - 1 = \frac{\alpha K_i + \alpha[I]}{\alpha K_i + \beta[I]} - 1 = \frac{\alpha K_i + \alpha[I] - \alpha K_i - \beta[I]}{\alpha K_i + \beta[I]} \tag{F-7a}$$

$$= \frac{[I](\alpha - \beta)}{\alpha K_i + \beta[I]} \tag{F-7b}$$

$$\Delta KV = \frac{K_M(\alpha - \beta)[I]}{V_{max}(\alpha K_i + \beta[I])} \tag{F-7c}$$

$$\frac{1}{\Delta KV} = \frac{V_{max}}{K_M} \cdot \frac{\alpha K_i}{(\alpha - \beta)} \cdot \frac{1}{[I]} + \frac{V_{max}}{K_M} \cdot \frac{\beta}{(\alpha - \beta)} \tag{F-7d}$$

The plot of $1/\Delta KV$ versus $1/[I]$ has the indicated slope and intercept. The calculations were discussed in Chapter 3.

Delta Slope. $\Delta V = (1/V_{max})(1/f_v - 1)$

$$\frac{1}{f_v} - 1 = \frac{\alpha K_i + [I]}{\alpha K_i + \beta[I]} - 1 = \frac{\alpha K_i + [I] - \alpha K_i - \beta[I]}{\alpha K_i + \beta[I]} \tag{F-8a}$$

$$= \frac{[I](1 - \beta)}{\alpha K_i + \beta[I]} \tag{F-8b}$$

$$\Delta V = \frac{1}{V_{max}} \cdot \frac{(1 - \beta)[I]}{(\alpha K_i + \beta[I])} \tag{F-8c}$$

$$\frac{1}{\Delta V} = V_{max} \frac{\alpha K_i}{(1 - \beta)} \cdot \frac{1}{[I]} + V_{max} \frac{\beta}{(1 - \beta)} \tag{F-8d}$$

The calculations based upon the slope and intercept of the plot of $1/\Delta V$ versus $1/[I]$ were discussed in Chapter 3.

Delta First-Order Rate. $\Delta' = (V_{max}/K_M)(1 - f_v/f_k)$

$$1 - \frac{f_v}{f_k} = 1 - \frac{\alpha K_i + \beta[I]}{\alpha K_i + \alpha[I]} = \frac{\alpha K_i + \alpha[I] - \alpha K_i - \beta[I]}{\alpha K_i + \alpha[I]} \tag{F-9a}$$

$$= \frac{[I](\alpha - \beta)}{\alpha K_i + \alpha[I]} \tag{F-9b}$$

$$\Delta' = \frac{V_{max}}{K_M} \cdot \frac{(\alpha - \beta)[I]}{(\alpha K_i + \alpha[I])} \tag{F-9c}$$

$$\frac{1}{\Delta'} = \frac{K_M}{V_{max}} \cdot \frac{\alpha K_i}{(\alpha - \beta)} \cdot \frac{1}{[I]} + \frac{K_M}{V_{max}} \cdot \frac{\beta}{(\alpha - \beta)} \tag{F-9d}$$

The application of the plot of $1/\Delta'$ versus $1/[I]$ was discussed in Chapter 3.

Pre-Steady State Transient Kinetics

If an enzyme catalyzes a reaction in two discrete steps and if the rate constant for the first step is much faster than the rate constant for the second step then in principle it is possible to "titrate" the active enzyme centers in the reaction mixture, i.e., to measure the molar concentration of the enzyme. The reaction model is:

$$E + S \underset{}{\overset{K_m}{\rightleftharpoons}} ES \xrightarrow{k_2} EA + P_1 \qquad EA \xrightarrow{k_3} E + P_2 \qquad \text{(G-1a)}$$

$$K_m = \frac{[E][S]}{[ES]} \qquad \text{(G-1b)}$$

$$\frac{d[P_1]}{dt} = k_2[ES] \qquad \text{(G-1c)}$$

$$\frac{d[P_2]}{dt} = k_3[EA] \qquad \text{(G-1d)}$$

$$\frac{d[EA]}{dt} = k_2[ES] - k_3[EA] \qquad \text{(G-1e)}$$

$$[E_t] = [E] + [ES] + [EA] \qquad \text{(G-1f)}$$

The rate constants k_1 and k_{-1} are for the forward and backward rates of the first step, i.e., $K_m = k_{-1}/k_1$. In addition an enzyme-substrate system such as this makes it possible to measure the rate constants for the reaction of a high-affinity inhibitor with the enzyme.

Enzyme Titration

Usually these titrations have been developed for hydrolytic enzymes (proteases, esterases) and EA is in fact an acylated enzyme. Also, for the most part, the experimental procedure has been to measure P_1, by either absorbance or fluorescence spectrophotometry, although with at least one titrant, N-*trans*-cinnamoylimidazole, it is the disappearance of S which is followed by absorbance decrease. Since we assume rapid equilibrium conditions for the analysis, the results are the same. In the following analysis it is assumed that $[S] >> [E_t]$ (Bender et al. 1967); the case where $[E_t] >> [S]$ will be discussed at the end of the analysis (Bender et al. 1962).

Substituting $[E] = [ES]K_m/[S]$ into equation G-1f and rearranging gives:

$$[ES] = \frac{[E_t] - [EA]}{1 + K_m/[S]} \qquad \text{(G-2a)}$$

which, with equation G-1e, yields:

$$\frac{d[EA]}{dt} = \frac{k_2[E_t]}{1 + K_M/[S]} - \frac{k_2[EA]}{1 + K_M/[S]} - k_3[EA] \qquad \text{(G-2b)}$$

or:

$$\frac{d[EA]}{dt} = \frac{k_2[E_t]}{1 + K_M/[S]} - \left(k_3 + \frac{k_2}{1 + K_m/[S]}\right)[EA] \qquad \text{(G-2c)}$$

For a particular experiment k_2, k_3, K_m, and $[S]$ are constant, so with:

$$a = \frac{k_2[E_t]}{1 + K_m/[S]} \quad \text{and} \quad b = k_3 + \frac{k_2}{1 + K_m/[S]} \qquad \text{(G-3a)}$$

we have:

$$\frac{d[EA]}{dt} = a - b[EA] \qquad \text{(G-3b)}$$

Integrating between the limits of $0,[EA]$ and $0,t$ gives:

$$[EA] = \left(\frac{a}{b}\right)(1 - e^{-bt}) \qquad \text{(G-3c)}$$

Substituting this value of $[EA]$ into equation G-2a results in:

$$[ES] = \frac{[E_t] - (a/b)(1 - e^{-bt})}{1 + K_m/[S]} \qquad \text{(G-4)}$$

This is now substituted into equation G-1c to give the expression for the rate of appearance of the first product P_1:

$$\frac{d[P_1]}{dt} = \frac{k_2[E_t] - k_2(a/b)(1 - e^{-bt})}{1 + K_m/[S]} \tag{G-5a}$$

This may be integrated between the limits $0,[P_1]$ and $0,t$ to give an equation which predicts the concentration of P_1 as a function of time:

$$[P_1] = \frac{k_2[E_t] - k_2(a/b)}{1 + K_m/[S]} \cdot t + \frac{k_2 a(1 - e^{-bt})}{b^2(1 + K_m/[S])} \tag{G-5b}$$

or, in a parametric form:

$$[P_1] = At + B(1 - e^{-bt}) = At + B - Be^{-bt} \tag{G-5c}$$

The curve which corresponds to this equation is shown in Figure G-1. Several points about this curve should be noted. At very large times $[P_1] = At + B$, a straight line with intercept B and slope A. When time is short, so that the plot is curved due to the exponential term in equation G-5c, then at any time t the *difference* between the actual plot and the extrapolated straight line is $\Delta P_1 = Be^{-bt}$. A plot of $\ln(\Delta P_1)$ versus t is a straight line with the intercept B and the slope $-b$. The experimental parameters obtained from this curve are A, B and b.

A is the steady-state rate of formation of product when the reaction system is turning over, i.e., the rate one would expect in a normal initial velocity determination. It is equivalent to v in the M-M equation, where V_{max} and K_M have the significance shown by equations 2-4a and 2-4b in Chapter 2.

The intercept B allows the calculation of $[E_t]$. From equation G-5b:

$$B = \frac{k_2 a}{b^2(1 + K_m/[S])} \tag{G-6a}$$

Substituting for a and b from G-3a, and noting that $K_M = K_m k_3/(k_2 + k_3)$:

$$B = \frac{[E_t]\{k_2/(k_2 + k_3)\}^2}{(1 + K_M/[S])^2} \tag{G-6b}$$

When $k_3 \ll k_2$ and $K_M \ll [S]$, conditions usually, but not always, found with titrating substrates, then G-6b reduces to $B = [E_t]$; the intercept of Figure G-1 is a direct measure of the concentration of active enzyme sites in the reaction mixture.

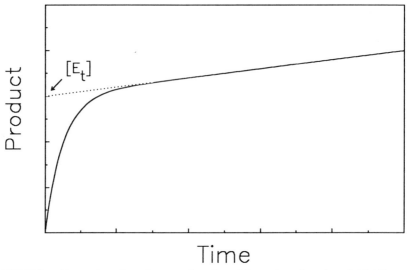

FIGURE G-1. Formation of product as a function of time when there is an initial "burst" followed by a slower turnover reaction.

If $[S]$ is not much greater than K_M then it is necessary to determine B at several values of $[S]$ and make a double reciprocal plot of G-6b:

$$\frac{1}{\sqrt{B}} = \frac{(k_2 + k_3)}{k_2} \cdot \frac{1}{\sqrt{[E_t]}} + \frac{K_M(k_2 + k_3)}{k_2\sqrt{[E_t]}} \cdot \frac{1}{[S]} \qquad \text{(G-6c)}$$

If $k_2 \gg k_3$ then $[E_t]$ is equal to $1/(\text{intercept})^2$ of this plot. If $k_3 \gg k_2$ then no "burst" is seen. EA deacylates as soon as it is formed, and a simple turnover reaction is seen. The value of K_M is nearly the same as K_m, and the rate constant for disappearance of substrate equals k_2.

The third parameter, the pseudo-first order rate constant b, is, from equation G-3a:

$$b = \frac{(k_2 + k_3)[S_0] + k_3 K_m}{K_m + [S_0]} \qquad \text{(G-7a)}$$

Since the experiment under consideration has $[S_0] \gg K_M$, $k_3 K_M \ll (k_2 + k_3)[S_0]$, so:

$$b \approx \frac{(k_2 + k_3)[S_0]}{K_m + [S_0]} \qquad \text{(G-7b)}$$

and, in the Hanes plot of b at several different values of $[S_0]$:

$$\frac{[S_0]}{b} = \frac{K_m}{k_2 + k_3} + [S_0]\frac{1}{(k_2 + k_3)} \tag{G-7c}$$

From the slope/intercept of this plot K_m is calculated. Knowing that $K_M = K_m k_3/(k_2 + k_3)$ and with the value of K_M from ordinary initial rate runs, k_3 may be calculated (more accurately obtained from $V_{\max} = k_3[E_t]$). Then k_2 is obtained from the slope of the plot.

The values of K_m and k_2 may be found in another way. When $[E_t] >> [S_0]$ and $k_2 >> k_3$, then, given the same equilibrium model as equation G-1a:

$$K_m = \frac{[E][S]}{[ES]} = \frac{([E_t] - [ES] - [EA])([S_0] - [P_1] - [ES])}{[ES]} \tag{G-8a}$$

$$[ES] = \frac{([E_t] - [EA])([S_0] - [P_1]) - [ES]([E_t] + [S_0] - [P_1] - [ES] - [EA])}{K_m} \tag{G-8b}$$

With the experimental conditions that $[E_t] >> [S_0]$, we also have $[E_t] >> [EA]$, so $[E_t] - [EA] \approx [E_t]$, and $[E_t] >> ([S_0] - [P_1] - [ES] - [EA])$, so equation G-8b becomes:

$$[ES] = \frac{[E_t]([S_0] - [P_1])}{K_m} - \frac{[ES][E_t]}{K_m} \tag{G-8c}$$

$$[ES]\left(1 + \frac{[E_t]}{K_m}\right) = \frac{[E_t]([S_0] - [P_1])}{K_m} \tag{G-8d}$$

Then, rearranging to get to the pre-steady state expression for rate of product formation:

$$\frac{d[P_1]}{dt} = k_2[ES] = \frac{k_2([S_0] - [P_1])}{1 + K_m/[E_t]} = k_{obs}([S_0] - [P_1]) \tag{G-9a}$$

where

$$k_{obs} = \frac{k_2}{1 + K_m/[E_t]} \tag{G-9b}$$

The formation of product P_1 is a pseudo-first order reaction, and the rate constant k_{obs} is obtained by plotting $\ln([P_1])$ versus t. Since $[E_t] >> [S_0]$

there is no turnover reaction; all the substrate is consumed in the initial formation of EA. The reaction is carried out at several different initial concentrations of enzyme, and from the Hanes plot of $[E_t]/k_{obs}$ versus $[E_t]$ (slope $= 1/k_2$, intercept $= K_m/k_2$), k_2 and K_m are obtained.

Rate Constants for Inhibitor-Enzyme Reaction

Leytus et al. (1983a,1984) used a titration substrate to monitor the rate of reaction of an enzyme with high-affinity inhibitors such as soybean trypsin inhibitor or p-nitrophenyl-guanidinobenzoate. The inhibitor may react with the enzyme via an EI complex, or directly in a second order reaction:

$$E + I \underset{}{\overset{K_i}{\rightleftharpoons}} EI \overset{k_i}{\longrightarrow} EI' \tag{G-10a}$$

$$E + I \overset{k''}{\longrightarrow} EI' \tag{G-10b}$$

The simple M-M kinetics for the reaction with substrate are assumed:

$$E + S \overset{K_m}{\rightleftharpoons} ES \overset{k_p}{\longrightarrow} E + P \tag{G-10c}$$

We use the general rate constant k_p in this analysis, because this development does not require $k_2 \gg k_3$. A somewhat unusual approach is taken for the experimental run; the inhibitor and the substrate are mixed in the buffer in the spectrophotometric cuvette, and enzyme is added to start the reaction. The situation is one of simultaneous competition between inhibitor and substrate for the active enzyme in a pre-steady state time frame.

We consider first the inhibition by the two step reaction, equation G-10a. We have $[E_t] \ll [S_0]$ and $[E_t] \ll [I_0]$, so the concentrations of substrate and inhibitor do not change significantly during the reaction, and the dissociation constants have their usual significance:

$$K_m = \frac{[E][S]}{[ES]} \text{ and } K_i = \frac{[E][I]}{[EI]} \tag{G-11a}$$

These can be rearranged and combined to give four expressions for the concentrations of the various enzyme species:

$$[E] = \frac{[ES]K_m}{[S]} \text{ and } [E] = \frac{[EI]K_i}{[I]} \tag{G-11b}$$

$$[ES] = [EI]\left(\frac{K_i}{[I]}\right)\left(\frac{[S]}{K_m}\right) \text{ and } [EI] = [ES]\left(\frac{K_m}{[S]}\right)\left(\frac{[I]}{K_i}\right) \tag{G-11c}$$

The total amount of enzyme present is:

$$[E_t] = [E] + [ES] + [EI] + [EI']$$ (G-12a)

Substituting for $[E]$ and $[EI]$, viz. $[E]$ and $[ES]$, gives:

$$[ES] = \frac{[E_t] - [EI']}{q} \text{ and } [EI] = \frac{[E_t] - [EI']}{r}$$ (G-12b)

where:

$$q = \frac{K_m}{[S]} + \left(\frac{K_m}{[S]}\right)\left(\frac{[I]}{K_i}\right) + 1$$ (G-12c)

$$r = \frac{K_i}{[I]} + \left(\frac{K_i}{[I]}\right)\left(\frac{[S]}{K_m}\right) + 1$$ (G-12d)

The rate of formation of inactive enzyme EI' is:

$$\frac{d[EI']}{dt} = k_i[EI]$$ (G-13a)

Substituting from G-12b gives:

$$\frac{d[EI']}{dt} = \frac{k_i[E_t]}{r} - \frac{k_i[EI']}{r} = a - b[EI']$$ (G-13b)

Integrating with the boundary conditions $0,[EI']$ and $0,t$ gives:

$$[EI'] = [E_t](1 - e^{-bt})$$ (G-13c)

Substituting this value of $[EI']$ into G-12b gives a second result:

$$[ES] = \left(\frac{[E_t]}{q}\right)e^{-bt}$$ (G-14a)

$$\frac{[dP]}{dt} = k_p[ES] = k_p\left(\frac{[E_t]}{q}\right)e^{-bt}$$ (G-14b)

Integrating between $0,[P]$ and $0,t$ gives:

$$[P] = \left(\frac{k_p[E_t]}{qb}\right)(1 - e^{-bt})$$ (G-14c)

At infinite time e^{-bt} is zero, so $[P_\infty] = k_p[E_t]/qb$, and:

$$[P_\infty] - [P] = \left(\frac{k_p[E_t]}{qb}\right)e^{-bt} \qquad \text{(G-14d)}$$

$$\ln\left([P_\infty] - [P]\right) = -bt + \ln\left(\frac{k_p[E_t]}{qb}\right) \qquad \text{(G-14e)}$$

The indicated plot of $\ln([P_\infty] - [P])$ versus t has a slope of $-b$. In some cases the experimental trace does not come to a constant value of $[P]$ within a reasonable time in which case the program EST1ST (Appendix E) may be applied. If the data can be gathered by a computer for good accuracy, it is possible to take the instantaneous velocity $d[P]/dt$ at several points along the curve, and a plot of $\ln(d[P]/dt)$ versus t (equation G-14b) has the slope $-b$.

From the definition of r (equation G-12d) and having $b = k_i/r$ (equation G-13b),

$$b = \frac{k_i[I]}{K_i(1 + [S]/K_m) + [I]} \qquad \text{(G-15)}$$

This is a rectangular hyperbola of the same form as the M-M equation. The pseudo first-order rate constant b is determined at several concentrations of inhibitor, and a Hanes plot of $[I]/b$ versus $[I]$ has the slope $1/k_i$ and intercept $K_i(1 + [S]/K_m)/k_i$. With K_m and $[S]$ known from runs made in the absence of inhibitor, K_i is readily calculated.

If the reaction between enzyme and inhibitor is strictly second order (equation G-10b) the experimental procedure and determination of the pseudo first-order rate constant is the same, but the dependence of the rate constant upon inhibitor concentration is linear rather than hyperbolic. With the same stipulation that $[E_t]$ is much less than either substrate or inhibitor concentration, from equation G-11a we have:

$$[ES] = \frac{[E][S]}{K_m} \qquad \text{(G-16a)}$$

$$\frac{d[ES]}{dt} = \left(\frac{d[E]}{dt}\right)\left(\frac{[S]}{K_m}\right) \qquad \text{(G-16b)}$$

The rate of formation of inactivated enzyme is:

$$\frac{d[EI']}{dt} = k''[E][I] \qquad \text{(G-16c)}$$

The total enzyme concentration is given by:

$$[E_t] = [E] + [ES] + [EI']$$ (G-17a)

and, differentiating with respect to time:

$$0 = \frac{d[E]}{dt} + \frac{d[ES]}{dt} + \frac{d[EI']}{dt}$$ (G-17b)

Substituting G-16b and G-16c into G-17b gives:

$$0 = \frac{d[E]}{dt} + \left(\frac{d[E]}{dt}\right)\left(\frac{[S]}{K_m}\right) + k''[E][I]$$ (G-17c)

Upon rearrangement this leads to:

$$\frac{d[E]}{[E]} = -hdt$$ (G-18a)

where:

$$h = \frac{k''[I]}{1 + [S]/K_m}$$ (G-18b)

Integration of equation G-18a with the limits of $0,t$ and $[E_t],[E]$ gives:

$$[E] = [E_t]e^{-ht}$$ (G-18c)

The rate of formation of product is $d[P]/dt = k_p[ES]$, and substituting from equations G-11a and G-18c we have:

$$\frac{d[P]}{dt} = \frac{k_p[S][E_t]e^{-ht}}{K_m}$$ (G-19a)

Integrating between the limits $0,t$ and $0,[P]$ yields:

$$[P] = \left(\frac{k_p[S][E_t]}{K_m h}\right)(1 - e^{-ht})$$ (G-19b)

The analysis of the experimental curve is exactly the same as for equation G-14c above. A plot of $\ln([P_\infty] - [P])$ versus t is a straight line with the slope $-h$. Likewise a plot of the instantaneous rate $\ln(d[P]/dt)$ versus t has a slope of $-h$.

The second-order rate constant for inactivation of enzyme, k'', is determined from equation G-18b. The pseudo first-order rate constant h is measured at several concentrations $[I]$; a plot of h versus $[I]$ is a straight line *passing through the origin* of the graph, with a slope of $k''/(1 + [S]/K_m)$. Knowing $[S]$ and K_m the rate constant k'' is readily calculated.

H Viscometric Measurements of Enzyme Activity

Viscometric measurements are primarily used to measure the activities of depolymerase enzymes. The group which has received the most theoretical attention in this regard is the endo-cellulases. A number of assays for amylases are based upon decreases in viscosity, but the non-random mode of hydrolysis of starch by α-amylase vitiates the theoretical analysis which would allow the conversion of the rate of change of solution viscosity into rate of change of product concentration (i.e., rate of enzyme action). Manning (1981) has developed the methodology for treating data from viscosimetric measurements of the action of endo-cellulase on carboxymethylcellulose solution to the extent that the actual rate of bond splitting by the enzyme can be determined; V_{max} and K_M can be calculated for the enzyme from this treatment.

Viscometric Constants

Viscosity measurements using capillary viscometers (Ostwald, Ubbelohde type) must be corrected for two systematic errors: kinetic energy deviations and end effects. The end effect error is avoided by having the proper viscometer design; the two ends of the capillary section must be trumpet-shaped rather than square-shouldered. Commercial viscometers incorporate this feature. The kinetic energy correction becomes important if the outflow time is less than 200 seconds. The correction is made by measuring the outflow time t of a series of liquids of known viscosity η and density ρ. (Sucrose solutions are convenient for this purpose; some values of η and ρ at 20°C are listed in Table H-1.) These values are used to determine the correction constants in the Hagenbach-Couette equation:

$$\frac{\eta}{\rho} = \alpha t - \frac{\beta}{t} \tag{H-1}$$

These constants are specific to a particular viscometer and solvent, and are used in calculating η from each experimental measurement of outflow

Table H–1.   Sucrose Solutions for Viscosity Corrections

Sucrose, wt. %	η , centipoise	ρ, kg/L
20	1.945	1.07991
25	2.447	1.10251
30	3.187	1.12594
35	4.323	1.15024
40	6.167	1.17541
45	9.383	1.20151
50	15.43	1.22854

time made with that viscometer. In the measurements described next, the calculations are based upon relative viscosity, and the densities of the solvent and of the solution of substrate are essentially identical, so the "kinematic viscosity," η/ρ, can be used as primary data.

The second deviation which must be considered is non-Newtonian flow, a usual characteristic of solutions of large polymers. The apparent viscosity η of a non-Newtonian solution is dependent upon the shear rate at which it flows through the capillary. This shear rate G depends upon the volume v of liquid which flows through a capillary of radius r in time t as follows:

$$G = \frac{8v}{3\pi r^3 t} \tag{H-2}$$

Outflow times are measured for a solution of the polymer in question, using several different viscometers of known capillary radius and efflux volume. For each measurement the specific viscosity $\eta_{sp} = \eta_r - 1 = (\eta/\rho)_{soln}/(\eta/\rho)_{buffer} - 1$ is calculated, and η_{sp} is plotted versus G. The dependence of η_{sp} on G is:

$$\eta_{sp} = \eta_{sp0} + \theta\eta_{sp0}^2 G \tag{H-3}$$

From the intercept of this plot the specific viscosity at zero shear rate is obtained, and θ may then be calculated from the slope of the plot. This is repeated at several different concentrations of polymer to find the best estimate of the value of θ. This parameter is characteristic of a specific batch of polymeric substrate. Thus it would only need to be determined each time a new lot of substrate is purchased for use.

Additionally from this series of measurements one obtains a set of data of η_{sp0} as a function of polymer concentration C_s. A plot of these data pairs may be extrapolated to zero polymer concentration; the intercept on the

y-axis is the intrinsic viscosity, $[\eta]$. The concentration dependence of this plot is best given by the Baker equation:

$$\eta_{sp0} + 1 = \eta_{r0} = (1 + [\eta]C_s)^n \qquad \text{(H-4)}$$

A plot of $\log\{\eta_{sp0} + 1\}$ versus $\log\{1 + [\eta]C_s\}$ gives a straight line with slope n.

Two more constants characteristic of the specific lot of substrate are required. The intrinsic viscosity is related to the viscosity average molecular weight of the polymer by $[\eta] = HM_v^x$. In turn, $M_v = KM_n$, where M_n is the number average molecular weight of the polymer, i.e., the total weight present divided by the number of moles (which may be determined for CM-cellulose by a chemical measurement of the reducing end groups). K remains constant as long as the shape of the polymer weight distribution does not change. (The non-random hydrolysis of starch molecules by amylases does change the distribution shape, hence this analysis will not apply in that case.) The experimental constants H and K may be combined in the equation:

$$[\eta] = K'M_n^x \qquad \text{(H-5)}$$

A series of CM-cellulose samples of varying M_n are made by incubating substrate with enzyme and stopping (by heating, adjusting pH, etc.) at various incubation times. The moles of reducing groups present are determined chemically (see Chapter 9), and $M_n = C_s/\text{moles}$. For each sample $[\eta]$ is obtained as described above. A plot of $\log\{[\eta]\}$ versus $\log\{M_n\}$ is linear, with an intercept of $\log\{K'\}$ and a slope of x.

From the experiments described one obtains the necessary viscometric constants for a batch of substrate, namely θ, n, x and K'. These need to be redetermined each time a new lot of substrate is put into use, hence it is desirable to buy a large amount of substrate so that the determination might only have to be performed once or twice a year. The viscometer constants α and β only have to be determined once for each new viscometer when purchased.

Enzyme Rate Calculations

The key concept in this analysis is to relate a change in product concentration to a change in solution properties, i.e., measured viscosity. The product of action of cellulase enzyme is a hydrolyzed glycoside linkage. The concentration of product equals the concentration of reducing groups, which is equal to C_s/M_n. The virtual substrate concentration equals the total weight divided by the molecular weight of the monomeric unit minus the initial product concentration. (For sodium carboxymethyl cellulose of degree of substitution of 0.8, the average monomer weight is 162, the glycosyl residue, plus 0.8 times 62, or 50, for a monomer weight of 212.) The rate of product formation is:

$$\frac{dP}{dt} = \frac{d(C_s/M_n)}{dt} \qquad \text{(H-6)}$$

Manning (1981) equated this to an activity A, as earlier put forth by Almin and Eriksson (1968), without realizing that A is a differential. This mistake on his part makes his subsequent analyses of the events related to enzymatic action invalid, although close.

From equation H-5 we have $1/M_n = K'[\eta]^{-1/x}$ which is substituted into H-6 to give:

$$\frac{dP}{dt} = K'C_s\left(\frac{d[\eta]^{-1/x}}{dt}\right) \qquad \text{(H-7)}$$

Integration between the conditions existing at the beginning of the reaction ($t = 0$) and the conditions at any later time t gives:

$$P_t - P_0 = K'C_s([\eta]_t^{-1/x} - [\eta]_0^{-1/x}) \qquad \text{(H-8)}$$

The next step is to relate the derived quantities, intrinsic viscosity $[\eta]$, to experimental measurements of relative viscosity η_r at the particular concentration of substrate C_s.

The relative viscosity at zero shear rate may be derived from equation H-3, by substituting $\eta_{sp} = \eta_r - 1$:

$$\eta_r - 1 = \eta_{r0} - 1 + (\eta_{r0} - 1)^2\theta G \qquad \text{(H-9a)}$$

and by expanding and collecting terms, to give:

$$\theta G\eta_{r0}^2 + (1 - 2\theta G)\eta_{r0} + \theta G - \eta_r = 0 \qquad \text{(H-9b)}$$

This is solved for the quadratic in η_{r0} as:

$$\eta_{r0} = f(\eta_r) = \frac{2\theta G - 1 + \sqrt{1 + 4\theta G(\eta_r - 1)}}{2\theta G} \qquad \text{(H-9c)}$$

(Two comments should be made here. θ is negative, although Manning mistakenly gave it a positive sign in his paper. The positive square root of the determinant is the one required to give values of η_{r0} which agree with actual data.)

The relationship between relative viscosity at zero shear rate and intrinsic viscosity is given by equation H-4. By substituting $f(\eta_r)$ for η_{r0} and rearranging we get:

$$[\eta] = \frac{f(\eta_r)^{1/n} - 1}{C_s} \qquad \text{(H-10a)}$$

and:

$$\frac{1}{[\eta]^{1/x}} = \frac{C_s^{1/x}}{[f(\eta_r)^{1/n} - 1]^{1/x}} \qquad \text{(H-10b)}$$

Substituting this expression into equation H-8, we obtain the final relationship between measured relative viscosity data and product concentration:

$$P_t - P_0 = K'C_s^{(1+1/x)}\left\{ \frac{1}{[f(\eta_r)_t^{1/n} - 1]^{1/x}} - \frac{1}{[f(\eta_r)_0^{1/n} - 1]^{1/x}} \right\} \qquad \text{(H-11)}$$

The experimental procedure would be as follows, in outline. To a solution containing C_s grams per liter of substrate, enzyme is added at $t = 0$ and the efflux (flow) times e_t are measured in the viscometer at various incubation times t. From a plot of $\log\{e_t\}$ versus t, the efflux time at $t = 0$ is obtained. For each e_t find η/ρ from equation H-1, η_r as the ratio of η/ρ for reaction mixture divided by η/ρ for solvent (buffer), calculate the shear rate G from equation H-2, then $f(\eta_r)_t$ from equation H-9c. Next, calculate the expression $[f(\eta_r)_t^{1/n} - 1]^{1/x}$, and plot the reciprocal of this versus incubation time (note that the incubation time for each measurement should be taken as the midpoint of the flow measurement time). It is obvious that these calculations will be expedited with a short program for the personal computer.

The plot will look like the usual plot of product versus time, given that $[P]$ does not equal 0 at $t = 0$. Conceptually this may be understood in this way. If all the polymeric substrate were present as one gigantic molecule, so that the moles of reducing groups per liter were zero, then $[P]_0$ would indeed be zero, and $f(\eta_r)_0$ would be essentially infinite. If all the bonds in the polymer were hydrolyzed, then $f(\eta_r)_\infty$ would be derived from that solution of sodium carboxymethyl glucose, and the value calculated above would equate to $[P]_\infty$ by the factor $K'C_s^{(1+1/x)}$. Further, $[P]_\infty = [S]_0$, just as in any other enzymatic reaction which is allowed to continue to completion. The value of any point on the plot of $1/[f(\eta_r)_t^{1/n} - 1]^{1/x}$ multiplied by the factor $K'C_s^{(1+1/x)}$ gives $[P]$ in the same absolute concentration terms as C_s, and may be used in data analysis according to standard enzyme kinetics. It must finally be re-emphasized that this entire procedure is only valid if the hydrolysis of bonds by enzyme is random, and the ratio M_v/M_n remains essentially constant.

A Simpler Assay

Hulme (1971) earlier had shown that by meeting certain restraints on the system the change in $1/\eta_{sp}$ with time could be related to the actual rate of action of cellulase in absolute terms, i.e., micromoles of glucosidic bonds hydrolyzed per minute. The substrate solution is prepared as described above. The specific viscosity η_{sp} is found at a number of concentrations of the substrate. From a plot of η_{sp}/c versus c, the intrinsic viscosity $[\eta]$ and Huggins' constant K is determined, according to the Huggins equation:

$$\frac{\eta_{sp}}{c} = [\eta] + K[\eta]^2 c \tag{H-12}$$

The intercept is intrinsic viscosity, and slope/(intercept)^2 gives K. This measurement is necessary, because for the approximations made by Hulme to be valid, $4K\eta_{sp}$ must be less than 3. Using literature values of 0.86 for x, 1.33×10^{-4} dl/g for H in the Mark-Houwink equation, and 2 for the value of M_v/M_n, Hulme arrived at an expression for the relationship between number average molecular weight of substrate M_n and the specific viscosity of the solution η_{sp}:

$$M_n = 0.5\left(\frac{\sqrt{1 + 4K\eta_{sp}} - 1}{2HKc}\right)^{1/0.86} \tag{H-13a}$$

For the substrate used, Hulme found $K = 0.24$, and with $c = 0.385$ g/dl (due to dilution with enzyme), over a range of 0.5 to 1.5 for η_{sp}, the relationship between M_n and η_{sp} was nearly linear, and:

$$M_n = S\eta_{sp}; \qquad S = 3.94 \times 10^4 \tag{H-13b}$$

For each batch of substrate it is necessary to determine K, calculate M_n at several (assumed) values of η_{sp}, and determine S.

Hulme developed the basic equation of Almin and Eriksson (1968) relating enzyme activity to change in viscosity, to arrive at:

$$A = \frac{P}{S[d(1/\eta_{sp})/dt]} \tag{H-14}$$

where A is the enzyme activity in micromoles bonds hydrolyzed per minute, P is the grams of substrate in the reaction mixture, S is the ratio M_n/η_{sp}, and

$d(1/\eta_{sp})/dt$ is the slope of a plot of the reciprocal of specific viscosity versus incubation time. (It should be pointed out that from the definition of η_{sp}, $1/\eta_{sp} = e_{t0}/(e_{ts} - e_{t0})$, where e_{t0} is the outflow time for solvent alone, and e_{ts} is the outflow time measured during the assay, assuming the densities of solvent and assay solution are equal.)   For the specific experimental conditions reported by Hulme A equaled 1.27 times the slope of the plot of $1/\eta_{sp}$ versus time. This proportionality factor must be evaluated for each batch of substrate.

At other substrate concentrations the proportionality factor changes as follows. To change from the substrate concentration c, e.g., 0.5 g/dl, to a concentration c', a factor $(c'/c)^{1 + 1/x}$ must be introduced where x is 0.86 (see above). For example, if c' were 1, the adjustment would be $(1/0.5)^{2.163}$, or 4.478, and the proper proportionality factor would be $5.69(4.478 \times 1.27)$ for Hulme's conditions.

References

AACC. *See* Am. Assoc. Cereal Chemists.

Acker, L. 1969. Water activity and enzyme reaction. *Food Techno.* 23:27-34.

Acker, L. 1985. Phospholipases of cereals. In *Advances in Cereal Science and Technology*, Vol 7, ed. Y. Pomeranz, p. 85, St. Paul, MN: Am. Assn. Cereal Chemists.

Acker, L., and J. Geyer. 1968. Uber die phospholipase B des gerstenmalzes. Mitt. I. Studium substrataktivierenden faktoren. *Z. Lebensm. Unters. Forsch.* 137:231-237.

Adams C. A., T. C. Robberts, and K. C. Butler. 1976. Automated determination of proteolytic enzymes and of amino nitrogen by use of trinitrobenzenesulfonic acid. *Anal. Biochem.* 70:181-186.

Adlers-Nissen, J. 1979. Determination of the degree of hydrolysis of food protein hydrolysates by trinitrobenzenesulfonic acid. *J. Agric. Food Chem.* 27:1256-1262.

Allain, C. C., C. P. Henson, M. K. Nadel, and A. J. Knoblesdorff. 1973. Rapid single-step kinetic colorimetric assay for lactate dehydrogenase in serum. *Clin. Chem.* 19:223-227.

Almin, K. E., and K-E. Eriksson. 1968. Influence of carboxymethlcellulose properties on the determination of cellulase activity in absolute terms. *Arch. Biochem. Biophys.* 124:129-134.

Alsen, C., and F. K. Ohnesorge. 1973. Determination of carbonic anhydrase (E. C. 4.2.1.1) activity by means of the pH-stat technique. *Z. Klin. Chem. Klin. Biochem.* 11:329-332.

Am. Assoc. Cereal Chemists (AACC). 1983. *Approved Methods of the American Association of Cereal Chemists.* 8th Ed. St. Paul, MN: Am. Assoc. Cereal Chemists.

Anastasi, A., M. A. Brown, A. Kembhavi, M. J. H. Nicklin, C. A. Sayers, D. C. Sunter, and A. J. Barrett. 1983. Cystatin, a protein inhibitor of cysteine proteinases. Improved purification from egg white, characterization, and detection in chicken serum. *Biochem. J.* 211:129-138.

Anson, M. L. 1938. The estimation of pepsin, trypsin, papain, and cathepsin with hemoglobin. *J. Gen. Physiol.* 22:79-89.

AOAC. *See* Assoc. Off. Anal. Chemists.

Appleby, C . A., and R. K. Morton. 1959. Lactic dehydrogenase and cytochrome b2 of bakers yeast. *Biochem. J.* 71:492-499.

Armstrong, J. McD. 1964. The molar extinction coefficient of 2,6-dichlorophenol indophenol. *Biochim. Biophys. Acta* 86:194-197.

Asp, E. and L. T. Midness. 1983. Measuring alpha-amylase activity in yeast dough and bread. *Cereal Foods World* 28:725-728.

Assoc. Off. Anal. Chemists (AOAC). 1980. Method 10.179 Free amino nitrogen in wort. In *Official Methods of Analysis of the Association of Official Analytical Chemists*. Washington, D. C.: Assoc. Off. Anal. Chemists.

Atkins, G. L., and I. A. Nimmo. 1980. Current trends in the estimation of Michaelis-Menten parameters. *Anal. Biochem.* 104:1-9.

Avigad, G. 1978. An NADH coupled assay system for galactose oxidase. *Anal. Biochem.* 86:470-476.

Axelrod, B., T. M. Cheesbrough, and S. Laakso. 1981. Lipoxygenase from soybeans. *Methods Enzymol.* 71:441-451.

Babson, A. L., and S. R. Babson. 1973. Kinetic colorimetric measurement of serum lactate dehydrogenase activity. *Clin. Chem.* 19:766-769.

Banauch, D., W. Brummer, W. Ebeling, H. Metz, H. Rindfrey, H. Lang, K. Leybold, and W. Rick. 1975. Eine glucose-dehydrogenase fur die glucose-bestimmung in korperflussigkeiten. *Z. Klin. Chem. Klin. Biochem.* 13:101-107.

Barman, T. E. 1969. *Enzyme Handbook*. 2 Volumes. Supplement 1 (1974). New York, Heidelberg, Berlin: Springer Verlag.

Barrett, A. J., and J. T. Dingle. 1972. The inhibition of tissue acid proteinases by pepstatin. *Biochem. J.* 127:439-441.

Beers, R. F., Jr., and I. W. Sizer. 1952. A spectrophotometric method for measuring the breakdown of hydrogen peroxide by catalase. *J. Biol. Chem.* 195:133-140.

Bender, M. L., G. R. Schonbaum, and B. Zerner. 1962. Mechanism of action of proteolytic enzymes. X. Spectrophotometric investigations of the mechanism of α-chymotrypsin-catalyzed hydrolyses-detection of acyl enzyme intermediates. *J. Am. Chem. Soc.* 84:2540-2550.

Bender, M. L., M. L. Begue-Canton, R. L. Blakely, L. J. Brubacher, J. Feder, C. R. Gunter, F. J. Kezdy, J. V. Killheffer, T. H. Marshall, C. G. Miller, R. W. Roeske, and J. K. Stoops. 1966. The determination of the concentration of hydrolytic enzyme solutions. α-chymotrypsin, trypsin, papain, elastase, subtilisin, and acetylcholinesterase. *J. Am. Chem. Soc.* 88:5890-5913.

Bender, M. L., F. J. Kezdy, and F. C. Wedler, Jr. 1967. α-Chymotrypsin: Enzyme concentration and kinetics. *J. Chem. Educ.* 44:84-88.

Benson, J. R., and P. E. Hare. 1975. o-Phthalaldehyde: Fluorogenic detection of primary amines in the picomole range. Comparison with fluorescamine and ninhydrin. *Proc. Natl. Acad. Sci. U.S.* 72:619-622.

Bergmeyer, H. U. 1974. *Methods of Enzymatic Analysis*. 4 Volumes. New York: Academic Press.

Bernfeld, P. 1955. Amylases, α and β. *Methods Enzymol.* 1:149-158.

Beyer, W. F., Jr., and I. Fridovich. 1987. Assaying for superoxide dismutase activity: Some large consequences of minor changes in conditions. *Anal. Biochem.* 161:559-566.

Bieth, J. 1974. Some kinetic consequences of the tight binding of protein-proteinase inhibitors to the proteolytic enzymes and their application to the determination of dissociation constants. In *Bayer Symposium V. Proteinase Inhibitors*, eds. H. Fritz, H. Tschesche, L. J. Greene, and E. Truscheit, pp. 463-469. New York: Springer-Verlag.

Birch, G. G., N. Blakeborough, and K. J. Parker, eds. 1981. *Enzymes and Food Processing.* London: Applied Science Publishers.

Borgen, J., and I. Romslo. 1977. Lysozyme determined in serum or urine by a simple nephelometric method. *Clin. Chem.* 23:1599-1601.

Bovaird, J. H., T. T. Ngo, and H. M. Lenhoff. 1982. Optimizing the o-phenylenediamine assay for horseradish peroxidase: Effects of phosphate and pH, substrate and enzyme concentration, and stopping reagents. *Clin. Chem.* 28:2423-2426.

Bowski, L., R. Saini, D. Y. Ryer, and W. R. Vieth. 1971. Kinetic modeling of the hydrolysis of sucrose by invertase. *Biotechnol. Bioeng.* 13:641-656.

Boyer, P. D. ed. *The Enzymes.* 3rd Ed. Vol. 1 (1970) -Vol. 17 (1986). New York: Academic Press.

Boyer, R. F. 1977. A spectrophotometric assay of polyphenoloxidase activity. *J. Chem. Educ.* 54:585-586.

Bradford, M. M. 1976. A rapid and sensitive method for the quantitation of microgram quantities of protein utilizing the principle of protein-dye binding. *Anal. Biochem.* 72:248-254.

Brewer, J. M., A. J. Pesce, and S. R. Anderson. 1974. Absorption and Fluorescence. In *Experimental Techniques in Biochemistry*, eds. J. M. Brewer, A. J. Pesce, and R. B. Ashworth, pp. 216-239. Englewood Cliffs, NJ: Prentice-Hall.

Brockerhoff, H., and R. G. Jensen. 1974. *Lipolytic Enzymes.* New York: Academic Press.

Burd, J. F., and M. Usategui-Gomez. 1973. A colorimetric assay for serum lactate dehydrogenase. *Clin. Chim. Acta* 46: 223-227.

Caillat, J. M., and R. Drapron. 1974. La Lipase du blé. Caracteristiques de son action en milieux aqueux et peu hydrate. *Ann. Technol. Agric.* 23: 273-286.

Campbell, J. A. 1980. Measurement of alpha-amylase in grains. *Cereal Foods World* 25:46-49.

Canevascini, G., and C. Gattlen. 1981. A comparative investigation of various cellulase assay procedures. *Biotechnol. Bioeng.* 23:1573-1590.

Carman, G. M., A. S. Fischl, M. Doughterty, and G. Maerker. 1981. A spectrophotometric method for the assay of phospholipase D activity. *Anal. Biochem.* 110:73-76.

Carmona, F. G., E. Pedreño, J. D. Galindo, and F. G. Canovas. 1979. A new spectrophotometric method for the determination of cresolase activity of epidermis tyrosinase. *Anal. Biochem.* 95:433-435.

Ceriotti, F., P. A. Bonini, M. Munrone, L. Barenghi, M. Luzzana, A. Mosca, M. Ripamonti, and L. Rossi-Bernardi. 1985. Measurement of lipase activity by a differential pH technique. *Clin. Chem.* 31:257-260.

Ceska, M. 1971. A new approach for quantitative and semi-quantitative determinations of enzymatic activity with simple laboratory equipment. Detection of α-amylase. *Clin. Chim. Acta* 33: 135-145.

Ceska, M., E. Hultman, and B. Ingleman. 1969. The determination of alpha-amylase. *Experientia* 25:555-556.

Chance, B. 1949. The properties of the enzyme-substrate compounds of horseradish peroxidase and peroxides. III. The reaction of complex II with ascorbic acid. *Arch. Biochem.* 24:389-409.

Chance, B., and A. C. Maehly. 1955. Assay of catalases and peroxidases. *Methods Enzymol.* 2:764-775.

Charney, J., and R. M. Tomarelli. 1947. A colorimetric method for the determination of proteolytic activity of duodenal juice. *J. Biol. Chem.* 171:501-505.

Chase, T. Jr., and E. Shaw. 1967. p-Nitrophenyl-p'-guanidinobenzoate HCl: A new active site titrant for trypsin. *Biochem. Biophys. Res. Commun.* 29:508-514.

Chen, P. S., Jr., T. Y. Toribara, and H. Warner. 1956. Microdetermination of phosphorus. *Anal. Chem.* 28:1756-1758.

Child, J. J., D. E. Eveleigh, and A. S. Sieben. 1973. Determination of cellulase activity using hydroxyethylcellulose as substrate. *Can. J. Biochem.* 51:39-43.

Ciucu, A., and C. Patroescu. 1984. Spectrometric method of determining the activity of glucose oxidase. *Anal. Letts.* 17:1417-1427.

Cleland, W. W. 1967. The statistical analysis of enzyme kinetic data. *Adv. Enzymol.* 29:1-32.

Cleland, W. W. 1970. Steady state kinetics. In *The Enzymes*, 3rd Ed. Vol. 2, ed. P. D. Boyer. pp. 1-65. New York: Academic Press.

Cleland, W. W. 1979. Statistical analysis of enzyme kinetic data. *Methods Enzymol.* 63:103-137.

Cleland, W. W. 1982. The use of pH studies to determine chemical mechanism of enzyme-catalyzed reaction. *Methods Enzymol.* 87:390-404.

Cohn, E. J., and J. T. Edsall. 1943. *Proteins, Amino Acids and Peptides.* New York: Reinhold Publishing Corp.

Cooper, J. R., and H. S. Gowing. 1983. A method for estimating phosphate in the presence of phytate and its application to the determination of phytase. *Anal. Biochem.* 132:285-287.

Cornish-Bowden, A. 1976. *Principles of Enzyme Kinetics.* London: Butterworths.

Cornish-Bowden, A., and R. Eisenthal. 1978. Estimation of Michaelis constant and maximum velocity from the direct linear plot. *Biochim. Biophys. Acta* 523:268-272.

Cornish-Bowden, A., and R. Eisenthal. 1974. Statistical considerations in the estimation of enzyme kinetic parameters by the direct linear plot and other methods. *Biochem. J.* 139:721-730.

Croxdale, J. G., and P. J. Vanderveer. 1986. Quantitative measurements of hexokinase activity in the shoot apical meristem, leaf primordia, and leaf tissues of *Dianthus chinensis L. Plant Physiol.* 81:186-191.

Cubadda, R., and E. Quatirucci. 1974. Separation by gel electrofocusing and characterization of wheat esterases. *J. Sci. Food Agric.* 25:417-422.

D'Appolonia, B. L., L. A. MacArthur, W. Pisesookbunterng, and C. F. Ciacco. 1982. Comparison of the grain amylase analyzer with the amylograph and falling number methods. *Cereal Chem.* 59:254-257.

Davidson, A. L., and W. J. Arion. 1987. Factors underlying significant underestimations of glucokinase activity in crude liver extracts: Physiological implications of higher cellular activity. *Arch. Biochem. Biophys.* 253:156-167.

Dawson, C. R., and R. J. Magee. 1955. Ascorbic acid oxidase. *Methods Enzymol.* 2:831-835.

DeGroot, H., H. DeGroot, and T. Noll. 1985. Enzymic determination of inorganic phosphates, organic phosphates and phosphate-liberating enzymes by use of nucleoside phosphorylase-xanthine oxidase (deydrogenase)-coupled reactions. *Biochem. J.* 229:255-260.

DelRio, L. A., M. G. Ortega, A. L. Lopez, and J. L. Gorge. 1977. A more sensitive modification of the catalase assay with the Clark oxygen electrode. *Anal. Biochem.* 80:409-415.

Delange, R. J., and E. L. Smith. 1971. Leucine aminopeptidase and other N-terminal exopeptidases. In *The Enzymes*, 3rd Ed. Vol. 3, ed. P. D. Boyer. pp. 81-89. New York: Academic Press.

Distler, J. J., and G. W. Jourdian. 1973. The purification and properties of β-galactosidase from bovine testes. *J. Biol. Chem.* 248:6772-6780.

Dixon, M. 1953. The effect of pH on the affinities of enzymes for substrates and inhibitors. *Biochem. J.* 55:161-179.

Dixon, M., and E. C. Webb. 1979. *Enzymes*. 3rd Ed. New York: Academic Press.

Doehlert, D. C., and S. H. Duke. 1983. Specific determination of α-amylase activity in crude plant extracts containing α-amylase. *Plant Physiol.* 71:229-234.

Draper, N. R., and H. Smith. 1981. *Applied Regression Analysis*. 2nd Ed. New York: John Wiley & Sons.

Drapron, R., and L. Sclafani. 1969. Methode de determination de l'activite lipolytique des produits biblogique solides ou pateaux. *Ann. Technol. Agric.* 18:5-16.

Dunn, M. S. 1949. Casein. *Biochemical Prepns.* 1:22-24.

Dygert, S., L. H. Li, D. Florida, and J. A. Thoma. 1965. Determination of reducing sugar with improved precision. *Anal. Biochem.* 13:367-374.

Eisenthal, R., and A. Cornish-Bowden. 1974. The direct linear plot. A new graphical procedure for estimating enzyme parameters. *Biochem. J.* 139:715-720.

Enari, T. M., and J. Mikola. 1977. Pepidases in germinating barley grain: Properties, localization, and possible functions. *CIBA Found. Symp.* 50:335-339.

Erlanger, B. F., A. G. Cooper, and A. J. Bendich. 1964. On the heterogeneity of three-times-crystallized α-chymotrypsin. *Biochemistry* 3:1880-1883.

Erlanger, B. F., N. Kowosky, and W. Cohen. 1961. The preparation and properties of two new chromogenic substrates of trypsin. *Arch. Biochem. Biophys.* 95:271-278.

Esterbauer, H., E. Schwarzl, and M. Hayn. 1977. A rapid assay for catechol oxidase and laccase using 2-nitro-thiobenzoic acid. *Anal. Biochem.* 77:486-494.

Feder, J. 1968. A spectrophotometric assay for neutral protease. *Biochem. Biophys. Res. Commun.* 32:326-332.

Fields, R. 1971. The measurement of amino groups in proteins and peptides. *Biochem. J.* 124: 581-590.

Fiske, C. H., and Y. Subbarow. 1925. The colorimetric determination of phosphorous. *J. Biol. Chem.* 66:375-400.

Ford, J. R., J. A. Nunley II, Y. Li, R. P. Chambers, and W. Cohen. 1973. A continuously monitored spectrophotometric assay with nitrophenyl glycosides. *Anal. Biochem.* 54:120-128.

Fossati, P. 1985. Phosphate determination by enzymatic colorimetric assay. *Anal. Biochem.* 149:62-65.

Foster, R. J., and C. Niemann. 1953. The evaluation of the kinetic constants of enzyme catalyzed reactions. *Proc. Natl. Acad. Sci. U. S.* 39:999-1003.

Fretzdorff, B. 1978. Bestimming der β-xylosidase-aktivitat in roggen. *Z. Lebensm. Unters-Forsch.* 167:414-418.

Fridovich, I. 1975. Superoxide dismutase. *Ann. Rev. Biochem.* 44:147-159.

Fromm, H. J. 1970. A simplified schematic method for deriving steady-state rate equations using a modification of the "Theory of Graphs" procedure. *Biochem. Biophys. Res. Commun.* 40:692-697.

Fromm, H. J. 1975. *Initial Rate Enzyme Kinetics*. New York, Heidelberg, Berlin: Springer-Verlag.

Fuwa, H. 1954. A new method for a microdetermination of amylase activity by the use of amylose as the substrate. *J. Biochem.* 41:583-603.

Gibian, M. J., and R. A. Galaway. 1976. Steady-state kinetics of lipoxygenase oxygenation of unsaturated fatty acids. *Biochemistry* 15:4209-4214.

Gibian, M. J., and P. Vandenberg. 1987. Product yield in oxygenation of linoleate by soybean lipoxygenase: The value of the molar extinction coefficient in the spectrophotometric assay. *Anal. Biochem.* 163:343-349.

Glick, D. 1944. Concerning the reineckate method for the determination of choline. *J. Biol. Chem.* 156:643-651.

Godfrey, T., and J. Reichelt, eds. 1983. *Industrial Enzymology: The Application of Enzymes in Industry.* New York: The Nature Press.

Goldstein, A. 1944. Mechanism of enzyme-inhibitor-substrate reactions. Cholinesterase-erine-acetylcholine system. *J. Gen. Physiol.* 27:529-580.

Good, N. E., G. D. Winget, W. Winter, T. N. Connolly, S. Izawa, and R. M. M. Singh. 1966. Hydrogen ion buffers for biological research. *Biochemistry* 5:467-477.

Gorin, G., S. F. Wang, and L. Papapavlou. 1971. Assay of lysozyme in its lytic action on *M. lysodeikticus* cells. *Anal. Biochem.* 39:113-127.

Grant, D. R. 1974. Studies on the role of ascorbic acid in chemical dough development. I. Reaction of ascorbic acid with flour-water suspension. *Cereal Chem.* 51:684-692.

Grossowicz, N., and M. Ariel. 1983. Methods for determination of lysozyme activity. *Methods. Biochem. Anal.* 29:435-446.

Grossowicz, N., M. Ariel, and T. Weber. 1979. Improved lysozyme assay in biological fluids. *Clin. Chem.* 25:484-485.

Guilbalt, G. G. 1975. Fluorometric determination of dehydrogenase activity using resorufin. *Methods Enzymol.* 41:53-56.

Guilbalt, G. G. 1976. *Handbook of Enzymatic Methods of Analysis.* New York: Marcel Dekker.

Guilbalt, G. G. ed. 1984. *Analytical Use of Immobilized Enzymes.* (Vol. 2. Modern Monographs in Analytical Chemistry.) New York: Marcel Dekker.

Guilbault, G. G., amd E. B. Rietz. 1976. Enzymatic fluorometric assay of α-amylase in serum. *Clin. Chem.* 22:1702-1704.

Guilbault, G. G., and J-M. Kauffmann. 1987. Enzyme-based electrodes as analytical tools. *Biotech. Appl. Biochem.* 9:95-113.

Hagberg, S. 1961. Note on a simplified rapid method for determining alpha-amylase activity. *Cereal Chem.* 38: 202-204.

Hagerman, A. E., and P. J. Austin. 1986. Continuous spectrophotometric assay for plant pectin methylesterase. *J. Agric. Food. Chem.* 34:440-444.

Hagihara, B., H. Matsubara, M. Nakai, and K. Okunuki. 1958. Crystalline bacterial proteinase. I. Preparation of crystalline proteinase of *Bacillus subtilis. J. Biochem* 45:185-194.

Hall, F.F., T. W. Culp, T. Hayakawa, C. R. Ratcliff, and N. C. Hightower. 1970. An improved amylase assay using a new starch derivative. *Am. J. Clin. Path.* 53:627-634.

Hamerstrand, G. E., L. T. Black, and J. D. Glover. 1981. Trypsin inhibitors in soy products:

Modification of the standard analytical procedure. *Cereal Chem.* 58:42-45.

Hardie, D. G., and D. J. Manners. 1974. A viscometric assay for pullulanase-type, debranching enzymes. *Carbohydrate Res.* 36:207-210.

Hartley, B. S., and B. A. Kilby. 1954. Reaction of p-nitrolphenyl esters with chromotrypsin and insulin. *Biochem. J.* 56:288-297.

Hartsuck, J. A., and W. N. Lipscomb. 1971. Carboxypeptidase. In *The Enzymes*, 3rd Ed. Vol. 3, ed. P. D. Boyer. pp. 1-14. New York: Academic Press.

Haworth, W. N., S. Peat, and P. E. Sagroff. 1946. A new method for the separation of the amylose and amylopectin components of starch. *Nature* 157:19.

Heller, M. 1978. Phospholipase D. *Adv. Lipid Res.* 16:267-326.

Hickey, M. E., P. P. Waymack, and R. L. Van Etten. 1976. pH-dependent leaving group effects on hydrolosis of phosphate and phosphonate esters catalyzed by wheat germ acid phosphatase. *Arch. Biochem. Biophys.* 172:439-448.

Hill, J. B., and D. S. Cowart. 1966. An automated colorimetric mutarotase assay. *Anal. Biochem.* 16:327-337.

Hill. R. L., D. H. Spackman, D. M. Brown, and E. L. Smith. 1958. Leucine aminopeptidase. *Biochem. Prepns.* 6:35-48.

Hockeborn, M., and W. Rick. 1982. Determination of the catalytic activity of lipase by a continuous titrimetric test. *J. Clin. Chem. Clin. Biochem.* 20:773-785.

Holm, K. H. 1980. Automatic spectrophotometric determination of amyloglucosidase activity in fermentation samples with a glucose dehydrogenase reagent. *Anal. Chim. Acta* 117:359-362.

Hopsu-Havu, V. K., P. Rintola, and G. G. Glenner. 1968. A hog kidney aminopeptidase liberating N-terminal dipeptides. Partial purification and characteristics. *Acta Chem. Scand.* 22:299-308.

Howell, B. F., S. McCune, and R. Schaffer. 1979. Lactate-to-pyruvate or pyruvate-to-lactate assay for lactate dehydrogenase: A re-examination. *Clin. Chem.* 25:269-272.

Hsu, E., and E. Varriano-Marston. 1983. Comparison of nephelometric and phadebas methods of determining alpha-amylase activity in wheat flour supplemented with barley malt. *Cereal Chem.* 60:46-50.

Huang, J. S., and J. Tang. 1976. Sensitive assay for cellulose and dextranase. *Anal. Biochem.* 73:369-377.

Hulme, M. A. 1971. Viscometric determination of carboxymethlcellulase in standard international units. *Arch. Biochem. Biophys.* 147:49-54.

Hummel, B. C. W. 1959. A modified spectrophotometric determination of chymotrypsin, trypsin and thrombin. *Can. J. Biochem. Physiol.* 37:1393-1399.

Ikeda, K., and T. Kusano. 1983. *In vitro* inhibition of digestive enzymes by indigestible polysaccharides. *Cereal Chem.* 60:260-263.

Ilany-Feigenbaum, J. 1966. A colorimetric method for the determination of proteolytic activity. *J. Food Sci.* 31:29-31.

Imamura, S., and Y. Horiuti. 1978. Enzymatic determination of phospholipase D activity with choline oxidase. *J. Biochem.* 83:677-680.

Interesse, F. S., P. Ruggiero, G. D'Avella, and F. Lamparelli. 1980. Partial purification and some properties of wheat (*Triticum aestivum*) o-diphenolase. *J. Sci. Food Agric.* 31:459-466.

Interesse, F. S., P. Ruggiero, F. Lamparelli, and G. D' Avella. 1982. Characterization of wheat o-diphenolase isoenzyme. *Phytochemistry* 22:1885-1889.

Iwai, H., F. Ishihara, and S. Akihama. 1983. A fluorometric rate assay of peroxidase using the homovanillic acid-o-dianisine-hydrogen peroxide system. *Chem. Pharm. Bull.* 31:3579-3582.

Iwasaki, T., K. Tokuyasu, and M. Funatsu. 1964. Determination of cellulase activity employing glycol cellulose as a substrate. *J. Biochem.* 55:30-36.

Jacks, T.J., and H. W. Kircher. 1967. Fluorometric assay for the hydrolytic activity of lipase using fatty acyl esters of 4-methylumbelliferone. *Anal. Biochem.* 21:279-285.

Jacobsen, N. E., and P. A. Bartlett. 1981. A phosphonamidate dipeptide analogue as an inhibitor of carboxypeptidase A. *J. Am. Chem. Soc.* 103:654-657.

Jensen, R. G. 1983. Detection and determination of lipase (acylglycerol hydrolase) activity from various sources. *Lipids.* 18:650-657.

John, M., G. Trevel, and H. Dellweg. 1969. Quantitative chromatography of homologous glucose oligomers and other saccharides using polyacrylamide gel. *J. Chromatog.* 42:476-484.

Johnston, K. J., and A. E. Ashford. 1980. A simultaneous-coupling azodye method for the quantitative assay of esterase using α-naphthyl acetate as substrate. *Histochemical J.* 12:221-234.

Kakade, M. L., N. Simons, and I. E. Liener. 1969. An evaluation of natural versus synthetic substrates for measuring the antitryptic activity of soybean samples. *Cereal Chem.* 46:518-526.

Kakade, M. L., J. J. Rackis, J. E. McGhee, and G. Puski. 1974. Determination of trypsin inhibitor activity of soy products: A collaborative analysis of an improved procedure. *Cereal Chem.* 51:376-382.

Kaplan, F., P. Setlow, and N. O. Kaplan. 1969. Purification and properties of a DPNH-TPNH diaphorase from *Clostridium kluyverii. Arch. Biochem. Biophys.* 132:91-98.

Kaufmann, R. A., and N. W. Tietz. 1980. Recent advances in measurement of amylase activity—a comparative study. *Clin. Chem.* 26:846-853.

Keyes, M. H., and F. E. Semersky. 1972. A quantitative method for the determination of the activities of mushroom tyrosinase. *Arch. Biochem. Biophys.* 148:256-261.

Kezdy, F. J, and E. T. Kaiser. 1970. Titration of the active site of chymotrypsin with

2-hydroxy-2-nitro-α-toluenesulfonic acid sultone. *Methods Enzymol.* 19:6-20.

Kezdy, F. J., and E. T. Kaiser. 1976. Active site titration of cysteine proteases. *Methods Enzymol.* 45:3-12.

Khalifah, R. G. 1971. The carbon dioxide hydration activity of carbon anhydrase. I. Stop-flow kinetics on the native human isoenzymes B and C. *J. Biol. Chem.* 246:2561-2573.

Klein, B., J. A. Foreman, and R. L. Searcy. 1969. The synthesis and utilization of Cibachron Blue-Amylose: A new chromogenic substrate for determination of amylase activity. *Anal. Biochem.* 31:412-425.

Klein, B., and F. Standaert. 1975. Fluorometric serum lipase assay: Evaluation of monodecanoyl fluorescein as a substrate. *Clin. Chem.* 21:1479-1485.

Koh, T. Y., and B. T. Khouw. 1970. A rapid method for the assay of dextranase. *Can. J. Biochem.* 48:225-227.

Kok, P. J., H. C. Koltkamp, and B. Seidel. 1978. Calibration of a turbidimetric assay of serum lipase activity. *Clin. Chim. Acta* 83:123-128.

Kolehmainen, K., and J. Mikola. 1971. Partial purification and enzymatic properties of an aminopeptidase from barley. *Arch. Biochem. Biophys.* 145:633-642.

Komiyama, T., H. Suda, T. Ayogi, T. Takeuchi, H. Umezawa, K. Fujimoto, and S. Umezawa. 1975. Studies on inhibitory effects of phorphoramidon and its analogs on thermolysin. *Arch. Biochem. Biophys.* 171:727-731.

Kroneck, P. M. H., F. A. Armstrong, H. Merkle, and A. Marchesini. 1982. Ascorbate oxidase: Molecular properties and catalytic activity. In *Ascorbic Acid: Chemistry, Metabolism, and Uses*, eds. P. A. Seib, and G. N. Tolbert (In Advances in Chemistry Series, 200) p. 223. Washington, D.C.: ACS Publishers.

Kruger, J. E. 1976. A note on the semi-automated determination of catalase in wheat. *Cereal Chem.* 53:796-801.

Kruger, J. E., and K. Preston. 1977. The distribution of carboxypeptidases in anatomical tissues of developing and germinating wheat kernels. *Cereal Chem.* 54:167-174.

Kruger, J. E., and K. R. Preston. 1978. Changes in aminopeptidases of wheat kernels during growth and maturation. *Cereal Chem.* 55:360-372.

Kruger, J. E., D. Lineback, and C. E. Stauffer, eds. 1987. *Enzymes and Their Role in Cereal Technology*. St. Paul, MN: Am. Assoc. Cereal Chemists.

Kuiper, J., J. A. Roels, and M. H. J. Zuidweg. 1978. Flow-through viscometer for use in the automated determination of hydrolytic enzyme activites: Application in protease, amylase and pectinase assays. *Anal. Biochem.* 90:192-203.

Kunitz, M. 1947. Crystalline soybean trypsin inhibitors. II. General properties. *J. Gen. Physiol.* 30:291-310.

Kwon, D. Y., and J. S. Rhee. 1986. A simple and rapid colorimetric method for determination of free fatty acids for lipase assay. *J. Am. Oil Chemists Soc.* 63:89-92.

Laidler, K. J., and B. F. Peterman. 1979. Temperature effects in enzyme kinetics. *Methods*

Enzymol. 63:234-256.

Lamkin, W. M., B. S. Miller, S. W. Nelson, O. D. Taylor, and M. S. Lee. 1981. Polyphenol oxidase activities of hard red winter, soft red winter, hard red spring, white common, club, and durum wheat cultivars. *Cereal Chem.* 58:27-31.

Larsen, K. 1983. α-Amylase determination using maltopentaose as substrate. *J. Clin. Chem. Clin. Biochem.* 21:45-52.

Laskowski, M., Jr., and I. Kato. 1980. Protein inhibitors of proteinases. *Ann. Rev. Biochem.* 49: 593-626.

Laufer, S. J. 1938. The proteolytic activity determined by a viscosimetric method. *J. Assoc. Off. Anal. Chem.* 21:160-164.

Lee, H.-J., and I. B. Wilson. 1971. Enzymatic parameters: Measurement of V and Km. *Biochim. Biophys. Acta* 242:519-522.

Leytus, S. P., S. W. Peltz, and W. F. Mangel. 1983a. Adaptation of acyl-enzyme kinetic theory and an experimental method for evaluating the kinetics of fast-acting irreversible protease inhibitors. *Biochim. Biophys. Acta.* 742:409-418.

Leytus, S. P., L. L. Melhado, and W.F. Mangel. 1983b. Rhodamine-based compounds as fluorogenic substrates for serine proteinases. *Biochem. J.* 209:299-307.

Leytus, S. P., W. L. Patterson, and W. F. Mangel. 1983c. New class of sensitive and selective fluorogenic substrates for serine proteinases. *Biochem. J.* 215:253-260.

Leytus, S. P., D. L. Toledo, and W.F. Mangel. 1984. Theory and experimental method for determining individual kinetic constants of fast-acting, irreversible proteinase inhibitors. *Biochim. Biophys. Acta* 788:74-86.

Leinhard, G. E. 1973. Enzymatic catalysis and transition-state theory. *Science* 180:149-154.

Lin, E. C. C., and B. Magasanik. 1960. The activation of glycerol dehydrogenase from *Aerobacter aerogenes* by monovalent cations. *J. Biol. Chem.* 235:1820-1823.

Lin, Y., G. E. Means, and R. E. Feeney. 1969. The action of proteolytic enzymes on N,N-dimethyl proteins. *J. Biol. Chem.* 244:789-793.

Linko, P., and J. Larinkari, eds. 1980. *Food Process Engineering.* Vol 2. Enzyme Engineering in Food Processing. London: Applied Science Publishers.

Lui, D. L., and C. C. Walden. 1969. A spectrophotometric assay for celloliase. *Anal. Biochem.* 31:211-217.

Liu, H. Y., G. A. Peltz, S. P. Leytus, C. Livingston, H. Brocklehurst, and W. F. Mangel. 1980. Sensitive assay for plasminogen activator of transformed cells. *Proc. Natl. Acad. Sci. U. S.* 77:3796-3800.

Lloyd, N. E., K. Khaleeluddin, and W. R. Lamm. 1972. Automated method for the determination of D-glucose isomerase activity. *Cereal Chem.* 49:544-553.

Lorentz, K. 1979. α-Amylase assay: Current state and future development. *J. Clin. Chem. Clin. Biochem.* 17:499-504.

Lowry, O. H., N. J. Rosebrough, A. L. Farr, and R. J. Randall. 1951. Protein measurement with the Folin phenol reagent. *J. Biol. Chem.* 193:265-275.

Lowry, O. H., N. R. Roberts, and J. I. Kapphahn. 1957. The fluorometric measurement of pyridine nucleotides. *J. Biol. Chem.* 224:1047-1064.

Maehly, A. C., and B. Chance. 1954. The assay of catalases and peroxidases. *Methods Biochem. Anal.* 1:357-424.

Manning, K. 1981. Improved viscometric assay for cellulase. *J. Biochem. Biophys. Methods* 5:189-202.

Marchylo, B., and J. E. Kruger. 1978. A sensitive automated method for the determination of α-amylase in wheat flour. *Cereal Chem.* 55:188-196.

Marshall, L. B., and G. D. Christian. 1978. A rapid spectrophotometric method for iodimetric α-amylase assay. *Anal. Chim. Acta* 100:223-228.

Marshall, M., and G. W. Chism. 1979. A comparison of the suitability of three hydrogen donors in the determination of peroxidase activity. *J. Food. Sci.* 44:942-943.

Martin, H. F., and F. G. Peers. 1953. Oat lipase. *Biochem. J.* 55:523-529.

Martin, P., M-N. Raymond, E. Bricas, and B. Ribadeau-Dumas. 1980. Kinetic studies on the action of *Mucor pusillus, Mucor meihei* acid proteases and chymosins A and B on a synthetic chromophoric hexapeptide. *Biochim. Biophys. Acta* 612:410-420.

Martin, P., J-C. Collin, G. Pascaline, B. Ribadeau-Dumas, and M. Germain. 1981. Evaluation of bovine rennets in terms of absolute concentration of chymosin and pepsin A. *J. Dairy Res.* 48:447-456.

Maruhn, D. 1976. Rapid colorimetric assay of β-galactosidase and N-acetyl-α-glucosaminidase in human urine. *Clin. Chim. Acta* 73:453-461.

Mathewson, P. R., B. S. Miller, Y. Pomeranz, G. D. Booth, and G. D. Farenholz. 1981. Colorimetric alpha-amylase, falling number, and amylograph assays of sprouted wheat: Collaborative study. *J. Assoc. Off. Anal. Chem.* 64:1243-1247.

Mathewson, P. R., and W. Seabourn. 1983. A new procedure for specific determination of β-amylase in cereals. *J. Agric. Food Chem.* 31:1322-1326.

Matlashewski, G. J., A. A. Urquhart, M. R. Sahasrabudhe, and I. Altosaar. 1982. Lipase activity in oat flour suspensions and soluble extracts. *Cereal Chem.* 59:418-422.

Matsubara, S., T. Ikenaka, and S. Akabori. 1959. Studies on Taka-amylase A. VI. On the α-maltosidase activity of Taka-amylase A. *J. Biochem.* 46:425-431.

Mayer, A. M., and E. Harel. 1979. Polyphenol oxidases in plants. A review. *Phytochemistry* 18:193-215.

Mazzocco, F., and P. G. Pifferi. 1976. An improvement of the spectrophotometric method for the determination of tyrosinase catecholase activity by Besthorn's hydrazone. *Anal. Biochem.* 72:643-647.

McClure, W. R. 1969. A kinetic analysis of coupled enzyme assays. *Biochemistry* 8:2782-2786.

McDonald, C. E., and A. K. Balls. 1957. Analogues of acetylchymotrypsin. *J. Biol. Chem.*

227:727-736.

Melhado, L. L., S. W. Peltz, S. P. Leytus, and W. F. Mangel. 1982. p-Guanidinobenzoic acid esters of fluorescein as active-site titrants of serine proteases. *J. Am. Chem. Soc.* 104:7299-7306.

Mestecky, J., F. W. Kraus, D. C. Hurst, and S. A. Voight. 1969. A simple quantitative method for α-amylase determinations. *Anal. Biochem.* 30:190-198.

Miller, B. S. 1947. A critical study of the modified Ayre and Anderson method for the determination of proteolytic activity. *J. Assoc. Off. Anal. Chem.* 30:659-669.

Miller, G. L., R. Blum, W. E. Glennon, and A. L. Benton. 1960. Measurement of carboxymethylcellulase activity. *Anal. Biochem.* 1:127-132.

Miwa, I., and J. Okuda. 1974. An improved mutarotase assay using β-D-glucose oxidase and an oxygen electrode. *J. Biochem.* 75:1177-1179.

Mokrasch, L. C. 1967. Use of 2,4,6-trinitrobenzenesulfonic acid for the coestimation of amines, amino acids and proteins in mixtures. *Anal. Biochem.* 18:64-71.

Monsigny, M., C. Kieda, and T. Maillet. 1982. Assay for proteolytic activity using a new fluorogenic substrate (peptidyl-3-amino-9-ethylcarbazole). *EMBO J.* 1:303-306.

Mörsky, P. 1983. Turbidimetric determination of lysozyme with *Micrococcus lysodeikticus* cells: Reexamination of reaction conditions. *Anal. Biochem.* 128:77-85.

Myrtle, J. F., and W. J. Zell. 1975. Simplified copper-soap method for rapid assay of serum lipase activity. *Clin. Chem.* 21:1469-1473.

Nelson, N. 1944. A photometric adaptation of the Somogyi method for the determination of glucose. *J. Biol. Chem.* 153:375-380.

Nelson, W. L., E. I. Ciacco, and G. P. Hess. 1961. A rapid method for the quantitative assay of proteolytic enzymes. *Anal. Biochem.* 2:39-44.

Ng, T. K., and J. G. Zeikus. 1980. A continuous spectrophotometric assay for the determination of cellulase solubilizing activity. *Anal. Biochem.* 103:42-50.

Ngo, T. T., and H. M. Lenhoff. 1980. A sensitive and versatile chromogenic assay for peroxidase and peroxidase-coupled reactions. *Anal. Biochem.* 105:389-397.

Nishino, N., and J. C. Powers. 1979. Design of potent reversible inhibitors for thermolysin. Peptides contining zinc coordinating ligands and their use in affinity chromatography. *Biochemistry* 18:4340-4347.

Nishino, N., and N. Izumiya. 1982. Anti-tryptic activity of a synthetic bicyclic fragment of soybean Bowman-Birk inhibitor. *Biochem. Biophys. Acta* 708:233-235.

Nolte, D., and L. Acker. 1975. Phospholipase D des getreides. *Z. Lebensm. Unters-Forsch.* 158:149-156.

Noltmann, E. A., C. J. Gubler, and S. A. Kuby. 1961. Glucose-6-phosphate dehydrogenase (zwischenferment). I. Isolation of the crystalline enzyme from yeast. *J. Biol. Chem.* 236:1225-1230.

Northrop, J. H. 1933. Pepsin activity units and method for determining peptic activity. *J.*

Gen. Physiol. 16:41-58.

Northrop, J. H., and R. G. Hussey. 1923. A method for the quantitative determination of trypsin and pepsin. *J. Gen. Physiol.* 5:353-358.

Nummi, M., P. C. Fox, M-L. Niku-Paavola, and T-M. Enari. 1981. Nephelometric and turbidometric assays of cellulase activity. *Anal. Biochem.* 116:133-136.

Orsi, B.A., and K. F. Tipton. 1979. Kinetic analysis of progress curves. *Methods Enzymol.* 63:159-182.

Osborne, B. G., S. Douglas, T. Fearn, C. Moorhouse, and M. J. Heckley. 1981. Collaborative evaluation of a rapid nephelometric method for the measurement of alpha-amylase in flour. *Cereal Chem.* 58:474-476.

Pantel, S., and H. Weisz. 1979. A U.V.-absorptiostat and its applications. *Anal. Chim. Acta* 109:351-359.

Paoletti, F., D. Aldinucci, A. Mocali, and A. Caparrini. 1986. A sensitive spectrophotometric method for the determination of superoxide dismutase activity in tissue extracts. *Anal. Biochem.* 154:536-541.

Parry, R. M., Jr., R. C. Chandan, and K.H. Shahani. 1965. A rapid and sensitive assay of muramidase. *Proc. Soc. Exp. Biol. Med.* 119:384-386.

Perten, H. 1964. Application of the falling number method for evaluating alpha-amylase activity. *Cereal Chem.* 41:127-140.

Perten, H. 1966. A colorimetric method for the determination of alpha-amylase activity (ICC Method). *Cereal Chem.* 43:336-342.

Perten, H. 1984. A modified falling-number method suitable for measuring both cereal and fungal alpha-amylase activity. *Cereal Chem.* 61:108-111.

Petit, L. 1974. Etude critique des methodes chimique de dosage des activite s proteolytiques. *Ann. Technol. Agric.* 23:223-231.

Petra, P. H. 1970. Bovine procarboxypeptidase and carboxypeptidase A. *Methods Enzymol.* 19:460-503.

Pfeilsticker, K., and S. Roeung. 1980. Isolierung und charakterisierung der L-ascorbinsaeureoxidase aus weizenmehl. *Z. Lebensm. Unters-Forsch.* 171:425-429.

Preston, K. 1975. An automated fluorometric assay for proteolytic activity in cereals. *Cereal Chem.* 52:451-458.

Proelss, H. F., and B. W. Wright. 1977. Lipoxygenic micromethod for specific determination of lipase activity in serum and duodenal fluid. *Clin. Chem.* 23:522-531.

Racker, E. 1952. Spectrophotometric measurements of the metabolic formation and degradation of thiol esters and enediol compounds. *Biochem. Biophys. Acta* 9:577-578.

Ranum, P. M., K. Kulp, and F. R. Agasie. 1978. Modified amylograph test for determining diastatic activity in flour supplemented with fungal alpha-amylase. *Cereal Chem.* 55:321-331.

Reed, G. 1975. *Enzymes in Food Processing.* 2nd Ed. New York: Academic Press.

Reimerdes, E. H., and H. Klostermeyer. 1976. Determination of proteolytic activities on casein substrates. *Methods Enzymol.* 45:26-28.

Reiner, J.M. 1969. *Behavior of Enzyme Systems.* 2nd Ed. New York: Van Nostrand Reinhold Co.

Richardson, M. 1981. Protein inhibitors of enzymes. *Food Chem.* 6:235-253.

Rick, W., and M. Hockeborn. 1982a. Zur bestimmung der aktivitat der lipase mit dem sogenannten turbidimetrischen test. *J. Clin. Chem. Clin. Biochem.* 20:735-744.

Rick, W., and M. Hockeborn. 1982b. Zur bestimmung der katalytischen activitat der lipase mit trilinolein als substrat. *J. Clin. Chem. Clin. Biochem.* 20:745-752.

Rinderknecht, H., and E. P. Marbach. 1970. A new automated method for the determination of serum α-amylase. *Clin. Chim. Acta* 29:107-110.

Riordan, J. F., and B. L. Vallee. 1963. Acetylcarboxypeptidase. *Biochemistry* 2:1460-1468.

Roberts, I. M. 1985. Hydrolysis of 4-methylumbelliferyl butyrate: A convenient and sensitve fluorescent assay for lipase activity. *Lipids* 20:243-247.

Robins, E., H. E. Hirsch, and S. S. Emmons. 1968. Glycosidases in the nervous system. I. Assay, some properties, and distribution of β-galactosidase, β-glucuronidase and β-glucosidase. *J. Biol. Chem.* 243:4246-4252.

Robinson, D. 1956. The fluorimetric determination of β-glucosidase: Its occurrence in the tissues of animals, including insects. *Biochem. J.* 63:39-44.

Robyt, J. R., and W. J. Whelan. 1965. Anomalous reduction of alkaline 3, 5-dinitrosalicylate by oligosaccharides and its bearing on amylase studies. *Biochem. J.* 95:10P-11P.

Rossi, A., M. S. Palma, F. A. Leone, and M. A. Brigliador. 1981. Properties of acid phosphatase from scutella of germinating maize seeds. *Phytochemistry* 20:1823-1826.

Rotman, B. 1961. Measurement of activity of single molecules of β-D-galactosidase. *Proc. Natl. Acad. Sci. U.S.* 47:1981-1991.

Rowlett, R. S., and D. N. Silverman. 1982. Kinetics of the protonation of buffer and hydration of CO_2 catalyzed by human carbonic anhydrase. II. *J. Am. Chem. Soc.* 104:6737-6741.

Rudolph, F. B., B. W. Baugher, and R. S. Beissner. 1979. Techniques in coupled enzyme assays. *Methods Enzymol.* 63:22-41.

Rungruangsak, K., and B. Panijpan. 1978. Absorbance change in the visible region should be reconsidered for assay of starch cleavage by α-amylases. *Clin. Chem.* 24:1085.

Salas, M., E. Viñuela, and A. Sols. 1963. Insulin-dependent synthesis of liver glucokinase in the rat. *J. Biol. Chem.* 238:3535-3538.

Saleemuddin, M., A. Hassan, and A. Hussain. 1980. A simple, rapid, and sensitive procedure for the assay of endoproteases using Coomassie Brilliant Blue G-250. *Anal. Biochem.* 105:202-206.

Sarda, L., and P. Desnuelle. 1958. Action de la lipase pancreatique sur les esters en emulsion. *Biochim. Biophys. Acta* 30:513-521.

Schonbaum, G. R., B. Zerner, and M. L. Bender. 1961. The spectrophotometric determination of the operational normality of an α-chymotrypsin solution. *J. Biol. Chem.* 236: 2930-2935.

Schwabe, C. 1973. A fluorescent assay for proteolytic activity. *Anal. Biochem.* 53:484-490.

Schwert, G. W., and Y. Takenaka. 1955. A spectrophotometric determination of trypsin and chymotrypsin. *Biochim. Biophys. Acta* 16:570-575.

Schwimmer, S. 1981. *Sourcebook of Food Enzymology.* Westport, CT: AVI Publishing Co.

Segel, I. H. 1975. *Enzyme Kinetics.* New York: John Wiley & Sons.

Seidl, D., and I. E. Liener. 1972. Isolation and properties of complexes of the Bowman-Birk soybean inhibitor with trypsin and chymotrypsin. *J. Biol. Chem.* 247:3533-3538.

Shaw, E., M. Mares-Guia, and W. Cohen. 1965. Evidence for an active-center histidine in trypsin through use of a specific reagent, 1-chloro-3-tosylamido-7-amino-2-heptanone, the chloromethylketone derived from Nα-tosyl-L-lysine. *Biochemistry* 4:2219-2224.

Shiga, M., M. Saito, K. Ueno, and K. Kina. 1984. Synthesis of a new tetrazolium salt giving a water-soluble formazan and its application in the determination of lactate dehydrogenase activity. *Anal. Chim. Acta* 159:365-368.

Shotton, D. M. 1970. Elastase. *Methods Enzymol.* 19:113-140.

Shukuya, R., and G. W. Schwert. 1960. Glutamic acid decarboxylase. I. Isolation procedures and properties of the enzyme. *J. Biol. Chem.* 235:1649-1652.

Silano, V. 1987. α-Amylase inhibitors. In *Enzymes and Their Role in Cereal Technology*, eds. J. E. Kruger, D. Lineback, and C. E. Stauffer. pp. 141-199. St. Paul, MN: Am. Assoc. Cereal Chemists.

Singh, B., and H. G. Sedeh. 1979. Characteristics of phytase and its relationship to acid phosphatase and certain minerals in triticale. *Cereal Chem.* 56:267-272.

Singleton, W. S., M. S. Gray, M. L. Brown, and J. L. White. 1965. Chromatographically homogenous lecithin from egg phospholipids. *J. Am. Oil Chemists Soc.* 42:53-56.

Sinha, A. K. 1972. Colorimetric assay of catalase. *Anal. Biochem.* 47:389-394.

Skelton, G. S., J. Orszagh, J. Gregoire, G. Parent, and L. Katombe. 1976. Rapid assay of proteinase activity in papain by measuring the increase in opalescence of a casein solution during hydrolysis. *Analyst* 101:992-995.

Skursky, L., J. Kovar, and M. Stachova. 1979. A sensitive photometric assay for alcohol dehydrogenase activity in blood serum. *Anal. Biochem.* 99:65-71.

Smith, P. K., R. I. Krohn, G. T. Hermanson, A. K. Mallia, F. H. Gartner, M. D. Provenzano, E. K. Fujimoto, N. M. Goeke, B. J. Olson, and D. C. Klenk. 1985. Measurement of protein using bicinchoninic acid. *Anal. Biochem.* 150:76-85.

Snyder, S.L., and P. Z. Sobocinski. 1975. An improved 2,4,6-trinitrobenzenesulfonic acid method for the determination of amines. *Anal. Biochem.* 64:284-288.

Sober, J. A., ed. 1970. *Handbook of Biochemistry*, 2nd Ed. Cleveland, OH: CRC Press.

Somogyi, M. 1952. Notes on sugar determination. *J. Biol. Chem.* 195:19-23.

Stauffer, C.E. 1971. The effect of pH on thermolysin activity. *Arch. Biochem. Biophys.* 147:568-570.

Stauffer, C. E. 1975. A linear standard curve for the Folin-Lowry determination of protein. *Anal. Biochem.* 69:646-648.

Stauffer, C. E. 1987. Proteases, peptides, and inhibitors. In *Enzymes and Their Role in Cereal Technology*, eds. J. E. Kruger, D. Lineback, and C. E. Stauffer, pp. 201-238. St. Paul, MN: Am. Assoc. Cereal Chemists.

Stauffer, C. E., and R. L. Glass. 1966. The glycerol ester hydrolases of wheat germ. *Cereal Chem.* 43:644-657.

Stauffer, C. E. and D. Etson. 1969. The effect of subtilisin activity of oxidizing a methionine residue. *J. Biol. Chem.* 244:5333-5338.

Stauffer, C. E. and R. S. Treptow. 1973. Inactivation of subtilisin carlsberg in surfactant and salt solutions. *Biochim. Biophys. Acta* 295:457-466.

Street, H. V., and J. R. Close. 1956. Determination of amylase activity in biological fluids. *Clin. Chim. Acta* 1:256-268.

Strothkamp, R. E., and C. R. Dawson. 1977. A spectroscopic and kinetic investigation of anion binding to ascorbate oxidase. *Biochemistry* 16:1926-1930.

Strumeyer, D. H. 1967. A modified starch for use in amylase assays. *Anal. Biochem.* 19:61-71.

Swinbourne, E. S. 1960. Method for obtaining the rate coefficient and final concentration of a first-order reaction. *J. Chem. Soc.* 1960:2371-2372.

Taylor, W. H. 1957. Formol titration: An evaluation of its various modifications. *Analyst* 82:488-498.

Teorell, T., and E. Stenhagen. 1938. Universal buffer over the pH range 2.0 to 12.0. *Biochem. Z.* 299:416-419.

Tipples, K.H. 1969. A viscometric method for measuring alpha-amylase activity in small samples of wheat and flour. *Cereal Chem.* 46:589-598.

Tipton, K. F., and H. B. F. Dixon. 1979. Effects of pH on enzymes. *Methods Enzymol.* 63:183-233.

Tomarelli, R. M., J. Charney, and M. L. Harding. 1949. The use of azoalbumin as a substrate in the colorimetric determination of peptic and tryptic activity. *J. Lab. Clin. Med.* 34:428-433.

Tono, T., and S. Fajita. 1982. Determination of ascorbic acid by spectrophotometric method based on difference spectra. II. Spectrophotometric determination based on difference spectra of L-ascorbic acid in plant and animal foods. *Agric. Biol. Chem.* 46:2953-2959.

Tschetkarov, M., D. Koleff, and S. Banikova. 1967. Viscosimetric determination of cellulase activity using a sodium carboxymethylcellulose substrate. *Monatsh. Chem.* 98:1916-1929.

Tschetkarov, M., and D. Koleff. 1967. Viscosimetric determination of the Michaelis-Menten constant of β-1,4-glucan-4-glucan hydrolase (EC 3.2.1.4) [Cx-cellulase enzymes]. *Monatsh. Chem.* 100:986-997.

Udenfriend, S. 1962. *Fluorescence Assay in Biology and Medicine.* 2 Volumes. New York: Academic Press.

Vallee, B. L., and F. L. Hoch. 1955. Zinc, a component of yeast alcohol dehydrogenase. *Proc. Natl. Acad. Sci. U.S.* 41:327-338.

Van Etten, R. L. 1982. Human prostatic acid phosphatase: A histidine phosphatase. *Ann N. Y. Acad. Sci.* 390:27-51.

van Leeuwen, L. 1979. New saccharogenic determination of α-amylase in serum and urine. *Clin. Chem.* 25:215-217.

Vanni, P., and G. M. Hanozet. 1980. Continuous optical assay of sucrase and other glucosidases. *Experientia* 36:1035-1037.

Veldink, G. A., J. F. G. Vliegenthaart, and J. Boldingh. 1977. Plant lipoxygenases. *Prog. Chem. Fats other Lipids* 15:131-166.

Verger, R. 1980. Enzyme kinetics of lipolysis. *Methods Enzymol.* 64:340-392.

Verhagen, J., G. A. Veldink, M. R. Egmond, J. F. G. Vliegenthaart, J. Boldingh, and J. Van der Star. 1978. Steady-state kinetics of the anaerobic reaction of soybean lipoxygenase-1 with linoleic acid and 13-L-hydroperoxylineoleic acid. *Biochim. Biophys. Acta* 529:369-379.

Visuri, K., J. Mikola, and T. M. Enari. 1969. Isolation and partial characterization of a carboxypeptidase from barley. *Eur. J. Biochem.* 7:193-199.

Walsh, K.A. 1970. Trypsinogens and trypsins of various species. *Methods Enzymol.* 19:41-63.

Walsh, K. A., and P. E. Wilcox. 1970. Serine proteases. *Methods Enzymol.* 19:31-41.

Weaver, L. H., W. R. Kester, and B. W. Matthews. 1977. A crystallographic study of the complex of phosphoramidon with thermolysin. A model for the presumed catalytic transition state and for the binding of extended substrates. *J. Mol. Biol.* 114:119-132.

Weibel, M. K. 1976. A coupled enzyme assay for aldose-1-epimerase. *Anal. Biochem.* 70:489-494.

Westerik, J. O., and R. Wolfenden. 1972. Aldehydes as inhibitors of papain. *J. Biol. Chem.* 247:8195-8197.

Whitaker, J. R. 1972. *Principles of Enzymology for the Food Sciences.* New York: Marcel Dekker.

Whitaker, J. R. ed. 1974. *Food Related Enzymes.* (Advances in Chemistry Series, 136.) Washington, DC: ACS Publishers.

Wohlgemuth, J. 1908. Uber eine neue methode zur quantitativen bestimmung des diastatischen ferments. *Biochem. Z.* 9:1-9.

Wolfenden R. 1977. Transition state analogs as potential affinity labeling reagents. *Methods Enzymol.* 46:15-28.

Wood, P. J., D. Paton, and I. R. Siddiqui. 1977. Determination of a β-glucan in oats and barley. *Cereal Chem.* 54:524-533.

Wood, P. J., and J. Weisz. 1987. Detection and assay of (1-4)-β-D-glucanase, (1-3)-β-D-glucanase, (1-3)(1-4)-β-D-glucanase, and xylanase based on complex formation of substrate with Congo Red. *Cereal Chem.* 64:8-15.

Zaks, A., and A. M. Klibanov. 1984. Enzymatic catalysis in organic media at 100°C. *Science* 224:1249-1251.

Index